EUCLIDEAN GEOMETRY

EUCLIDEAN GEOMETRY

(An Uptodate Account of Plane Geometry of 2 Dimensions as Known Today)

M.S. Karmalkar

KALPAZ

Euclidean Geometry

M.S. Karmalkar

ISBN: 978-81-7835-886-4

Published in 2011 in India by
Kalpaz Publications
C - 30 Satyawati Nagar
Delhi - 110052
Phone : 9812729499
E-mail: kalpaz@hotmail.com
Laser Typy Setting : Lucky Graphic, Delhi
Printed at : Chawla Offset Press, Delhi

गणित हे विभूतिमत्वाला जाणत नाही

ते स्वयंसिद्ध असते

Mathematics does not depend on Super naturalism.

It can prove itself

Late Shri S M Karmalkar

Contents

Preface

My father, Late Shri S.M. Karmalkar who passed away on 18.12.72 had dedicated his life to the teaching of Mathematics to the high school students in the state of Maharashtra particularly in the Vidarba from 1921 to 1951. In addition he wrote a number of books in Mathematies and Astronomy. One of the books was Bhumity Shastrachtri Multatwe (भूमिति शास्त्राची मूलतत्वे) which means Fundamental Principles of geometry. It was a very popular book and was used as a text-book for geometry by the Matriculation students of those years.

2. There has been considerable change in the syllabus in the 10th, 11th and 12th classes. Calculus has now been introduced in the 12th class and computer courses have been introduced in the early stages of the school in the 5th class itself. Algebra and geometry are being taught simultaneously from 5th class to the 12th class. In view of this there is no single book that caters to the requirement of geometry alone. Besides this the fine points of geometry are lost in the confusion and geometry is not taught in full perspective, I therefore, thought of writing this book on geometry alone. It contains geometry as we know it today and nothing else.

3. Euclid (322 B.C. - 275 B.C.) was founder of the school of Mathematics at the University of Alexandria in Egypt. His text book 'The Elements' was used in schools for 2000 years and forms the basis of geometry even today. He is also credited to proving that there are infinite prime numbers. He also developed the unique factorisation theorem and devised an algorithm for the greatest common denominator.

4. Basically, I had three reasons to write this book.

(i) The first is that, I wanted to pay a tribute to my father and appreciate his efforts in teaching of Mathematics and popularising it among the students.

(ii) The second, perhaps the most important reason, was to provide an English translation of my father's original work which was in 'Marathi Language'. His work was needed to be presented and made available to the wider readership especially the non-Marathi speaking population of our country.

(iii) The third was that there are several difficult examples in my father's work which to my knowledge are not given anywhere. It was necessary to bring these to the notice of Mathematical fraternity of our country.

5. Obviously the book is not suitable for any school curriculum But for those aspiring for I.I.T. entrance examination, for which deeper knowledge of the subject is necessary, this book is bound to be is useful. And somebody who likes Mathematices may like to keep it on his shelf. He may like to add to or subtract from the book for which the author will feel grateful.

6. (a) The book has five chapters. Of this the first three cater to the requirement of tenth class curriculum as follows:

   1. St lines, triangles and Quadrilaterals
   2. Areas, proportionality and pythagoras theorems
   3. Circles and tangents

   It however, contains a detailed exposition of the subject much more than what is required for the present tenth class course.

   (b) The fourth chapter contains advanced topics as mentioned below

   1. Extension of Pythagoras theorems
   2. Theorems on intersecting chords of Circles.
   3. Nine pt. circle and Simpson's line
   4. Special propertes of isosceies Δ with angles 72°, 72° and 360°
   5. Contains special theorems like Ceva's theorem, Plolemy's theorem and Moreley's theorem
   5. Miscellaneous examples - P and Q Series.

   (c) In addition to the above, I have added a fifth chapter which contains solution of difficult problems in the book.

7. (a) Special thanks are due to Dr. V.G. Tikekar, Retired Professor and Chairman of the Dept of Mathenmatics I.I. of Science Bangalore for his extremely useful guidance in bringing out the book. His wide experience of Mathematics and suggestions have been very useful.

   (b) Thanks are also due to Shri Mukul Kumar Jain my son-in law for his general guidance in the conduct of the book and also in finding a suitable publisher for the book.

8. In the end though every care has been taken to avoid any mistakes some would have crept in. The same may be pointed out to the undersigned for correction.

9. I am also thankful to my wife Smt Nirmala Karmalkar for her cooperation in bringing out the book.
10. I have provided asterix mark to those examples which are important and difficult to solve.

– M.S. Karmalkar

Introduction

The word "Geometry" is derived from two Greek words, "Geo" meaning earth and "Metro" meaning measurement. Hence "geometry" means "measuring of the earth".

1. Our forefathers encountered certain difficulties when they set out to measure the earth for construction of buildings, drawing maps of cities, districts and countries. Therefore they invented certain propositions called 'theorems' and drew conclusions there from and that is how the science of geometry was born.

2. Euclidean geometry consists 2 parts - Part I of Plane Geomety and .Part-II of Solid Geometry.

In the first part figures drawn in planes are discussed with their properties while in the second part solid figures and their properties are elaborated. In the present book we deal with Part I only, i.e. properties of plane figures.

3. Importance of geometry

(a) It is the basis of modern mathematics. Hence it is learnt before advanced Mathematics is learnt.

(b) It is useful in day to day life e.g. in drawing maps of fields, buildings and so on.

(c) It is an exact science and hence does not have another meaning to a statement.

(d) It teachers you the method of reasoning and is very useful in daily life. A real geometrician will not make a wrong statement and is expected to do well-in-life.

4. Plane geometry of one or two dimension composed of two sections.

Section I Consists of theorems and conclusions derived from them.

Section II Consists of drawing figures with the help of compass and straight edge.

While the Ist one is called theoretical geometry the IInd one is called Practical gcometry. It is easy to distinguish them as problems of theoretical geometry and problems of practical geometry and are discussed in this book side by side.

5. It is undertood that in order to prove any statement in geometry, all statements made and proved earlier can be used. but not the statements made or proved later. Then to prove a statement in theorem No. 12, all the statements made upto and including theorem No11 and proved earlier can be used, but not the statements made in theorem No. 13 or after that. Further a statement has to be proved in order to take it for granted. For example, an angle has to be proved to be 90° in order to take it to be 90° not because it appears to be 90°. It has to be so from what is given in the example or from the statement made or proved earlier. But what about theorem No. 1. It has no statements made earlier. It has to be proved from statement which are self evident and require no proof. Such self evident statements are called *AXIOMS*

Definition

An *Axiom* is a self evident statement which requires no proof.

The following axioms are true in general in all branches of Mathematics.

(a) Things which are equal to the same thing are equal to one another

If $a = b$ and $b = c$, then $a = c$.

(b) Things which are equal to one another can be added and the sums shall be equal. If $a = b$ and $c = d$, then $a + c = b + d$.

(c) Things which are equal to one another can be subtracted and the remainders shall be equal.

If $a = b$ and $c = d$, then $a - c = b - d$.

(d) Things which are equal to one another can be multiplied and the products shall be equal

If $a = b$ and $c = d$, then $a . c = b . d$.

(e) Things which are equal to one another can be divided and the quotients shall be equal

If $a = b$ and $c = d$, then $\frac{a}{c} = \frac{b}{d}$.

6. Fundamental Definitions

Everybody knowns what is meant by a point, a line, a surface and a solid but in geometry these terms are used in a strict sense, which needs some explanation.

Definition

A *point* has position but no length, breadth ar thicknees.

(a) In other words it has no dimension. It means that we are to attach to a point no ideas as to its size i.e. no length or no breadth or no thicknees. It has only a position. A dot made by a sharp pencil may be taken as roughly representing a point but howsoever small it may be it has some length or breadth, is therefore not a true geometrical point. The smaller a dot is, the more nearly it represents a point. Actually, a point has only a position but no dimension.

Definition

A *line* has length and direction but no breadth or thickness

(b) When a point moves in a particular direction it traces out a straight line. But such a trace howsoever finely drawn has some degree of breadth and is therefore not a true geometrical line. The finer the trace the more true it is to the geometrical line. Straight line has one direction only.

Straight line ————————

Definition

When a line changes its direction it becomes a *curved* line.

curved line.

(c) In a similar manner from the idea of a curved line, we proceed to the idea of surface.

Definition

A surface has length and breadth but no thickness.

(d) A *solid* has length, breadth or thickness.

We may say that moving a point in one direction traces out a line. Moving a line in a plane traces out a surface. Moving a plane of two dimensions, in the third dimension traces out a solid. We thus see that a point, a straight line, a surface and a solid are basic elements of geometry.

7. Line and Ray

Definition:- A straight line starts at a point and it ends in another point it is called a segment of a line. When it does not end in a point it is called a ray. A ray is indefinitely long.

Segment AB   Ray AB

8. The following are the axioms of a straight line

(a) Any two points can be joined by a st. line.

(b) Any straight line segment can be extended indefinitely to form another segment.

(c) A line contains indefinitely many points.

(d) Through a given point their pass indefinitely many lines.

(e) Every portion of a st. line is straight and has one direction only. The same as the main st. line.

(f) Throught 2 points there is one and only one st. line that contains them both.

Definition

Three or more points are said to be *collinear points* when the is one line that contains them all.

Definition

Three or more lines are said to be *concurrent* if there is a point that contains them all

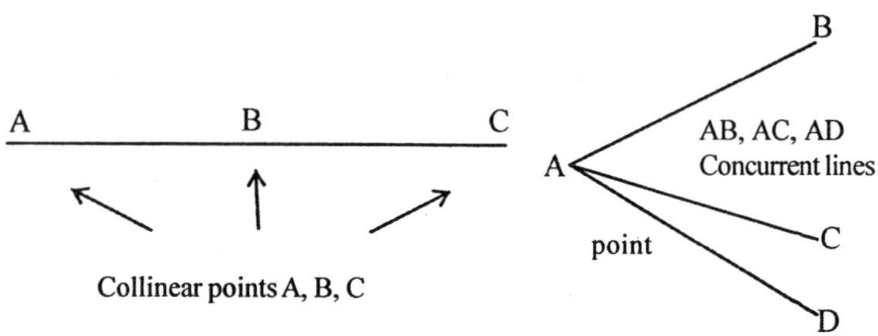

Collinear points A, B, C

9. Measurement of a Line

For measuring a line, we require a standard. In this case standard is the contimeter. Ordinary scale in a compass box is graduated in cm. The centimeter is divided into ten parts. Each part is called a millimeter. 10 centimeters is called a decimeter. 10 decimeters is called a metre. A cm is the unit for small measurement. Meter is a unit for larger measurement. Still larger unit is the kilometer.

Complete table of measurement is given below:

10 millimeter = 1 centimeter (cm)

10 centimeter = 1 decimeter (dm)

10 decimeter = 1 meter (m.)

10 meters = 1 Dekameter (Dm.)

10 dekameters = 1 Hectometer (Hm.)

10 Hectometers = 1 Kilomster (1 Km.)

Prior to 1957, Govt. of India changed into metric system of measurments which has now come into vogue, we used to measure in inches, feet, yards and mile.

The complete table which was in use is given below.

12 inches = 1 foot   220 yards = 1 furlong

3 feet = 1yard   8 furlongs = 1 mile

The table is cumbersome and is no longer used we use the metric system in which

8 kms = 5 miles

10. Definitions

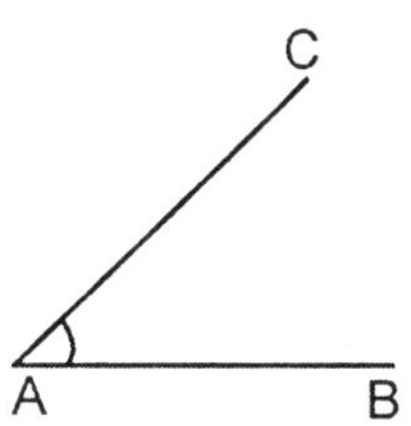

When two straight lines meet in a point they are said to form an *angle.*

The points where they meet is called the vertex. The two lines which form the angle are called the arms of the angle. Thus in the figure shown here with $\angle CAB$ is formed at the vertex A and AB and AC are the arms of the angle.

When 2 angles are dissimilar they are expressed as

$\angle C'A'B' > \angle CAB$

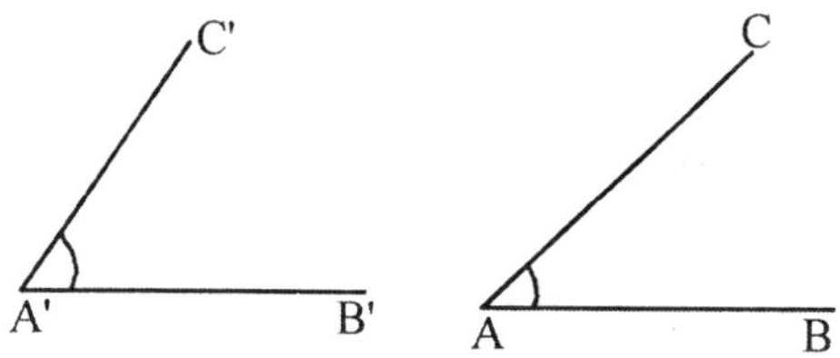

It is easy to see that the size of angle does not upon the length of the arms but upon the inclination of one arm to the other.

11. Measurement of the angle

To measure the angle we have to fix one arm OA and rotate the second arm OB from OA till it reaches the position OB. The turning is called the measure of the angle. Further we have a standard of measurement. If we rotate OA about O till it comes back to the position OA again, we call it rotation through 360°. Half the rotation is called 180° and one-fourth of the rotation is called 90°. Three fourth is called 270°.

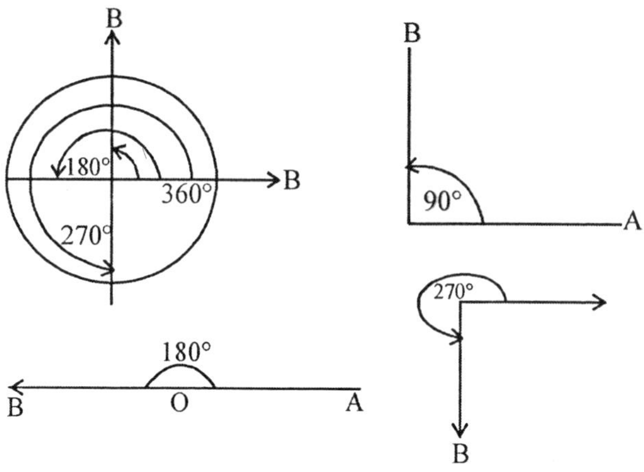

12. Definitions

(a) Angle less than 90° is called an **acute** angle.

(b) Angle more than 90° but less than 180° is called an **obtuse** angle.

(c) Angle more than 180° but less than 360° is called a **reflex** angle.

(d) When the arm is rotated about O till it makes a straight line with $\overline{OA}$ on the opposite side it is called a straight angle.

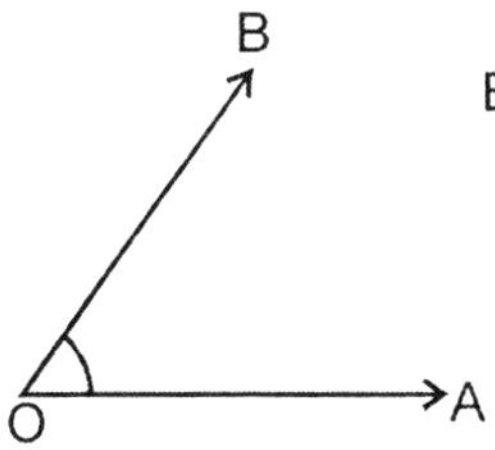

(1) An acute angle

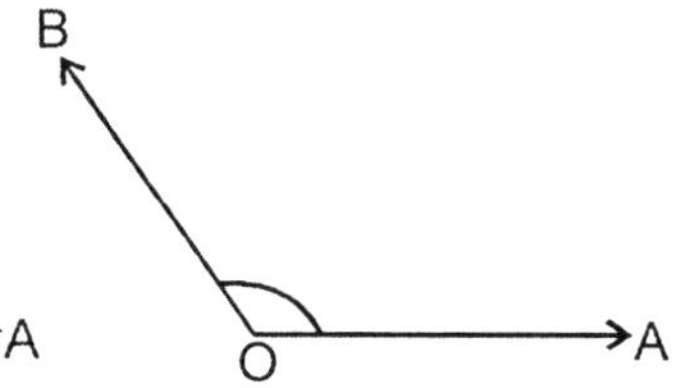

(2) An obtuse angle

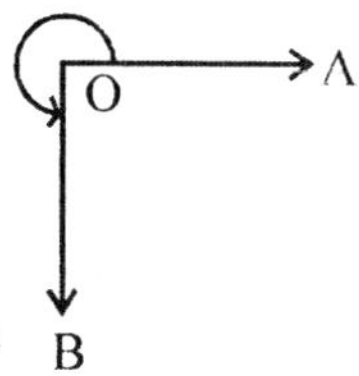

(3) A Reflex angle

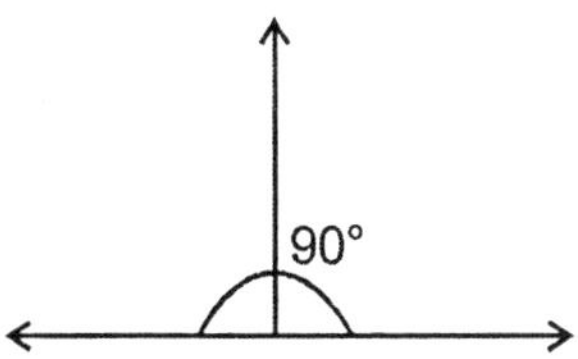

(e) When 2 st. lines meet each other, 2 angles are formed one less than 90° and the other more than 90°. We will take the angle less than 90° as the angle between two lines.

(f) If both $\angle$s are = 90°, each is called a right angle.

Axiom :- All right angles are equal.

13. Theoretical geometry consists of theorems and examples or riders theorem. A general statement is made at the beginning and is proved. It is followed by what is given and what is to be proved. Sometimes a construction is necessary. It is shown in dotted lines. It is customary to write (H. P.) at the end meaning there by that what was to be proved has been proved.

14. Definition

Angles which are on a common side of one common arm are called *Adjacent Angles.*

Thus $\angle AOB$ and $\angle COB$ are angles on a common side OB. They are called adjacent $\angle$s .

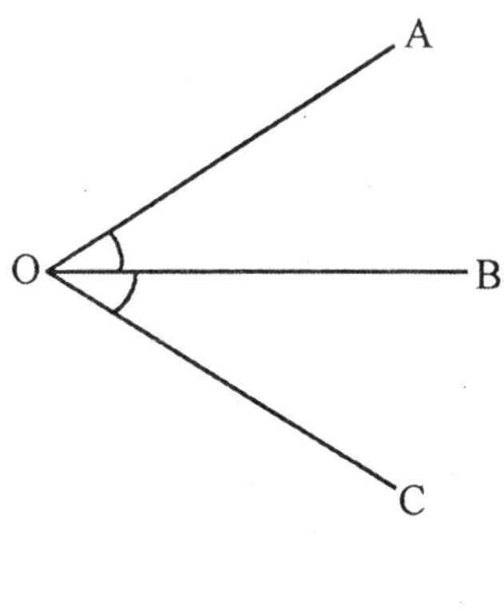

15. If a line AB is drawn and meeting it is a line OC at O. Two $\angle s$ are formed. $\angle AOC$ and $\angle BOC$. They are together = 180°

$\angle AOC + \angle BOC = 180°$

This is known as Linear Pair Axiom (L.P.A.)

Converse of L.P.A. is also true.

Converse means just the opposite as detailed below.

1. Theorem : All Brahmins are Hindus (True)
   Converse : All Hindus are Brahmins (False)
2. Theorem : The month of July has 31 days (True)
   Converse : If a month has 31 days it should be called July (False)
3. Theorem : All right angles are equal (True)
   Converse : If 2 angles are equal it should be called a right angle (False)

This shows that converse of a theorem is not necessarily true, but in the case of L.P.A. converse is true. This is proved in Thorem No. 2.

1

Straight Lines, Triangles and Quadrilaterals

Theorem No. 1

Two distinct straight lines meet at most in one point.

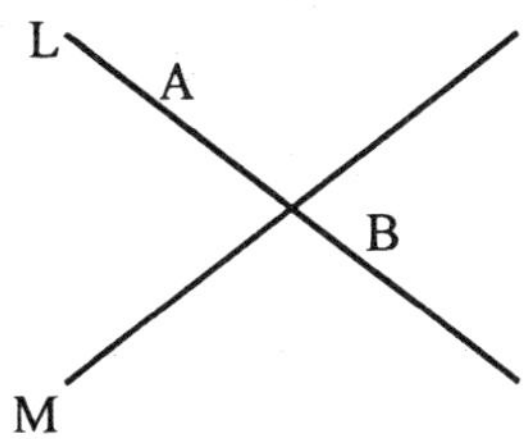

Given :- Two distinct straight lines 'L' and 'M'.

To Prove:- 'L' and 'M' intersect at most in one point only.

Proof:- Suppose 'L' and 'M' intersect in 2 points and let these be 'A' and 'B'. Then 'L' contains A and B and 'M' contains 'A' and 'B' i.e. A and B are points of lines L and M. But AB is one direction and L and M are 2 directions. It is not possible that one direction will contain two directions. Hence our supposition that 'L' and 'M' intersect in 2 points must be incorrect. It must intersect in 1 point only. (H.P)

Note:- In this type of proof we assume the opposite is possible, but this will lead to impossible result. Hence it is proved that the opposite is not possible.

Theorem No. 2

If two adjacent angles are together equal to 180°, then their uncommon arms are a pair of opposite rays.

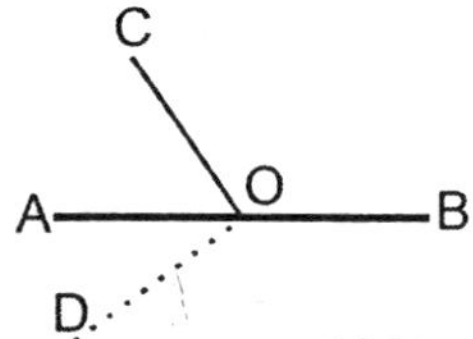

Given:- $\angle AOC + \angle BOC = 180°$

To prove :- The uncommon arms OA and OB are a pair of opposite rays.

Proof:- Let $\overrightarrow{OA}$ and $\overrightarrow{OB}$ be not a pair of opposite rays and ray $\overrightarrow{OD}$ be opposite to ray $\overrightarrow{OB}$

$\therefore \angle AOC + \angle BOC = 180°$ (given) ... (1)

and as $\overrightarrow{OD}$ is a opposite to $\overrightarrow{OB}$

$\angle DOC + \angle BOC = 180°$ ...(2)

From (1) and (2) we get

$\angle AOC = \angle DOC$

But OC is a common arm That means OA must fall along OD.

That means OA and OB are a pair of opposite rays. (H.P.)

Definitions

(1) If the sum of 2 angles is equal to 90°, each is said to be **complement** of the other.

(2) If the sum of 2 angles is equal to 180° each is said to be **supplement** of the other.

(3) If 2 straightlines intersect each other they form 4 angles.

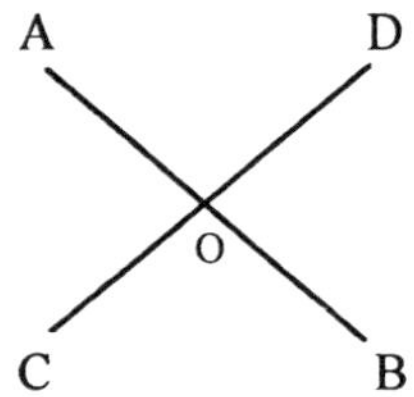

(1) $\angle AOC, \angle AOD, \angle BOD, \angle BOC$

(2) $\angle DOA$ and $\angle COB$ and $\angle AOC$ and $\angle BOD$ are opposite pair of angles.

Theorem No. 3

When 2 lines intersect each other the two pairs of opposite angles so formed are equal to one another.

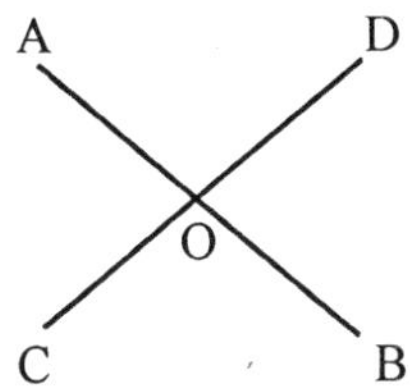

Given:- Two lines AB and CD intersect each other at O.

To Prove:- $\angle AOC = \angle DOB$

and $\angle AOD = \angle BOC$

Proof:- Because CD is a st line

$\angle AOC + \angle AOD = 180°$ (L.P.A.) (1)

Because AB is a straight line

$\angle BOD + \angle AOD = 180°$ (L.P.A.) (2)

From (1) and (2)

$\angle AOC = \angle BOD$

Similarly we can prove

$\angle AOD = \angle BOC$ (H.P.)

Exercise (a)

1. In the figure of Theorem 2 if $\angle AOC = 47°$ then what is $\angle BOC$.
2. If $\angle AOC = 55°$ what should be $\angle BOC$ so that OA and OB are opposite rays.
3. If $\angle AOB = 88°$ and $\angle AOC = 92°$ and B and O are an opposite sides of OA show that BC is a st. line.
4. Which angle is supplement to itself?
5. Which angle is complement to itself?
6. Which angle is supplement to $\frac{1}{3}$ itself?

7. An angle is 16° more than its complement then what is its measure?

8. Find the complement of the following angles.

   (i) 22° (ii) 63° (iii) 45° (iv) 75° (v) 35°

9. Find the supplement of the following $\angle s$.

   (i) 75° (ii) 90° (iii) 100° (iv) 115° (v) 133°

Definition

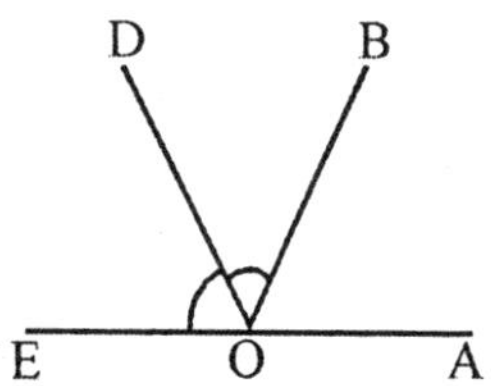

If one side of the angle is produced and the exterior angle so formed is bisected by a line. That line is called the external bisector of that angle.

OD is called the external bisector of BOA.

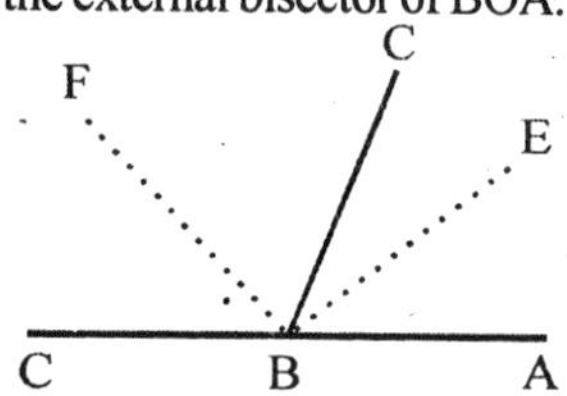

10. If $\angle ABC = 40°$ and BE and BF are the internal and external bisectors of $\angle ABC$ what is $\angle FBE$ equal to

11. In the previous example if $\angle ABC = 50°$, 60° and 92° show that $\angle FBE$ is always = 90°.

12. Prove that the angle between internal and external bisectors of an angle = 90°.

13. $\angle ABC$ is bisected by BD is DB is produced to E show that $\angle ABE = \angle CBE$.

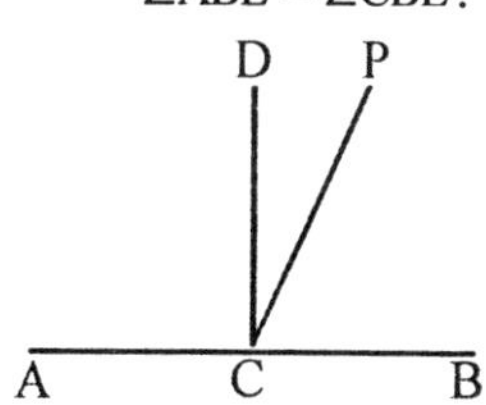

14. In the figure below CD $\perp$ AB and CP is any line show that

$$\angle DCP = \frac{1}{2}(\angle ACP - \angle PCB)$$

15. Several $\angle s$ are formed at 0. If $\angle AOB = 40°$, $\angle BOC = 65°$, $\angle COD = 85°$, $\angle DOE = 115°$, what is $\angle EOA$?

16. The measurce of an angle is twice the angle of its supplement. Find the angle.

17. If AB and CD are 2 st lines interseting in O and $\angle AOC = 90°$. Prove that $\angle COB$, $\angle BOD$, $\angle DOA$, are each = 90°.

18. In the figure of theorem No. 3 $\angle AOC = 76°$. What are the other $\angle s$ in the figure.

19. If AB and CD are intersecting in O the bisector of $\angle AOC$ and $\angle BOD$ are in the same st. line.

20. If $\angle AOB = 60°$ and the angle is bisected by OC. Which is produced backwards to D. DO bisects the reflex angle AOB each of which $\angle = 150°$.

Parallel Lines

Let us look at theorem No. 1 a little more closely. It states that two lines meet each other in at most one point. If meets the other line in another point, it can only be coincident line, but can it meet in zero point that means in no point at all. Yes we have the example of the railway lines. Two railway lines do not meet each other how so ever they are produced. Or consider the example of two lines in a note-book these also do not meet each other if they are produced both ways. Such lines are called parallel lines.

Definition

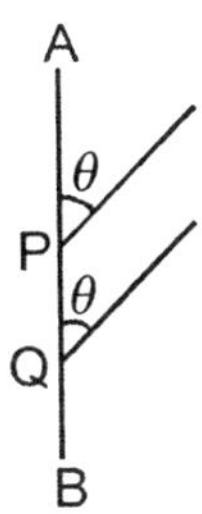

1. *Parallel lines* are two or more straight lines which do not meet each other even if they are produced both ways as much as we like. Suppose we have a line AB and take two points P and Q on it. Then we draw 2 lines at P and Q making the same angle with AB. These lines do not meet each other howsoever we produce them. They are therefore parallel lines. Parallel lines are those lines which make equal angles with any arbitrary line AB.

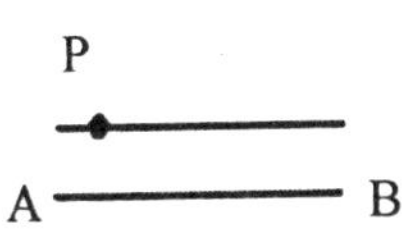

2. **Play-Fairs Axiom** : From a point P outside Play AB only one parallel line can be drawn to AB

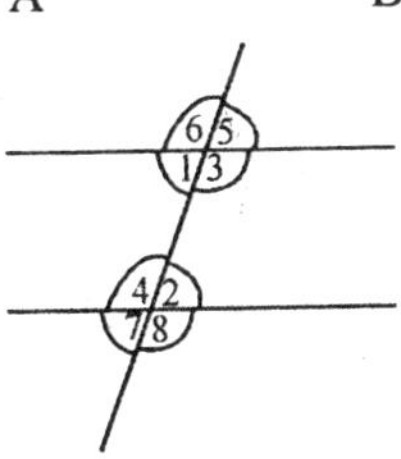

3. When we draw two parallel lines and draw one line intersecting them In all eight $\angle s$ are formed. They are named as below.

(a) 1 and 2, and 3 and 4 are called alternate interior angles

(b) 4 and 6, 2 and 5, 1 and 7, 3 and 8 are called corresponding angles.

(c) 2 and 3 and 1 and 4 are called interior angles on the same side of the transversal

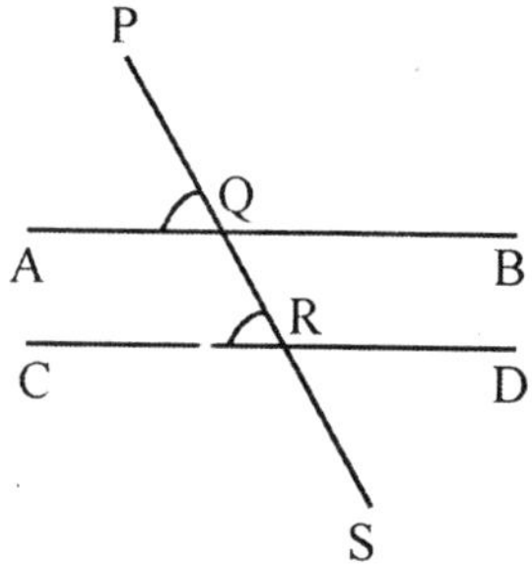

4. **Corresponding Angles Axiom (C.A.A.)**

If a transveresal meets two parallel lines, each pair of corresponding angles are equal Conversely, If a transversal meets two lines in such a way that the corresponding angles are equal, then the two lines are parallel.

Theorem No. 4

Two straight lines which are parallel to the same straight are parallel to one another.

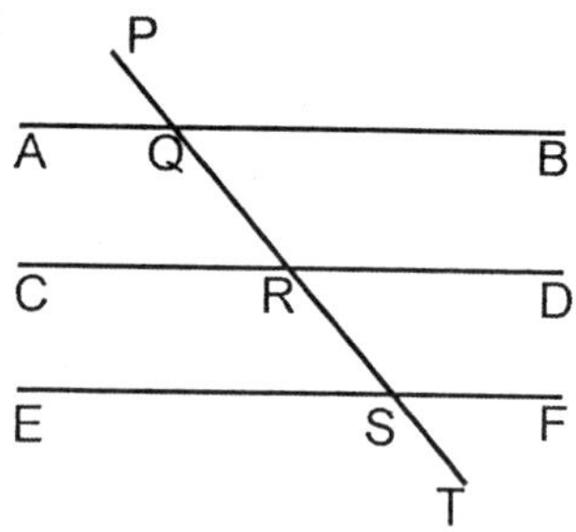

Given :- AB $\parallel$ EF and CD $\parallel$ EF

PQRST is any transversal

To Prove:- AB $\parallel$ CD

Proof:- AB $\parallel$ EF

$\therefore$ $\angle PQB = \angle PSF$ (C.A.A.) ...(1)

CD $\parallel$ EF

$\therefore$ $\angle PRD = \angle PSF$ (C.A.A.) ...(2)

From (1) and (2) we get

$\angle PQB = \angle PRD$

But there are corresponding angles

$\therefore$ AB $\parallel$ CD (C.A.A.) (H.P.)

Theorem No. 5

If 2 parallel lines are intercepted by a transversal then,

(1) *Both pairs of alternate ∠s are equal*

(2) *Both pairs of interior ∠s on the same side of the transversals are together equal to 180°*

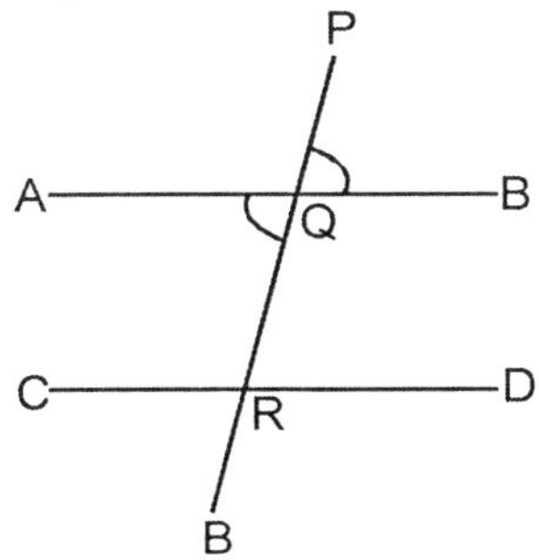

Given :- AB || CD PQRS is a transversal. Cutting parallel lines AB and CD in Q and R.

(1) **To Prove** $\angle AQR = \angle QRD$

$\angle BQR = \angle QRC$

Proof : AB and CD are parallel lines and PQRS is a transversal.

$\angle PQB = \angle QRD$ (C.A.A.)

But $\angle PQB = \angle AQR$ (Th. 3)

$\therefore$ $\angle AQR = \angle QRD$

Similarly we can prove (H.P.)

$\angle BQR = \angle QRC$

(2) **To prove** $\angle AQR + \angle QRC = 180°$

and $\angle BQR + \angle QRD = 180°$

Proof: AB is a st. line and QR meets it

$\therefore$ $\angle AQR + \angle RQB = 180°$ (L.P.A.)

But $\angle RQB = \angle QRC$ (Proved above)

∴ $\angle AQR + \angle QRC = 180°$

Similarly $\angle QRD + \angle RQB = 180°$ (H.P.)

Theorem No. 6

If 2 lines are cut by a transversal and if (1) Alternate ∠s are equal or if (2) Interior ∠s on the same side of the transversal are together = 180° then the 2 lines are parallel.

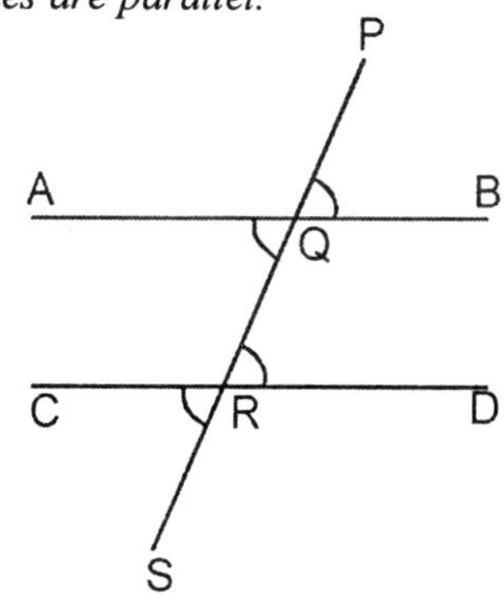

Given:- PQRS is a transversal that intersects AB and CD in points Q and R, besides

$\angle AQR = \angle QRD$

$\angle BQR = \angle QRC$

To Prove: AB || CD

Proof: $\angle AQR = \angle PQB$ (Th 3)

$\angle AQR = \angle QRD$ (given)

∴ $\angle PQB = \angle QRD$

AB || CD (C.A.A.) (H.P.)

Similarly $\angle AQP = \angle BQR$ (Th 3)

$\angle BQR = \angle QRC$ (given)

∴ $\angle AQP = \angle QRC$

∴ AB || CD (C.A.A.)

Given :	PQRS cuts two lines AB and CD in Q and R.	
and	$\angle AQR + \angle QRC = 180°$	
or	$\angle BQR + \angle QRD = 180°$	
To Prove:	AB ‖ CD	
Proof :	$\angle AQR + \angle AQP = 180°$	(L.P.A.)
	$\angle AQR + \angle QRC = 180°$	(given)
∴	$\angle AQP = \angle QRC$	
	AB ‖ CD	(C.A.A.)
Similarly	$\angle BQR + \angle QRD = 180°$	(given)
	$\angle BQR + \angle PQB = 180°$	(L.P.A.)
∴	$\angle PQB = \angle QRD$	
∴	AB ‖ CD	(H.P.)

Exercise (b)

1. Two intersecting lines cannot both be parallel to the same line.
2. AB, AC, AD, AE are parallel to a line l. What can be said about the points A, B, C, D, E.

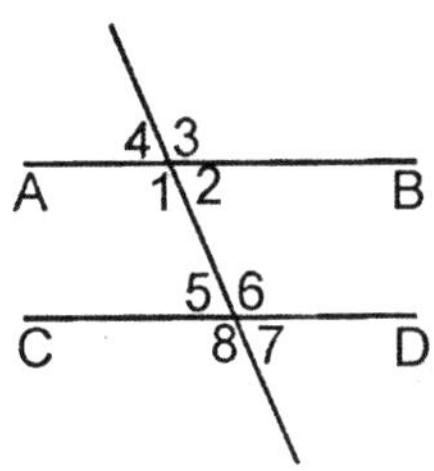

3. In the adjoining figure AB ‖ CD a st. line meets it if $\angle 5 = 45°$. Find $\angle 2, \angle 7, \angle 8, \angle 6$.

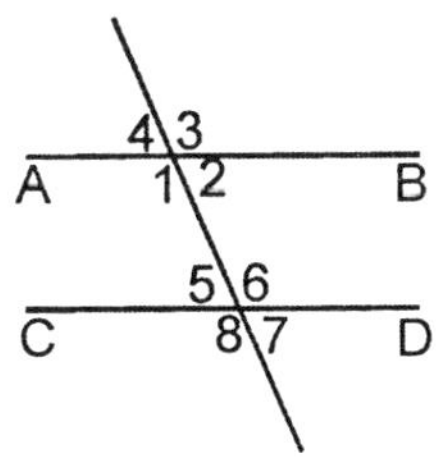

4. In the adjoining figure $\angle 1 = 118°$, $\angle 7 = 61°$ Is AB ‖ CD ?
5. In the adjoining figure $\angle 6 = 120°$ $\angle 4 = 60°$ Is AB ‖ CD ?

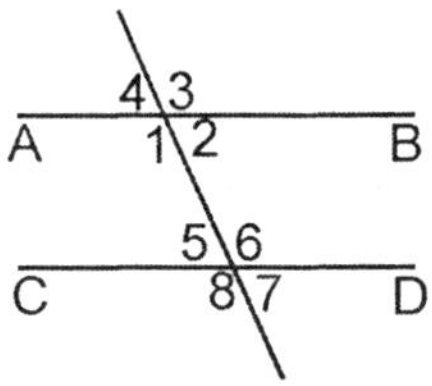

6. (a) AB || CD

(b) ∠2, ∠3 are in the ratio of 4 : 5.

Determine all the ∠s in the figure.

7. AB ⊥ CD. They intersect in O show that all the four ∠s at O are = 90°.

8. If 2 parallel lines are intersected by a third line then the lines bisecting the altenate ∠s are parallel.

9. If any 2 lines are intersected by a third line and if the lines bisecting the alternate interior angles are parallel then the lines are parallel.

10. In a Δ ABC, BA is produced to D if ∠DAC is bisected and the bisector is || BC then prove that ∠ACB = ∠ABC

11. AB || CD. The transversal EGHL meets the || lines at G and H. The corresponding ∠s EGB and GHD are bisected by GM and HN. Prove that GM || HN.

Triangles

Definition

A *triangle* is a figure bounded by three sides viz. AB, BC and CA. It has 3 angles ∠ABC, ∠BCA and ∠CAB. It is denoted by the symbol Δ.

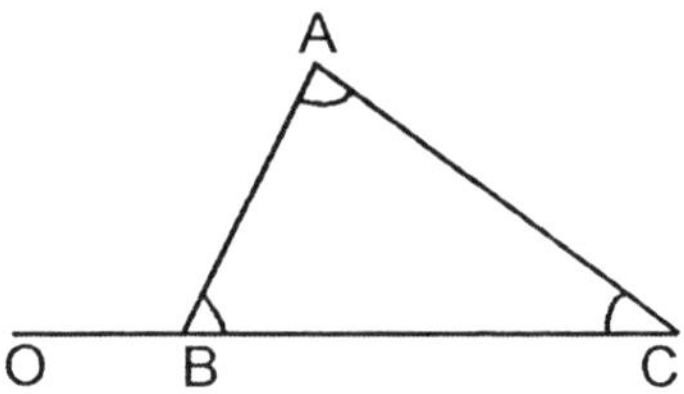

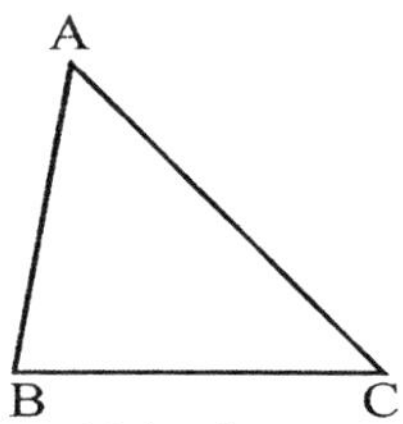

(a) A scalene Δ

(1) When the three sides of a Δ are unequal it is called a *Scalene triangle*. The 3 ∠s are also unequal.

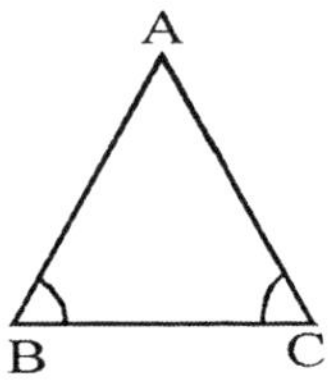

AB = AC

∠B = ∠C

(2) When two sides are equal it is called an *isosceles* Δ. It can be proved that the two angles of an isosceles Δ are equal.

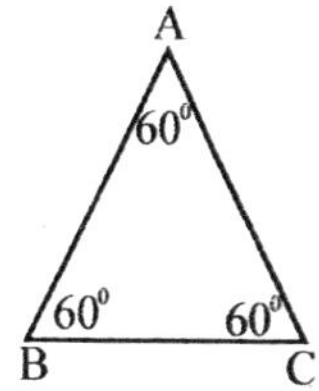

(c) Equilateral

∠ AB=BC=CA

$\angle A = \angle B = \angle C = 60^\circ$

(3) **When all the 3 sides of a Δ are equal to each other it is called an Equilateral Δ. All 3 $\angle$s can be proved to be equal and equal to 60°.**

Theorem No. 7

The sum of 3 angles of a Δ is equal to 180°

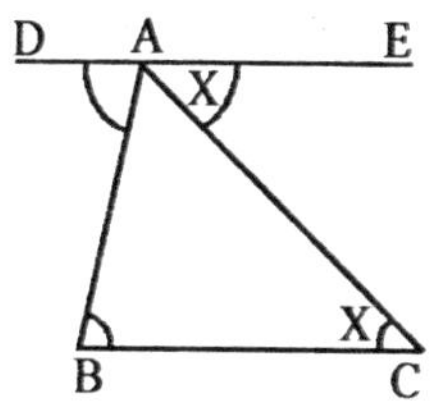

Given :- A triangle ABC

To Prove : - $\angle A + \angle B + \angle C = 180^\circ$

Construction :- Draw a line DE through $A \parallel BC$

Proof :- $DE \parallel BC$ and AB meets it

$\therefore$ $\angle DAB = \angle ABC$ ....(1)

$DE \parallel BC$ and AC meets it

$\therefore$ $\angle EAC = \angle ACB$ .....(2)

Adding (1) and (2)

$\angle ABC + \angle ACB = \angle DAB + \angle EAC$

Further adding $\angle BAC$ to both sides

$\angle ABC + \angle ACB + \angle BAC = \angle DAB + \angle EAC + \angle BAC = 180^\circ$

$\therefore$ The sum of the 3 $\angle$s of a $\Delta = 180^\circ$ (H.P.)

Corollary 1 :- When one side of a Δ is produced, the exterior $\angle$ so formed is equal to sum of 2 interior opposite $\angle$s

Proof:- $\angle A + \angle B + \angle C = 180°$ ....(1)

Produce BC to D. $\angle ACD$ is the exterior $\angle$ formed.

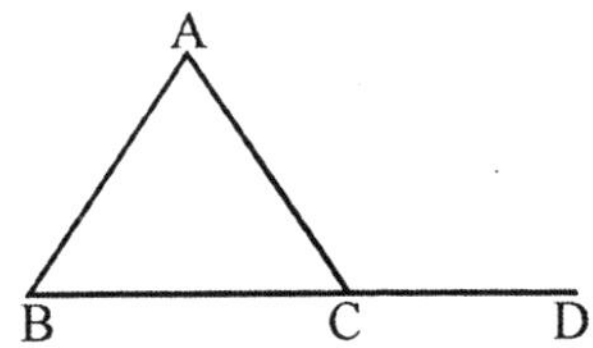

$\angle ACD + \angle C = 180°$ .....(2)

$\therefore$ From (1)+(2)

$\angle ACD = \angle A + \angle B$

Definition :- A quadrilateral is a figure bounded by four sides. It has four angles. ABCD is quadrilateral. AC and BD are the diagonal.

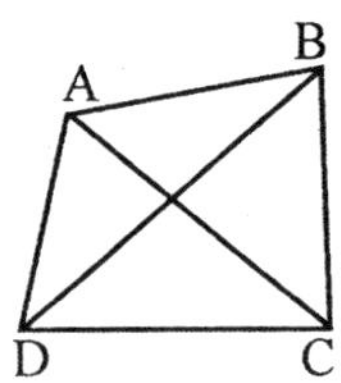

Corollary 2 :- The sum of the $\angle$s of a quadrilateral is 360°.

By drawing one diagonal say AC, a quadrilateral can be divided into 2 Δs. Obviously the sum of the $\angle$s of the quadrilateral is 360°.

Exercise (C)

1. In a ΔABC, $\angle B = 65°$, $\angle C = 45°$, Find $\angle A$.
2. In a ΔABC, $\angle A = 4 \times \angle C$ and $\angle B = 5 \times \angle C$, find all angles of the Δ.
3. The angles of a Δ are in the ratio of 3 : 5 : 7, what are they?
4. $AB \parallel CD$ and is cut by a transveral at Q and R. The bisectors of the angles AQR, BQR, QRD and QRC form a rectangle. Prove this.
5. Can a Δ have

(a) Two right angles (b) Two obtuse angles (c) Two acute angles
(d) All angles more than 60° (e) All angles $\angle$ 60° (f) All angles = 60°

6. In a ΔABC $\angle C = 90°$, CD is drawn $\perp$ AB. Show that $\angle ABC = \angle DCA$ and $\angle CAB = \angle DCB$

7. The bisectors of the base $\angle s$ of a ΔABC meet at O. Show that

   $\angle BOC = 90 + \frac{A}{2}$.

8. The sides AB and AC of a ΔABC are produced to D and E respectively and external $\angle s$ DBC and ECB are bisected to meet at G. Show that $\angle CGB = \left(90 - \frac{A}{2}\right)$

9. In a ΔABC, AD is the bisector of $\angle A$, AE $\perp$ BC, show that

   $\angle EAD = \frac{1}{2}(\angle B - \angle C)$

10. In M and N are plane mirrors at right angles to each other. In AB is the incident ray on mirror M at B and is reflected from mirror at C. Show that the resulting ray CD is $\parallel$ to AB.

11. Side BC of a ΔABC is produced to D. $\angle ACD = 100°$. The sides AB and CA are produced to P and Q. $\angle QAP = 30°$. Find all $\angle s$ of the Δ.

12. Sides BC, CA and AB are produced to D, E, F. Show that

    $\angle ACD + \angle BAE + \angle CBF = 360°$

13. Side BC of a ΔABC is produced to D.

    The bisector of $\angle A$ meets BC in $\angle$?

    Prove that $\angle ABC + \angle ACD = 2\angle ALC$.

14.(a) If a point is taken on the base of a Δ and perpendiculars are dropped on the sides then the sum of the angles which these perpendiculars make with the base is equal to the angle at the vertex.

   (b) If however one of these perpendicular meets outside the Δ, then sum of the angles which these perpendiculars make with the base are equal to the difference of base angles.

15. In a ΔABC the angle between the bisector of interior $\angle B$ and the bisector of the exterior angle to $\angle C$ is equal to half the angle A.

Polygons

Definitions

A closed figure, which has more than four line segments is called a *Polygon*

A polygon with five sides, is called a *Polygon*

A polygon with six sides, is called a *Heptagon*

A polygon with seven sides, is called a *Heptagon*

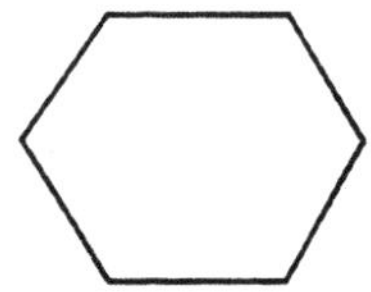

Convex Polygon

A polygon with eight sides, is called a *Octagon*

A polygon with nine sides, is called a *Nonagon*

A polygon with ten sides, is called a *Decagon*

A polygon with all its vertices on one side of all its line segments it is called a convex *Polygon.*

Concave Polygon

A polygon with at least two of its certion an either side of a line segment it is called a *Concave Polygon.*

In the book we shall deal with convex polygon only.

A polygon is called a regular polygon if all its sides are equal. It can be proved that all its angles are equal.

Theorem No. 8

The sum of the angles of a convex ploygon + 4 right angles is equal to twice right angles as the polygon has sides.

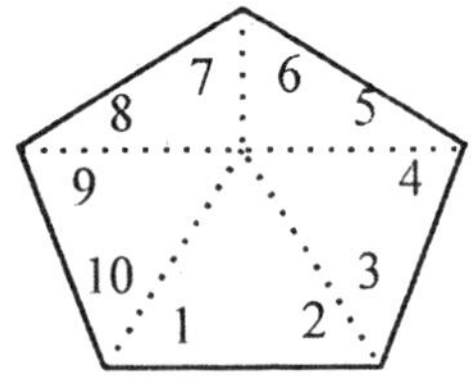

Given: We will take case of a pentagon - a polygon with five sides.

To Prove: ∠A + ∠B + ∠C + ∠D + ∠E + 4st angles = (2 x 5) rt. angles.

Construction : Take a point O inside the polygon and join it to the vertices.

Proof: There will be as many Δs as the polygon has sides.

In the above case there will be five Δs.

$\therefore$ $(\angle 1 + \angle 2 + \angle 3 + \angle 4 + \angle 5 + \angle 6 + \angle 7 + \angle 8 + \angle 9 + \angle 10)$ + 4rt angles at O = (2 x 5) rt. angles

$\therefore$ $(\angle A + \angle B + \angle C + \angle D + \angle E)$ +4rt angles at O= 10 rt. angles

$\therefore$ Total of Angles of polygon + 4rt angles = 10 rt. angles

The same will be true for polygon of any side. (H.P.)

Exercise (D)

1. The angles of a quadrilateral are 110°, 38°, 102°. What is the fourth angle?
2. In a quadrilateral the four angles are in the ratio of 1 : 2 : 3 : 4. Find the measure of each angle?
3. What is the measure of each angle of a regular (a) Pentagon (b) Hexagon (c) Heptagon (d) Octagon (e) Nonagon (f) Decagon (g) Dodecagon.
4. What is no. of sides of a regular polygon which has one angle equal to (a) 160° (b) 175° (c) 135°.
5. ABCDE is a regular pentagon. The bisector of $\angle A$ of the pentagon meets one side of the pentagon at right angles. Prove this.
6. ABCDE is a regular Pentagon. To draw the diagonals of the pentagon we have to join the angular points to other angular points except the adjacent ones. Hence only 2 diagonals from each of the 5 points can be drawn. Prove that each angle in tri-sected by the diagond
7. ABCDEF is a regular Hexagon. Each end point has 3 diagonals. Each diagonal divides the angle into 4 equal parts.
8. Similarly in an octagon there are 5 parts each equal to $22\frac{1}{2}°$.
9. In a nonagon, 6 diagonals divide the angle into 7 parts each =20°.

Theorem No. 9

If the sides of a polygon are producd in order, the exterior angles ∠c so for formed are together equal to 4 rt angles.

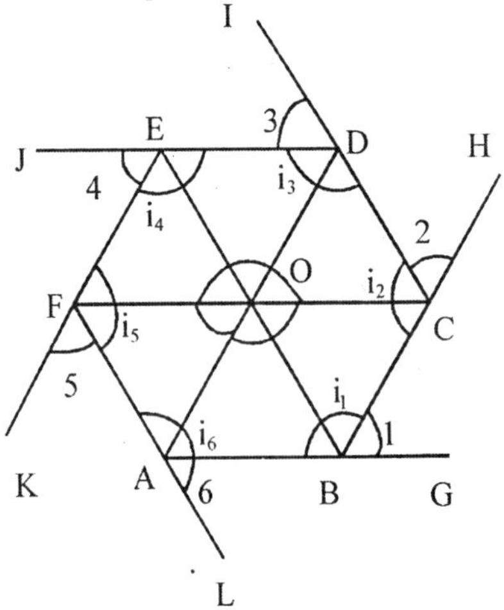

Given : A polygon (six sides) is given.

Side AB is produced to G

Side BC is produced to H

Side CD is produced to I

Side DE is produced to J

Side EF is produced to K

Side FA is produced to L

The exterior ∠s are marked 1, 2, 3, 4, 5, 6.

To Prove :- $\angle 1 + \angle 2 + \angle 3 + \angle 4 + \angle 5 + \angle 6 = 360°$

Proof :- Dividing the polygon into six △ s at O it is easy to see that

$(i_1 + i_2 + i_3 + i_4 + i_5 + i_6)$ + (4 rt angles at O)= 6Triangle's angles

= (2 x 6) rt angles.... (1)

$(i_1 + i_2 + i_3 + i_4 + i_5 + i_6) + (1 + 2 + 3 + 4 + 5 + 6) = 2 \times 6$ rt angles

(Considering St angles at A, B, C, D, E.) .... (2)

Comparing (1) and (2)

We get 1 + 2 + 3 + 4 + 5 + 6 = 4 rt angles. (H.P.)

Altenative Proof

Suppose a person goes round the polygon starting from A. He goes along AB. At B he turns through $\angle 1$ and then he goes along BC. At C he turns through $\angle 2$. Then he goes along CD. At D he turns through $\angle 3$. Then he goes along DE. At E he turns through $\angle 4$ and proceeds along EF. At F he turns through $\angle 5$ and proceeds along FA. At A he turns through $\angle 6$ and reaches the same point where he started. He faces along AB.

He has turned through 4 rt. angles

$\therefore \angle 1 + \angle 2 + \angle 3 + \angle 4 + \angle 5 + \angle 6 = 4$ rt. angles (H.P.)

* **Example:-** The sides of a polygon are produced both ways till they meet each other and a star-shaped figure is formed. Prove that sum of the angles at the vertices of the figure = (2s–8) right angles.

Congruency of Triangles

Congruency means equality in all respects :

1. If a st. line AB is congruent to another line CD, they must be equal in length to each other.

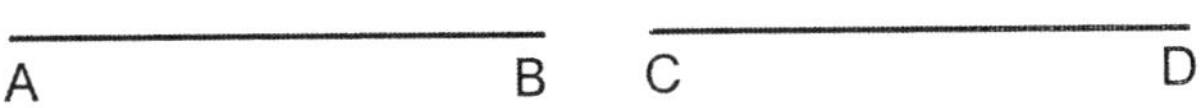

Suppose AB is congruent to CD. It means that if we place AB along CD. A must fall on C and B must fall on D. Since AB = CD. Therefore AB is Congruent to CD.

2. If measure of $\angle A$ = measure of $\angle B$ it means $\angle A$ is congruent $\angle B$. Suppose $\angle CAD$ is bodily lifted and placed on $\angle EBF$ in such a way that A must fall on B, AD along BF. Since $\angle A = \angle B$. AC must along BF and the 2 $\angle$s are congruent to each other.

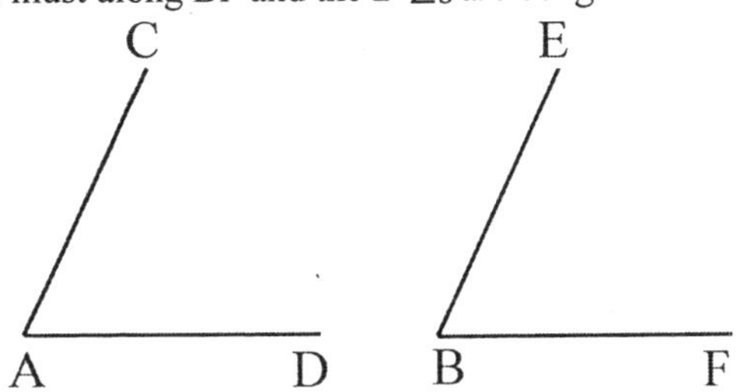

3. Let us now think of congruency of triangles. A triangle has 6 parameters - 3 sides and 3 angles. To draw a Δ we require 3 parameters.

   (1) Two sides and angle included in these sides.

   (2) One side and the 2 angles adjacent to these sides.

   (3) All the three sides.

In all the above cases we can draw a single Δ and there is no ambiguity.

4. Ambiguous Cases

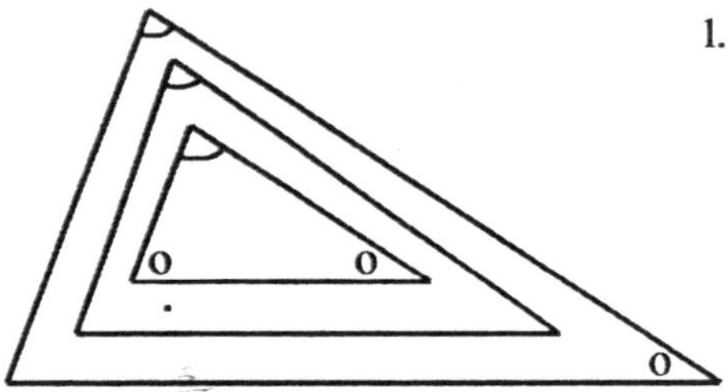

1. If we are given all the 3 angles, we cannot draw a single Δ.

   Because if we draw a Δ with 3 angles, we con draw another Δ' whose sides are parallel to the one drawn which we will have exacly the same angles. We can further draw another Δ whose sides are $||$ to the two Δs already drawn. Thus any no. of triangles can be drawn with identied $\angle$s and there cannot be congruency.

2. Similarly if a Δ has 2 sides and angle opposite the smaller of the 2 sides are given, there cannot be a single Δ.

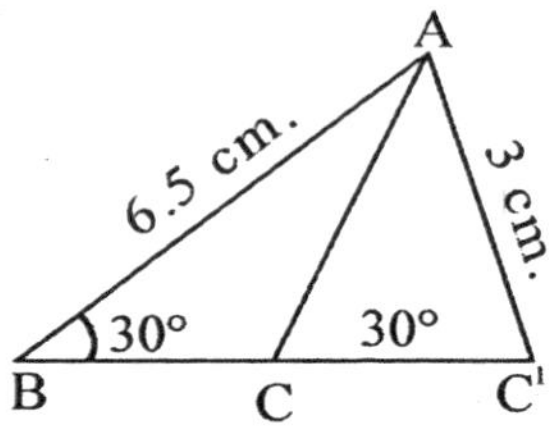

   Suppose we draw a Δ with AB = 6.5 cms. AC = 3 cms and $\angle B = 30°$. There are 2 positions of the Δ. We can draw a ΔABC as well a ΔABC'. Both satisfy the given condition.

   But for the above 2 cases we can draw a single Δ in the cases of 3 cases given in para 3 above.

Hence to draw a single Δ we require one of the following.

1. Two sides and angle included in them.
2. Side and 2 angles adjacent to the side.

3. All the 3 sides.

S.A.S. Congruency Axiom

Two triangles are congruent if 2 sides of the one are congruent to 2 sides of the other and the angle included in these sides are also congruent to each other.

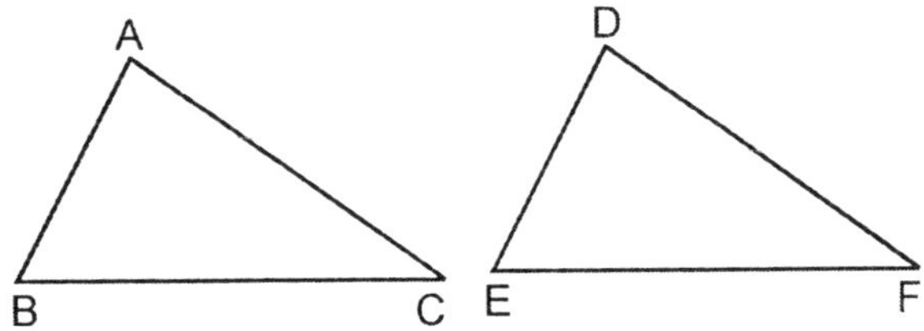

In the above 2 Δs, Δ ABC is congruent to Δ DEF if AB = DE and BC=EF and $\angle B = \angle E$.

Suppose Δ ABC is lifted and placed upon Δ DEF so that B falls on E and because BC = EF, C falls on F. Then because $\angle B = \angle E$ AB falls along DE and because AB = DE, A falls on D. Thus A and D coincides with Δ DEF.

That means Δ ABC = Δ DEF.

Theorem No. 10

Angles opposite equal sides of a Δ are equal.

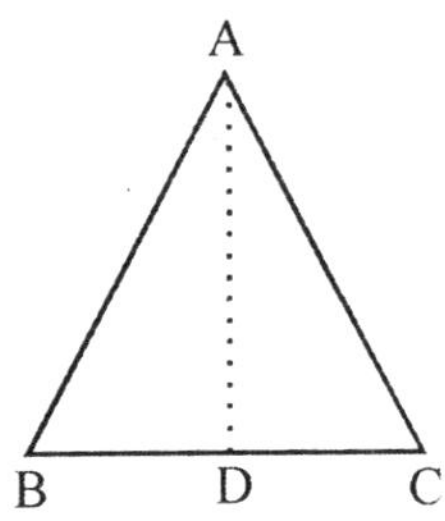

Given : ABC is a Δ in which AB = AC

To Prove : $\angle B = \angle C$

Construction : Draw AD bisector of $\angle A$

Proof : In Δs ABD and ACD

because $\begin{cases} AB = AD & \text{given} \\ \angle BAD = \angle CAD & \text{construction} \\ AD = AD & \text{common} \end{cases}$

$\therefore$ Δs are congruent (S.A.S Congruen Axiom)

$\therefore$ $\angle B = \angle C$ (H.P.)

Theorem No. 11

If 2 angles of one Δ are congruent to 2 angles of the other and the sides common to these sides are equal to one another, then the Δs are congruent. *(ASA)*

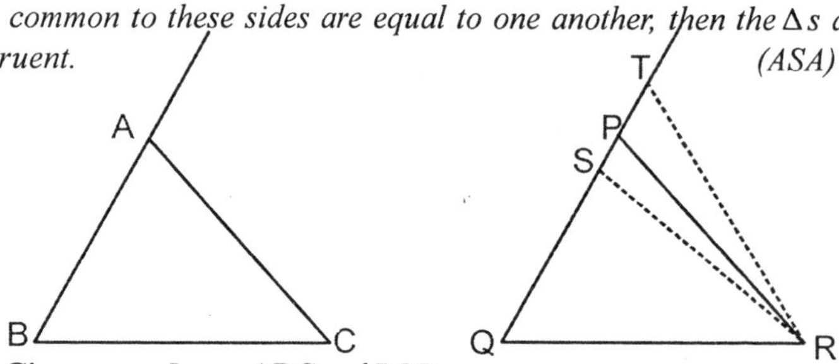

Given:- In Δs ABC and PQR

$\angle B = \angle Q$ and $\angle C = \angle R$

and BC = QR

To Prove :- $\Delta ABC \equiv \Delta PQR$

Proof:- There are 3 possibilities

(i) PQ > AB (ii) PQ < AB (iii) PQ = AB

We shall prove that (i) and (ii) are not true.

If PQ > AB then let AB = SQ

Join SR

In Δs ABC and SQR

because $\begin{cases} AB = SQ & \\ BC = QR & (\textit{given}) \\ \angle B = \angle Q & (\textit{given}) \end{cases}$

$\therefore$ $\Delta ABC \equiv \Delta SQR$ (SAS Axiom)

$\therefore$ $\angle ACB = \angle SRQ$ ....(1)

But we are given that

$$\angle ACB = \angle PRQ \qquad \text{....(2)}$$

From (1) and (2) we get

$$\angle SRQ = \angle PRQ$$

This means an angle = its part which is possible only if S coincides with P

$\therefore$ AB $\neq$ SQ

AB = PQ

Similarly it can be proved that PQ < AB is not true

If PQ < AB

Let AB = QT

Join TR

In Δs ABC and TQR

$$\text{because} \begin{cases} BA = QT & \\ \angle B = \angle Q & (given) \\ BC = QR & (given) \end{cases}$$

$\therefore$ Δ ABC $\equiv$ Δ TQR (SAS Axiom)

$\therefore$ $\angle$ ABC = $\angle$ TRQ

But we are gvien that

$$\angle ACB = \angle PRQ$$

$\therefore$ $\angle TRQ = \angle PRQ$

But this is not possible unless T conicides with P

$\therefore$ our supposition that

$\therefore$ QT = BA is not correct

QP = BA

$\therefore$ Δ PQR = Δ ABC (H.P.)

Corollary (A.A.S.)

If 2 $\angle$s and a non-included side of one Δ is equal to 2 $\angle$s and a non-included side of another Δ, the 2 Δs are congruent.

Any two $\angle$s of one Δ = 2 angles of another Δ. That means the third angle of both the Δs must be equal. It means 2 $\angle$s and included side of one Δ is equal to 2 $\angle$s and included side of the other triangle. Hence the Δs are congruent.

Theorem No. 12

(Converse of Theorem No. 10)

If 2 angles of a Δ are equal, then the siders opposite the equal angles are equal.

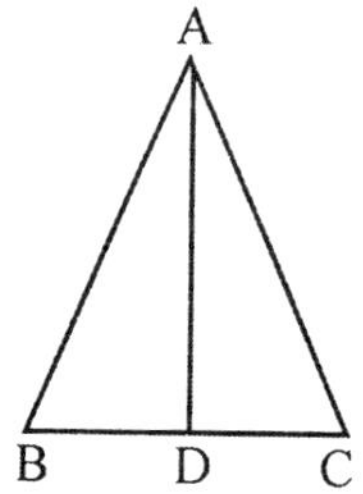

Given: A ΔABC in which $\angle B = \angle C$

To Prove: AB = AC

Construction : Lets AD be the bisector of $\angle BAC$

Proof: In Δ ABD and ACD

because $\begin{cases} AD = AD \\ \angle BAD = \angle CAD \\ \angle B = \angle C \end{cases}$

Δ ABD $\equiv$ ΔACD (AAS)

$\therefore$ AB = AC (H.P.)

Another Method to prove the above

Suppose AB $\neq$ AC But AB > AC

Cut off AD=AC from AB and join CD.

If in Δs ADC and ABC

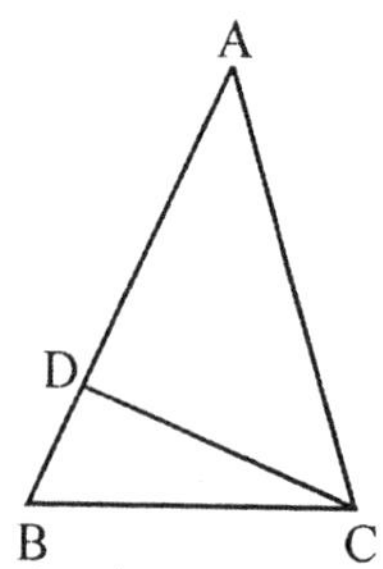

because $\begin{cases} AC = AC \\ AD = AD \\ \angle A \text{ is Common} \end{cases}$ *(Construction)*

$\therefore \quad \triangle ADC \equiv \triangle ABC$

But it cannot be as it is part of $\triangle ABC$

$\therefore$ Our supposition that $AB \neq AC$ made in the beginning must be wrong.

Similarly AB cannot be less than AC

$\therefore \quad$ AB is neither $> AC$ nor $< AC$

$\therefore \quad AB = AC.$

Note:- Here we employ an indirect method of proof frequenly used in geometry. It consists in showing that the theorem cannot be untrue since if it were it should lead to us some impossible conclusion. This form of proof is known as reductio ad- absurdom or proof by contradiction and is most commonly to demonstrate the converse of some foregoing theorem.

Theorem No. 13

If three sides of one $\triangle$ are equal to three sides of another $\triangle$, each to each, then the $\triangle$s are congruent.

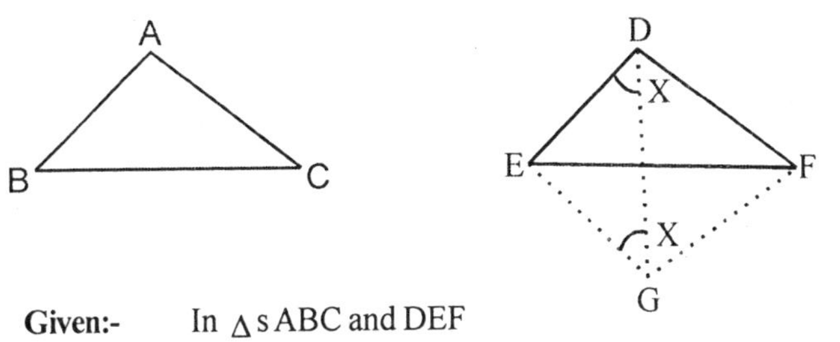

Given:- In $\triangle$s ABC and DEF

$BC = EF, AB = DE,$ and $AC = DF$

To Prove: $\triangle ABC \equiv \triangle DEF$

Proof: Apply $\triangle ABC$ to $\triangle DEF$, so that B falls on E and BC along EF and because $BC = EF$, C must fall an F and A falls on the opposite side of D.

Let G E F be the new position of $\triangle ABC$. Join D G..

Because $\quad ED = AB = EG$

$\therefore \quad \angle EDG = \angle EGD \quad$ (Th 10) ...(1)

and because $\quad AC = DF = FG$

$\therefore \quad \angle FDG = \angle FGD \quad$ (Th 10) ...(2)

Adding (1) and (2)

$$\angle EDF = \angle EGF$$

But $\angle EGF = \angle BAC$

$\therefore$ $\angle EDF = \angle BAC$

Now in Δs ABC and DEF

because $\begin{cases} AB = DE & (given) \\ \angle BAC = \angle EDF & (proved\ above) \\ AC = DF & \end{cases}$

$\therefore$ $\Delta ABC \equiv \Delta DEF$ SAS Axiom (H.P.)

Exercise (e - 1)

Example 1

Given:-	To Prove
(a) BD bisects $\angle ABC$	
AB = BC	$\angle DAM = \angle DCN$
(b) AB = BC	
AD = DC	$\angle ADB = \angle CDB$
(c) $\angle ABD = \angle CBD$	AD = DC
$\angle ADB = \angle CDB$	AB = BC

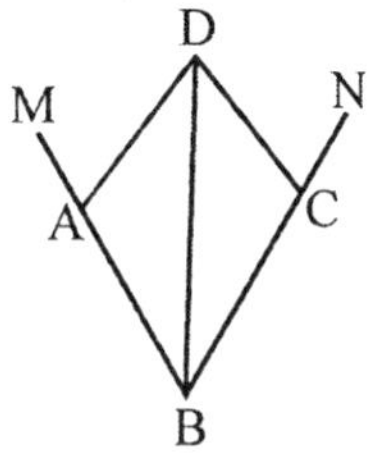

Example 2

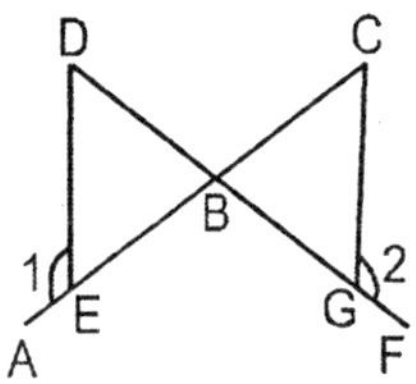

AC and DF are st lines

Given:-	To Prove
(a) BD = BC	ED = CG
EB = BG	
(b) BE = BG	$\angle D = \angle C$
$\angle BED = \angle BGC$	
(c) BE = BG	BD = BC
$\angle EDB = \angle GCB$	

(d) EB = BG	$\angle AED = \angle FGC$
BD = BC	
(e) DE = CG	$\angle BDE = \angle BCE$
$\angle E = \angle G$	

Example 3

Given:-	To Prove
(a) BE = BG	$\angle E = \angle G$
$\angle 3 = \angle 4$	$\angle 2 = \angle 5$
(b) $\angle 1 = \angle 2$	BE = BG
$\angle 3 = \angle 4$	
(c) EC = CG	BE = BG
$\angle 2 = \angle 5$	$\angle 3 = \angle 4$
(d) $\angle 3 = \angle 4$	$\angle E = \angle G$
$\angle 2 = \angle 5$	

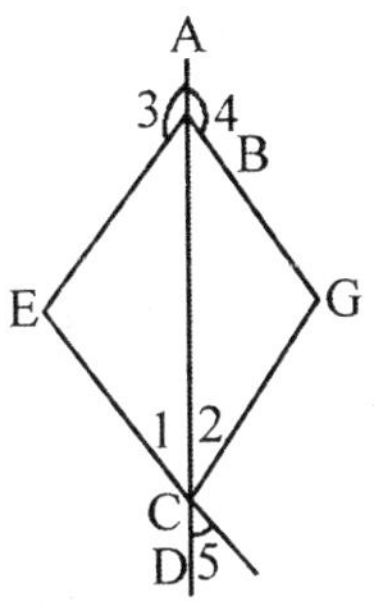

Example 4

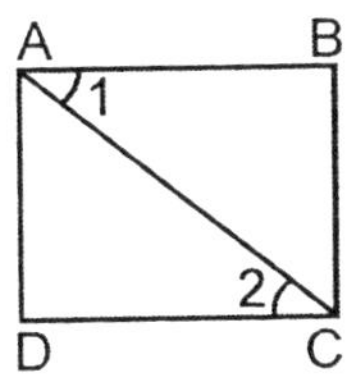

Given:-	To Prove
(a) $\angle 1 = \angle 2$	$\angle B = \angle D$
AD ⊥ AB	
BC ⊥ DC	
(b) $\angle 1 = \angle 2$	AD = BC
AB = CD	
(c) $\angle BAD = \angle BCD$	AB = DC
$\angle 1 = \angle 2$	

Example 5

ABCDE is a st. line

Given:-	**To Prove**
(a) AB = DC	AH = DG
$\angle ACH = \angle DBG$	
CH = BG	
(b) AB = CD	HC = BG
AH = GD	
$\angle 1 = \angle 3$	
(c) $\angle 3 = \angle 1$	
ABCDE and FDG are st. lines BG = CH	
CD = AB	
$\angle DBG = \angle ACH$	
(d) DG = AH	
$\angle G = \angle H$	CD = AB
$\angle 3 = \angle 1$	
(e) DG = AH	$\angle DBG = \angle ACH$
DC = AB	$\angle G = \angle H$
BG = CH	$\angle ABG = \angle DCH$

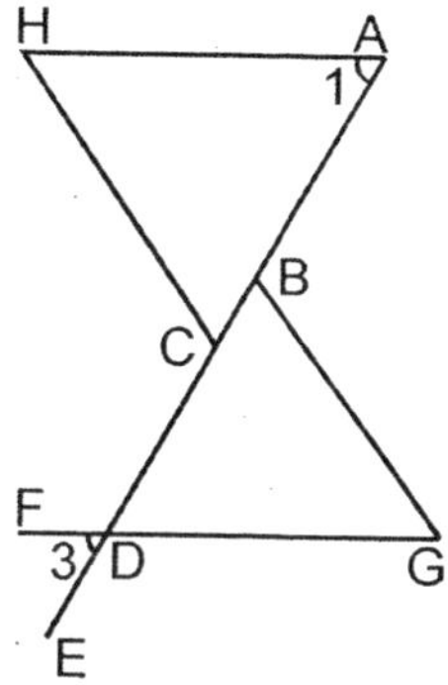

Example 6

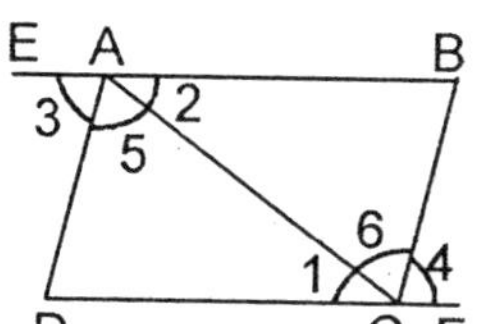

Given:-	**To Prove**
(a) $\angle EAC = \angle ACF$	AD = BC
$\angle 5 = \angle 6$	
(b) $\angle 3 = \angle 4$	AB = CD
$\angle 5 = \angle 6$	AD = BC
AB ‖ DC	
BE and FD are st. line.	

Example 7

A B C D E F is a st. line

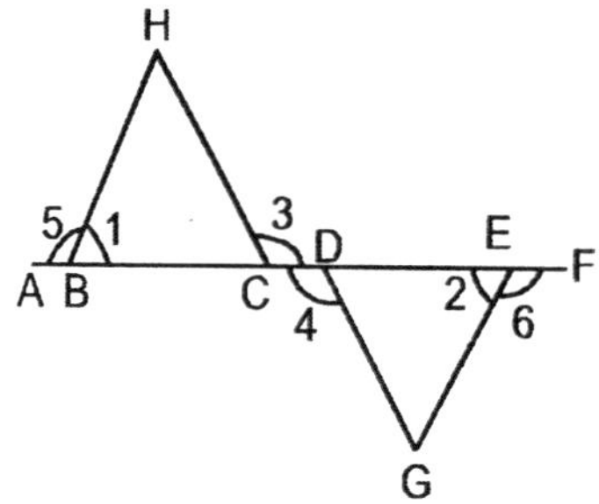

Given:-	To Prove
(a) $\angle 5 = \angle 6$	
$\angle 3 = \angle 4$	$\angle G = \angle H$
BD = CE	
(b) BD = CE	
HB = EG	HC = DG
$\angle 5 = \angle 6$	
(c) BD = CE	
HB = EG	$\angle 3 = \angle 4$
$\angle 5 = \angle 6$	

Example 8

E A B F is a st. line

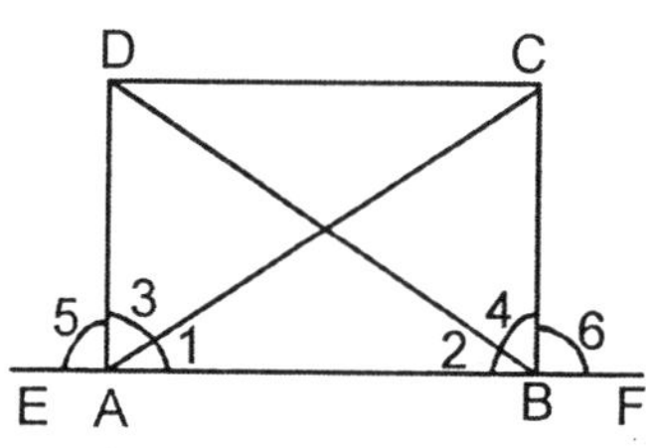

Given:-	To Prove
(a) DA $\perp$ EF	
CB $\perp$ EF	DB = AC
DA = CB	
(b) DA = CB	$\angle 1 = \angle 2$
$\angle 5 = \angle 6$	
(c) $\angle 5 = \angle 6$	AC = BD
$\angle 1 = \angle 2$	
(d) $\angle 5 = \angle 6$	AD = BC
$\angle 3 = \angle 4$	

Example 9

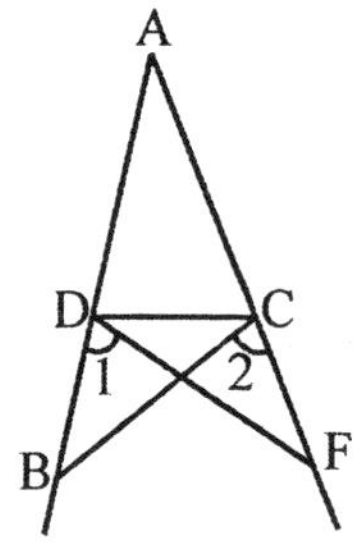

	Given:-	To Prove
(a)	AD = AC	DE = BC
	DB = CE	
(b)	AD = AC	DE = BC
	$\angle 1 = \angle 2$	
(c)	AB = AF	$\angle B = \angle F$
	BD = CF	
(d)	AB = AF	DB = CF
	AC = AD	

Example 10

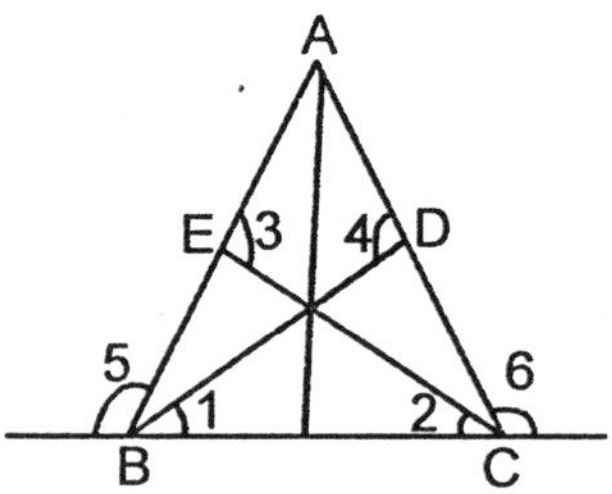

	Given:-	To Prove
(a)	$\angle 1 = \angle 2$	BE = CD
	$\angle 3 = \angle 4$	
(b)	$\angle ABC = \angle ACB$	BE = CD
	$\angle 1 = \angle 2$	
(c)	$\angle 5 = \angle 6$	BE = CD
	$\angle 1 = \angle 2$	
(d)	$\angle ABC = \angle ACB$	CE = BD
	$\angle 3 = \angle 4$	
(e)	$\angle 5 = \angle 6$	$\angle 1 = \angle 2$
	BE = CD	
	$\angle 3 = \angle 4$	
(f)	AB = AC	AE = AD
	$\angle 3 = \angle 4$	

Example 11

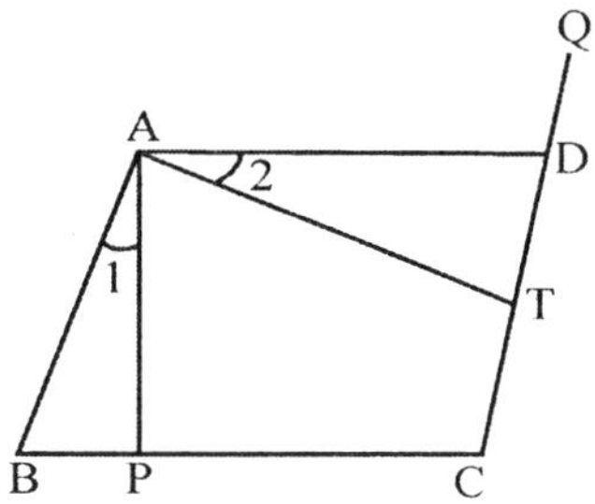

	Given:-	To Prove
(a)	AB = AT	BT = PD
	AP = AD	
	$\angle 1 = \angle 2$	
(b)	BP = DT	AP = AD
	AB = AT	$\angle APC = \angle ADQ$
	$\angle ABP = \angle ATD$	
(c)	AB = AP = AT = AD	$\angle APB = \angle ADT$; $\angle ABD = \angle ADB$
	$\angle 1 = \angle 2$	$\angle ABP = \angle ATD$; $\angle APT = \angle ADQ$

Exercise (e-2)

1. In an isosceles Δ the line joining the vertex to the midpoint of the base, bisects the angle at the vertex and is at right angles to the base, prove this.
2. Two isosceles Δs ABC and DBC are drawn on the same base BC but are on opposite sides. Show that AD bisects the base BC and is at right angles to it.
3. In the above example what will be the difference if the Δs are drawn on the same side.
4. In a hexagon opposite side are equal. If one pair of opposite angles is equal, prove that the other pairs of angles are also equal.
5. In an equilateral Δ, the points joining the mid points of the three sides form a Δ which is also equilateral.
6. If the bisector of the vertical angle meets the midpoints of the base prove that it is isosceles Δ.
7. Prove that Δ ABC will be isosceles if one of the following is true:

(a) Altitude bisects the base.

(b) Median AD is $\perp$ to the base.

(c) Bisector of exterior $\angle A$ is $\perp$ to the base.

(d) Altitude AD is bisector of $\angle A$ at the vertex.

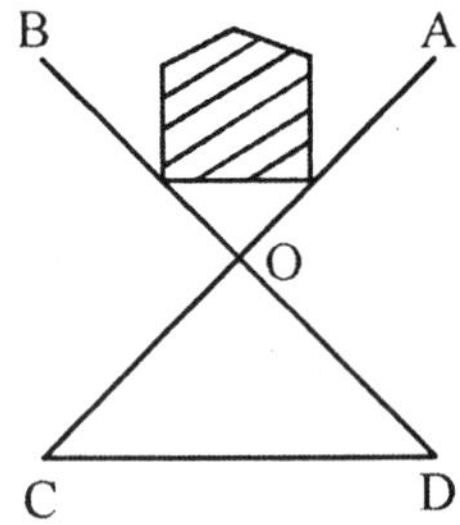

8. A man wishes to determine the distance between 2 objects A and B but there an obstacle between these objects which prevents from measuring the distance directly. He therefore devices an ingenious method to over come the difficulty. First he fixes a pole at a convenient point O from where both A and B are visible. Then he fixes another pole at a point D from where he can see both B and A and measures BO = OD.

In a similar manner he measures CO = OA and fixes a pole at a point C. Then he measures the distance CD, which is equal to the distance between AB.

9. In a similar manner a man wants to find out the breadth of a river. First he locates a tree on the bank of the river opposite to where he is. Then he fixes a stone opposite the tree and walks a certain distance AB along the river and fixes another stone at B and walks a distance BC along the river equal to AB. Then he walks ⊥ to the river and walks upto the point D so that he sees the tree on the other side and the stone B in the same straight line. Prove that he walks a distance = breadth of the river.

10. A B C D is a quadrilateral with all sides equal. Prove that AC and BD bisect each other at rightangles.

11. A B C D is a quadrilateral whose sides are all equal. E is a midpoint on AC. BE and ED are joined. Show that BED is a straight line.

12. In the adjoining figure AC = BC

$\angle DCA = \angle ECB$ and

$\angle DBC = \angle EAC$

Prove that EC = DC

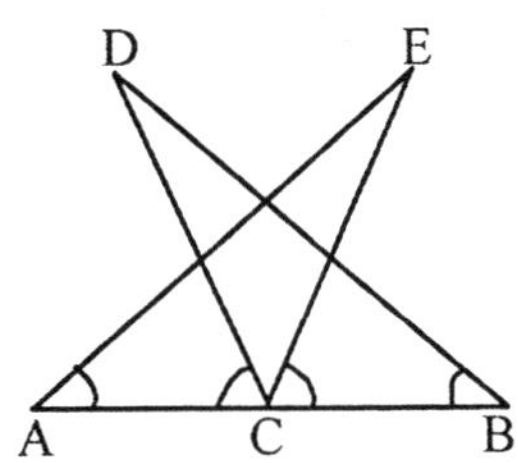

13. On one base are drawn two congruent triangles on opposite side of the base. Show that the line joining their vertices is either at right angles to the base or passes through the centre of the base.

14. On one base are drawn two Δs which are congruent to each other. They are on the same side of the base. Show that either the Δs are coincident or the line joining their vertices is parallel to the base.

Practical Geometry I

(Basic Constructions)

The figures that you have drawn so far were for proving a theorem or solving an example. They were drawn by hand with a pencil and at most by a scale. They were not accurate figures and could aften lead to erroneous results. On the other hand figures drawn with the help of scale and compass box and with actual measurement would be more accurate figures. A compass box contains in addition to scale and compass, a divider, a protractor and setsquares. The protractor is for measuring angles, while the set square are two right angled Δs with angle 45° - 45° and 30° - 60° respeclivsly. The divider is for comparing and dividing sides. In addition a good erasure and two sharpend pencils make up the whole compass box.

When the compass box is ready, you are also ready to draw the diagrams. First the simplest constrution is to draw a straight line. This can be drawn with the scale and pencil.

Construction No. 1

To draw the perpendicular bisector of a straight line of given length

Given:- A line AB of any length

To draw :- The perpendicular bisector of the line

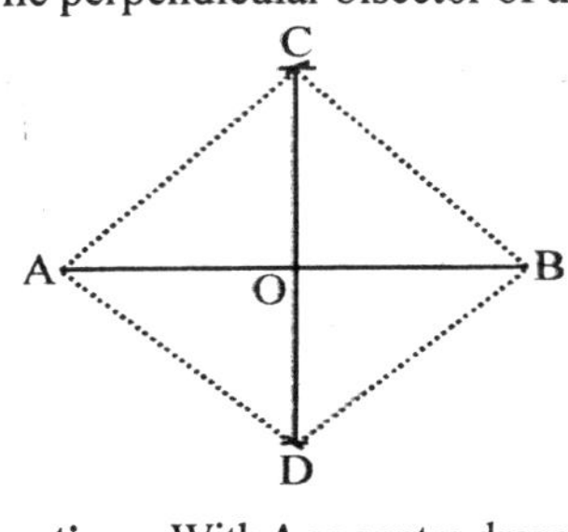

Steps of Construction:- With A as centre draw 2 arcs of circles on both sides of the line AB. The distance between the legs of compass should be more than half the line AB. Then with B as centre draw 2 more arcs intersecting the arcs drawn with A as centre at C and D. Join CD. It meets AB at O. O is the point of division.

Proof:- (1) Prove $\Delta ACD \equiv \Delta BCD$

(2) This gives $\angle ACD = \angle BCD$

(3) Then prove $\Delta ACO \equiv \Delta BCO$

(4) That given AO = BO and $\angle AOC = \angle BOC$

(5) That means $\angle AOC = 90°$

Construction No. 2

To Bisect an angle

Given:- An angle BOA

To draw :- A line bisecting the angle BOA.

Steps of Construction:- With O as centre draw an are of circle cutting the two arms of the angle at C and D. Then taking a distance more than $\frac{1}{2}$ CD in the compass draw 2 arcs of circles with C and D as centres cutting each other in point E. Join OE which is the required bisector.

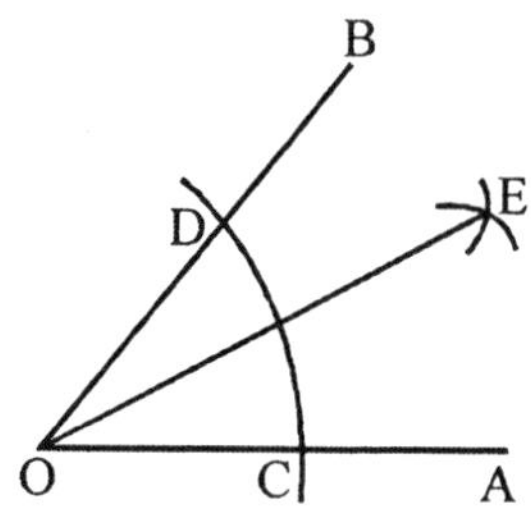

Proof:- Prove $\Delta ODE = \Delta OAE$

Construction No. 3

To draw a perpendicular to a line at a given point in it.

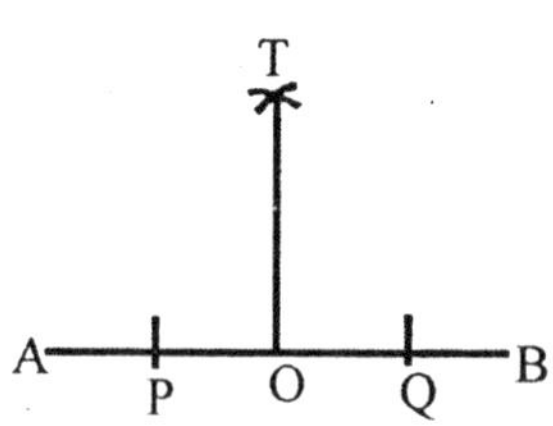

Given:- A line AB with a point in it.

To draw :- A perpendicular to AB at O.

Steps of Construction:- With O as centre and any distance in the compass as radius mark 2 points P and Q on AB so that OP = OQ. Then taking more distance in the compass than OP draw 2 arcs of circles on one side of AB cutting each other at T. Join OT. That is the perpendicular.

Construction No. 4

To draw a perpendicular to a line from a point outside the line.

Given:- A line AB and a point P out side it.

To draw :- A perpendicular from P to AB.

Steps of Construction:- With P as centre draw an arc of a circle cutting the line at C and D. With C and D as centres and more than $\frac{1}{2}$CD as radius draw 2 arcs of circles cutting each other at T opposite AB. Join PT. is the required perpendicular to AB.

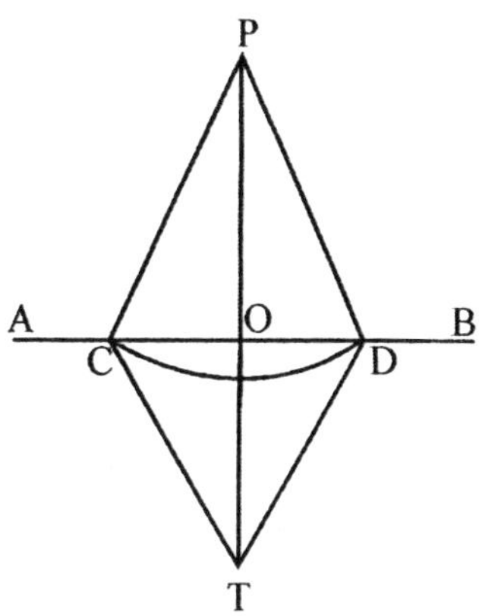

Proof:- First prove $\Delta PCT \equiv \Delta PDT$

$\therefore$ $\angle CPT = \angle DPT$

Now Prove $\Delta POC \equiv \Delta PDT$

But $\angle POC + \angle POD = 180°$

$\therefore$ $\angle POD = \angle POC = 90°$

Construction No. 5

To Construct an angle equal to a given angle

Given:- A angle BAC.

To Construct :- An angle QPR at PR=angle BAC

Steps of Construction:- With centre A draw an arc of a circle

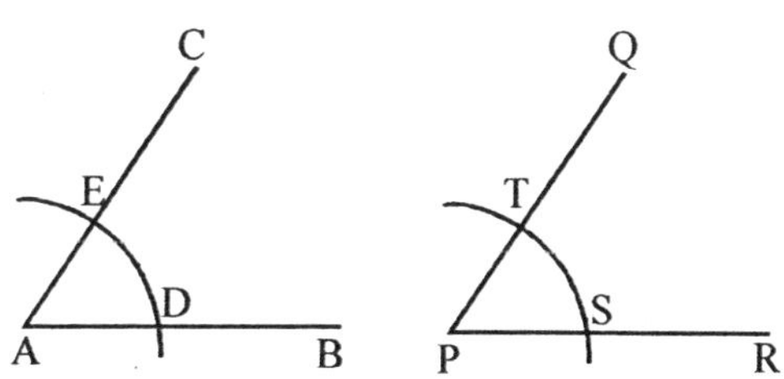

cutting the line AB at 'D' and AC at E. With the same distance in the compass draw another arc with P as centre cut PR at S. From this are cut off a distance ST=DE. Join T to P. PTS is the angles.

Proof:- Prove $\Delta EAD \equiv \Delta TPS$

Construction No. 6

To draw a line parallel to a given line at a point outside it.

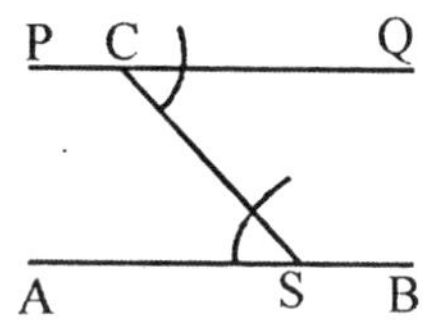

Given:- A line AB and a point C outside AB.

To draw :- A line PQ through C parallel to AB.

Steps of Construction:- Draw any line CS to any point S in AB. At C make $\angle QCS = \angle CSA$. Join QC and extend to P. PQ is the line.

Proof:- $\angle QCS = \angle CSA$

but these are alternate $\angle s$

$\therefore$ PQ || AB.

Construction No. 7

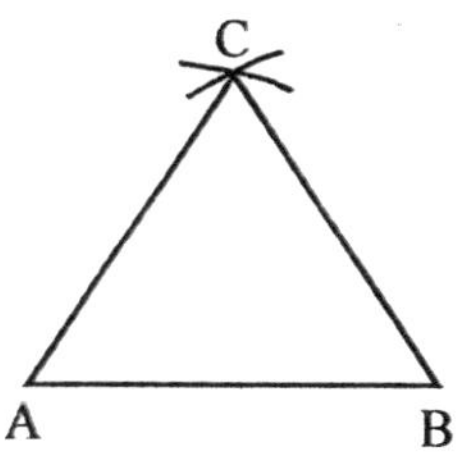

To construct an angle of 60°.

Given:- A segment of a line AB.

To draw :- An angle of 60° with AB.

Steps of Construction:- With A as centre and radius AB draw an arc of circle. Then with B as centre and the same radius draw another arc of circle cutting the earlier arc at C. Join A to C and B to C.

$\angle ACB = \angle CBA = \angle ACB = 60°$

Construction No. 8

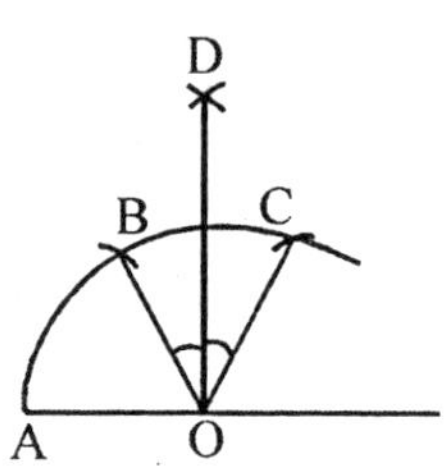

To construct an angle of 90° on a line.

Steps of Construction:- Draw OA any line. With O as centre and OA as radius draw a semi circle. Then with A as centre and the same distance in the legs of the compass, Draw an arc of circle cutting the semicircle at B. Then with B as centre and the same distance in the compass draw another arc of circle cutting the semi circle at C.

$\angle BOC$ =60°. Bisect BOC by a line OD

OD is the required perpendicular to OA

$\angle AOB = 60^\circ \quad \angle BOD = 30^\circ$

$\therefore \quad \angle AOB + \angle BOD = 90^\circ$

$\therefore \quad \angle AOD = 90^\circ$

Construction No. 9

To divide a line into a no. of equal parts. *(Say 4)*

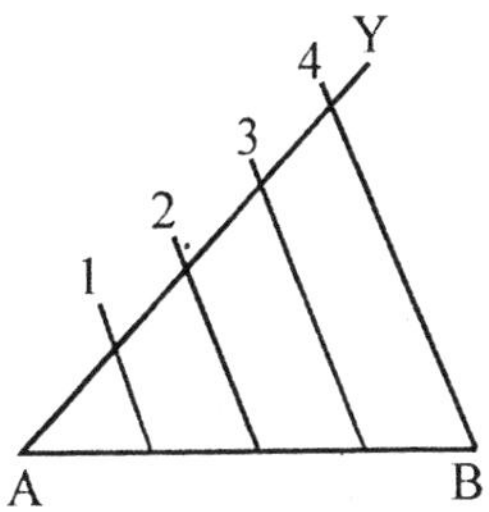

Given:- A line AB.

To draw :- Line AB into 4 equal parts.

Steps of Construction:- Draw any line AY making an acute angle with AB. This line is divided in 4 convenient equal parts. Join the last part to B. To this line draw parallel lines from intermediate parts i.e. parts 3, parts 2, parts 1 as per construction No. 6.

The line AB is divided into 4 equal parts.

Proof:- If parallel lines make equal intercepts on one line, they make equal intercepts on another also.

Construction No. 10

(i) To draw a Δ with sides a, b, c.

First draw side 'b' then in compass take a distance with side 'a' and draw an arc of circle. Then take a distance with side 'c' and draw an arc of a circle cutting the arc drawn earlie. Then join the ends of side 'b' with the point where the arc interscet.

(ii) To draw a Δ with side BC and $\angle B$ and $\angle C$.

Side BC is drawn first and then with protractor make $\angle B$ and then at the other end make $\angle C$.

The above 10 constructions are basic for all the problems in geometry.

Exercise (f)

1. Draw a ΔABC in which AB = 2.6 cms., BC = 2.6 cms and CA = 4.2 cms. Measure the $\angle$s of the Δ. Add them. This must be 180°.

2. In a Δ ABC, AB = 7.5 cms, BC = 7.00 cms, CA = 6.5 cms. Draw the Δ. From A drop AD $\perp$ BC. Measure it.

3. Draw any Δ. Draw the 3 bisectors of the three sides. They should be concurrent.

4. Draw any Δ. Bisect $\angle B$ and $\angle C$. Let them meet at D. Join AD. It will bisecet $\angle A$ verify the same.

Definition: A line drawn from any of the three vertices to the middle point of opposite side is called the median of the Δ. There are 3 medians of any Δ.

5. Draw any Δ. Draw the three medians. They must be concurrent.

6. Draw any Δ. Then drop perpendiculars from each vertex to the opposite side and verify that they are concurrent.

7. Draw any Δ at random then produce AB and AC. Bisect the exterior $\angle s$ so formed. Let them meet at D. Verify that AD bisects $\angle A$.

8. Draw a ΔABC where a = 3.9 cms b = 2.5 and c = 2.75 cms. Bisect $\angle B$. Let the bisector meet AC in point D. Measure AD and CD.

   Find the ratio $\dfrac{AD}{CD}$.

9. Draw an angle of $22\frac{1}{2}^\circ$ and another of $67\frac{1}{2}^\circ$.

Definition: Angle of Sun's elevation = Angle made by the Sun to the horizontal.

10. The sun's elevation = 30°. A man's shadow is 8 meters. What is the height of the man.

11. Two sides of a Δ in the shape of a field is 300 m x 300 m. The angle between is 45°. What is the measure of the third side.

12. Draw a ΔABC. When $\angle B = 30^\circ$, AB = 5.00 cms, AC = 4.00 cms. How many Δs can be drawn?

Sides of Triangles

Theorem No. 14

If one side of a Δ is longer than the other then the longer side has the greater angle opposite to it.

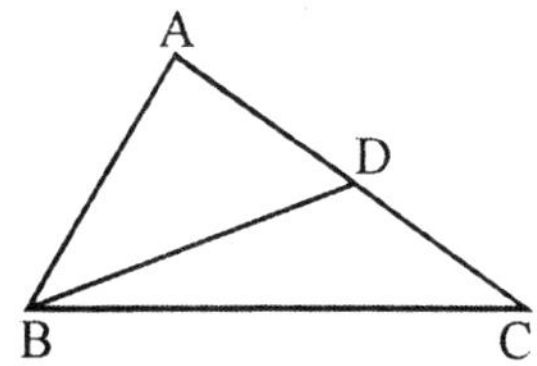

Given: A triangle ABC in which AC>AB.

To Prove: $\angle B > \angle C$.

Construction: From AC cut off AD = AB.

Proof: Because AD = AB

$\angle ABD = \angle ADB$ (Th. 10)

$\angle ADB = \angle DBC + \angle DCB$ (cor 1 Th. 7)

$\angle ADB > \angle DCB$

$\therefore$ $\angle ABD > \angle DCB$

But $\angle ABC > \angle ABD$

$\angle ABC > \angle DCB$

$\angle B < \angle C$ (H.P.)

Theorem No. 15

If one angle of a Δ is larger than another, then the side opposite the larger angle is longer than the side opposite the smaller.

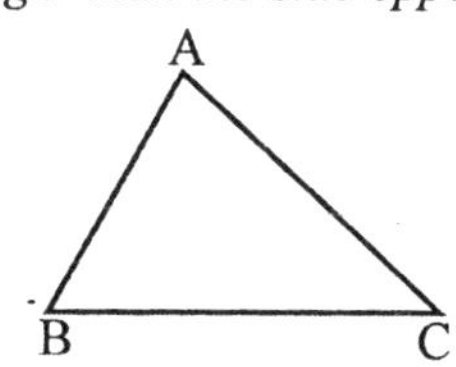

Given: A ΔABC in which $\angle B > \angle C$

To Prove: AC > AB.

Proof: Either AB > AC

or AB = AC or else AB < AC.

if AB > AC

then $\angle C > \angle B$

But this is not true as we are given that $\angle B > \angle C$

if AB = AC

$\angle B = \angle C$

This is also not true as we are given that $\angle B > \angle C$

Hence the third possibility must be true

$\therefore$ AC > AB (H.P.)

The method of proof used in this theorem is known as the method of Exhaution. It is used in cases where in one supposition is necessarily true when other supposition are untrue.

Theorem No. 16

The sum of lengths of any two sides of a Δ is more than the third side.

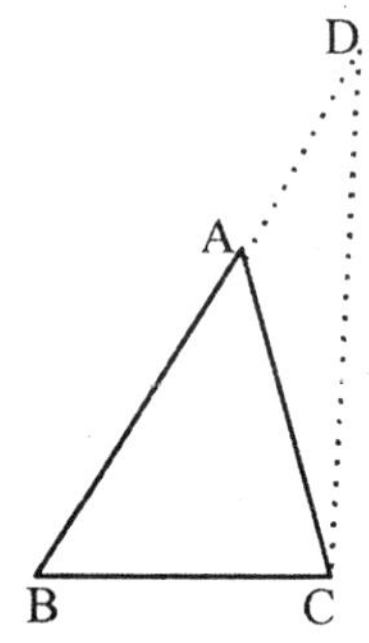

Given: A ΔABC

To Prove: AB + AC > BC

Construction: Produce BA to D so that AC = AD. Join DC.

Proof: AC = AD

$\therefore$ $\angle ACD = \angle ADC$

But $\angle BCD > \angle ACD$

$\therefore$ $\angle BCD > \angle ADC$

$\therefore$ BD > BC

$\therefore$ BA + AD > BC

BA + AC > BC (H.P.)

Theorem No. 17

Of all the lines from a points outside a line, the perpendicular is the shortest.

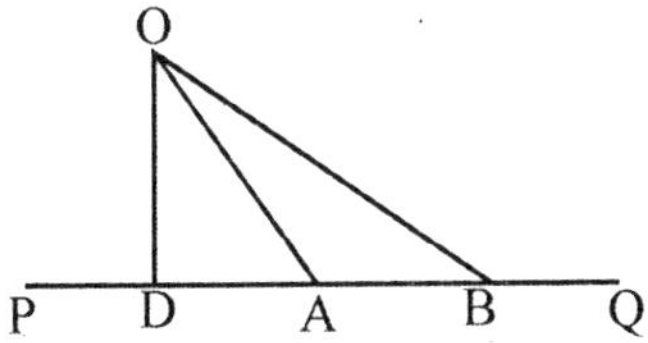

Given: PQ is a line and O is any points outside PQ.

$OD \perp PQ$, OA, OB are other lines

To Prove: OD is the shortest of all lines like OA, OB etc.

Proof: OD is perpendicular to PQ. OA, OB are other lines inclined to PQ at various acute angles.

Let us consider one such line say OA. In Δ ODA

$\angle D = 90° \; \angle OAD < 90°$

$\therefore$ OAD is acute angle

$\therefore$ OA > OD

All other lines like OA, OB are longer

OB > OD

$\therefore$ OD is shortest.

Definition:- D is called the foot of perpendicular OD.

Exercise (g)

1. Bisect the angle at the vertex and prove that the two sides of a triangle are greater than the third side.
2. Prove that any side of a Δ is greater than the difference of the other 2 sides.
3. In a ΔABC, AB > AC. If BD and CD are the bisectors of $\angle B$ and $\angle C$. Show that BD > CD.

4. If the sides AB and AC of a ΔABC are produced to D and E respectively and if $\angle DBC > \angle ECB$. Show that AB > AC.

5. In a ΔPQR, PQ = PR. S is a point on QR produced. Show that SQ> PQ or PR.

6. In a ΔABC, AB > AC and a point D is taken on the side BC. Show that AB > AD.

7. In a rt. angled Δ show that hypotsenuse is the longest side.

8. In a ΔPQR, PS is the bisector of $\angle P$. Show that if PR > PQ, $\angle PSR > \angle PSQ$.

9. In a quadrilateral (figure with 4 sides) the sum of the 2 diagonals is greater than the semi perimeter but less than the perimeter.

10. Prove that the sum of all the three altitudes of a Δ is less than the sum of the 3 sides.

11. A points D is taken in the interior of the ΔABC. Show that AB + AC > DB + DC.

12. Show that if a point D is taken in the interior of the ΔABC then,

$$AB + BC + CA > AD + BD + CD$$

and $$AD + BD + CD > AB + BC + CA$$

13. In a quadrilateral ABCD (figure with 4 sides,) AB is the longest side and CD is the shortest side then prove that each angle made with the shortest side is more than the opposite angle.

14. In a quadrilateral the distance of any point from the vertices is more than sum of the diagonal with one exception. Point out the exception.

15. In a quadrilateral prove that the sum of the diagonals is more than the semiperimeter but less than the perimeter of the quad.

16. In a Δ the semi perimeter is always greater than one side.

17. In a Δ ABC, AD is the median. Show that AB + AC > 2AD

    Hence show that the 3 medians are together less than the pericmeter.

18. In an equilateral ΔABC, point P is taken. Prove that you can make a Δ with PA, PB and PC.

19. In a Δ ABC side BA is produced. The external angle so formed is bisected and point P is taken on the bisector. Prove that (PB+PC) is greater than (AB+AC).

20. In a Δ ABC, AB > AC. A points P is taken on the bisector of $\angle A$ prove that AB - AC > PB - PC.

21. In a Δ ABC, AB < AC; BD = DC show that AD is located between the bisector of $\angle A$ and AB or AC whichever is greater.

22. In a quadrilateral ABCD. AC and BD intersect in O show that

    $AB + BC + CD + DA > AC + BD < 2(AC + BD)$.

23. In a Δ ABC, A is the maximum angle then you cannot draw of a Δ with AB, AC and 2 BC as sides.

24. In a Δ the sum of 3 medians is smaller than the perimeter of the Δ.

Theorem No. 18

If in 2 Right angled Δs the diagonal and one side of one Δ is respectively equal to the diagonal and a side of the other, the Δs are congruent.

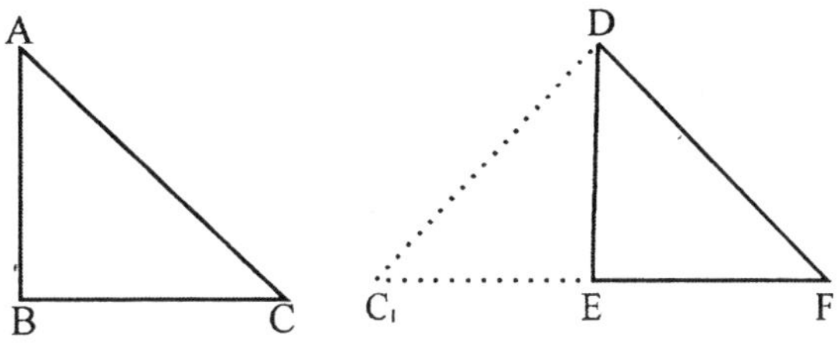

Given: In two right angled Δs ABC and DEF

$\angle B = \angle E = 90°$

AC = DF

AB = DE

To Prove: $\Delta ABC \equiv \Delta DEF$

Construction: Keep ΔABC along ΔDEF so that AB falls along DE and as AB = DE, A falls on D and B falls on E and the ΔABC falls on opposite side to ΔDEF, so that ΔABC falls along DEC_1

Proof: $\angle DEC' = \angle ABC = 90^\circ$

$\angle DEC' + \angle DEF = 90^\circ + 90^\circ = 180^\circ$

$\therefore$ C^1EF is a st line ... (2)

$\therefore$ DC^1F can be considered to be a Δ.

$DC^1 = AC = DF$

$\therefore$ $\angle DC'E = \angle DFE$

and $\angle DC'E = \angle ACB$

$\therefore$ $\angle ACB = \angle DFE$ ... (1)

Now in Δs ABC and DEF

because $\left\{\begin{array}{l} AC = DF \\ \angle ABC = \angle DEF \\ \angle ACB = \angle DFE \end{array}\right.$ (both 90°)

$\therefore$ These Δs are congruent. (H.P.)

Corollary: In 2 obtuse angled Δs if the obtuse angle, and the side opposite the obtuse angle and one of the sides that enclose the obtuse angle are equal the Δs are congru ent.

Theorem No. 19

If in Δs two sides of the one are respectively equal to two sides of the other but the angle included in these two sides are unequal then the third side of the Δ in which the angle is greater is longer than the third side in the other Δ.

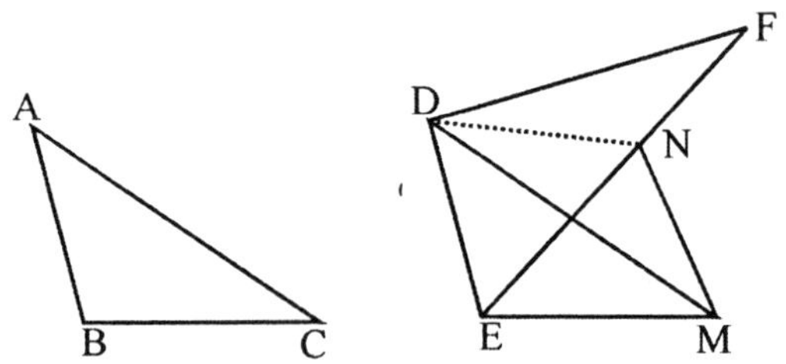

Given: In ΔABC and ΔDEF

AB= DE and AC = DF

But $\angle EDF > \angle BAC$

To prove: EF > BC

Construction: Make $\angle EDM = \angle BAC$

and Make DM = AC

Join EM

Bisect $\angle MDF$ and $\angle MDN = \angle FDN$

Let the bisector DN meet EF in N. Join NM

Proof: In Δs BAC and EDM

$$\text{because} \begin{cases} AB = DE & (given) \\ AC = DM & (const.) \\ \angle BAC \doteq \angle EDM & (const.) \end{cases}$$

These Δs are congruent (SAS Axiom)

$\therefore$ BC = EM ...(1)

Now in Δs MDN and FDN

$$\text{because} \begin{cases} DM = DN & (const) \\ \angle MDN = \angle FDN & (const.) \\ DN \text{ is Common Side} \end{cases}$$

$\therefore$ These Δ s are congruent

$\therefore$ MN = NF ...(2)

In Δ ENM

EN + NM > EM (Th. 16)

EN + NF > EM > BC using (1) and (2)

EF > BC (H.P.)

Corollary :- In 2 Δ s, two sides of the one are equal to 2 sides of the other and the third side of the one is longer the third side of the other then the angle opposite the third side in which it is more is larger than the corresponding angle of the other.

(This is converse of the above theorem)

Quadrilaterals

Quadrilateral is already defined. It is a union of four sides neither of which is parallel nor equal to each other. None of the angles are equal to each other. In short form it is written as quad.

Definitions

1. *A quad.* is called a *Parallelogram* if both pairs of its opposite sides are parallel

(1) Parallelogram

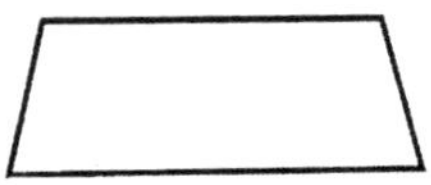

(2) Trapezium

2. *A quad.* is called a *Trapezuim* if only pair of opposite sides is parallel.

3. *A quad.* is called a *Rhombus* when its four sides are equal.

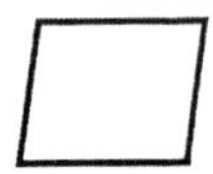

(3) Rhombus

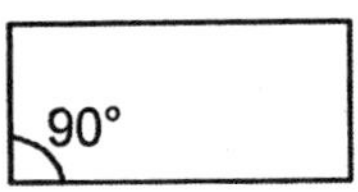

(4) Rectangle

4. If one angle of parallelogram is 90° it is called a *Rectangle.*

5. If the adjacent sides of a rectangle are equal it is called a *square.*

(5) Square

Theorem No. 20

In a parallelogram

(a) opposite sides are equal,

(b) opposite $\angle s$ *are equal,*

(c) the diagonal bisect each other but not at right angles.

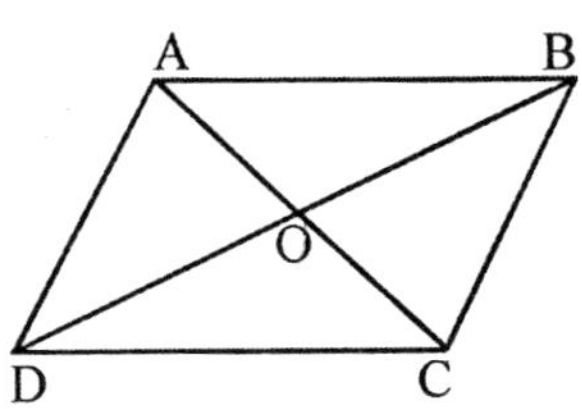

Given: ABCD is a parallelogram

AB || CD and AD || BC

(a) To prove: AB = CD, AD = BC

Construction: Join BD

Proof: (1) AB || DC and DB is the transversal

$\therefore \quad \angle ABD = \angle BDC$

(2) AD || BC and DB is the transversal

$\therefore \quad \angle ADB = \angle DBC$

Now in Δs ABD and CBD

because $\begin{cases} \angle ABD = \angle BDC \\ \angle ADB = \angle DBC \\ BD \text{ is common side} \end{cases}$

$\therefore$ These Δs are congrunt (A.S.A.)

$\left.\begin{array}{ll} & AB = DC \\ \text{and} & AD = BC \\ \text{But} & AB \neq BC \end{array}\right\}$ (H.P.)

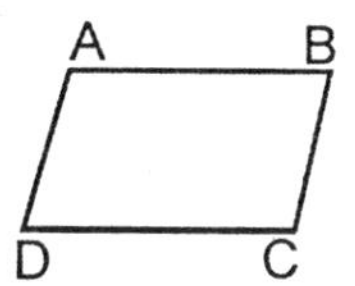

(b) To prove: $\angle A = \angle C, \angle B = \angle D$

Because AB || DC and AD is the transversal

$\angle A + \angle D = 180°$ (Th 5) ...(1)

Similarly AD || BC and DC is the trensversal

$\angle C + \angle D = 180°$ (Th 5) ...(2)

On comparing (1) and (2) we get

$\angle A = \angle C$

Similarly AB || DC and BC is the transversal

∴ $\angle B + \angle C = 180°$ (Th 5) ... (3)

AD || CB and DC is the transversal

∴ $\angle D + \angle C = 180°$ (Th 5) ... (4)

On comparing (3) and (4) we get

$\angle B = \angle D$ (H.P.)

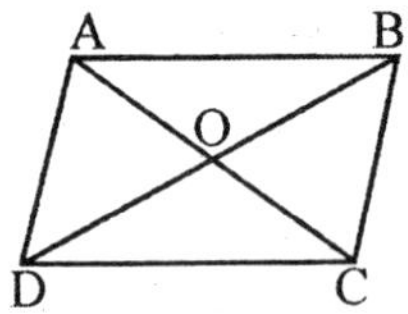

(c) To prove: AO = CO and DO = BO and $\angle AOB \neq 90°$

Proof:- AD || BC and AC is the transversal

$\angle DAO = \angle BCO$ (Th 5)

Similarly AD || BC and DB is the transversal

$\angle ADO = \angle CBO$

and AD = BC (Proved above)

Therefore Δ ADO ≡ Δ BCO

∴ AO = OC

and BO = DO

In Δs ABO and CBO

$$\begin{cases} AO = OC \\ BO \text{ is the common side} \\ But\ AB > BC \end{cases}$$ *see (1) above*

$\angle BOA > \angle BOC$

But these are together ≡ 180°

∴ $\angle BOA > 90°$ and $\angle BOC < 90°$

The diogonals bisect each other but not at 90° (H.P.)

Exercise (h)

1. Prove the converse of Th. 20 i.e. if in a quad.

   (a) If the opposite sides are equal

or (b) If the opposite $\angle s$ are eqnal

or (c) If the diagonals bisect each other but not at right angle then it is a parallelogram.

2. ABCD is a parallelogram. AX and CY bisect $\angle s$ A and C respectively, X and Y being points on DC and AB. Show AX $\parallel$ CY.

3. In a parallelogram the line joining the middle points of opposite sides is parallel to the parallel sides and equal to them.

4. Prove that each diagonal divider the parallelogram into 2 identical Δ s.

5. Through the vertices of a Δ parallel sides are drawn to the opposite sides. Show that resulting Δ is equal 4 times the original Δ.

6. AN and CP are perpendiculars drawn to the diagonal BD of a parallelogram from A and C respectvely. Prove that

   (i) $\Delta ADN \equiv \Delta CPB$

   (ii) AN = CP

7. Prove that if the diagonal of a parallelogram bisects one of the angles, it bisects the other angles also.

8. ABCD is a parallelogram and X and Y are the midpts of the side AB and DC respectively. AY and BX are joined and they intersect in P. CX and BY are joined. They intersect in Q. Prove that PXQY is a parallelogram.

9. ABCD is a parallelogram and X and Y are points on the diagonal BD such that DX = BY. Prove that AXCY is a parallelogram.

10. That parallelogram which has equal diagonals is a rectangle.

11. That parallelogram which has its diagonals at right angles is a rhombus.

12. If all the four angles of a parallelogram are bisected it forms a rectangle.

13. If a line is drawn from the centre of a parallelogram then

   (i) The centre of the parallelogram is its centre.

(ii) It divides the parallelogram into 2 equal parts.

14. BD and CE are medians of a ΔABC. BD and CE are produced to F and G respectively so that BD = DF and CE = EG. Show that F, G, A are collinear.

15. In a trapezium if the inclined sides are equal then

(i) Two pairs of equal angles.

(ii) The angles made by the inclined sides with the base are equal.

(iii) The line joining the mid pts of the parallel lines is perpendicular to the parallel lines.

Theorem No. 21

Two lines segments which join the terminal ends of two equal and parallel lines are themselves equal and parallel.

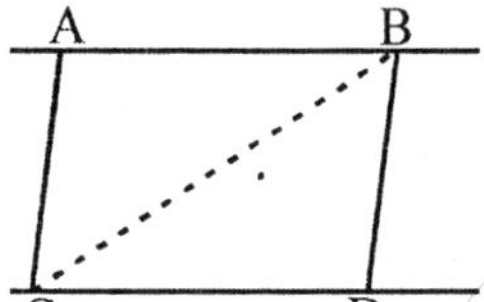

Given: AB and CD are two equal and parallel lines.

To Prove: $AC \parallel BD$ and $AC = BD$.

Construction: Join BC.

Proof: $AB \parallel CD$ and BC is a transversal

$\therefore$ $\angle ABC = \angle BCD$

In Δs ABC and DBC

because $\begin{cases} AB = CD & \text{(given)} \\ \angle ABC = \angle ABD & \text{(Proved above)} \\ BC & \text{is common} \end{cases}$

$\therefore$ $\Delta ABC \equiv \Delta DBC$

$\therefore$ $\angle ACB = \angle DBC$

$\therefore$ $AC \parallel DB$ (Th 6)

and $AC = BD$ (H.P.)

Theorem No. 22

The diagonals of a rhsombus are perpendicular to each other.

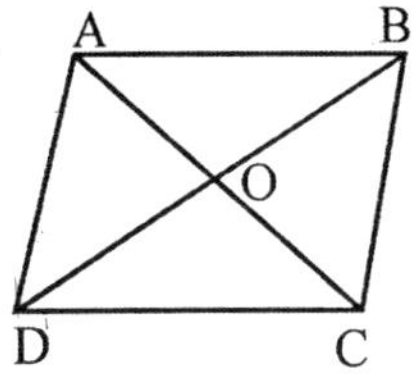

Given: AB = BC = CD = DA

Proof: In Δs ADC and ABC

because $\begin{cases} AD = AB & (given) \\ DC = BC & (given) \\ AC & is\ common \end{cases}$

The Δs are congruent

$\therefore$ $\angle DAO = \angle BAO$

Now in Δs DAO and BAO

because $\begin{cases} DA = AB & (given) \\ AO & is\ common \\ \angle DAO = \angle BAO & Proved\ above \end{cases}$

DO = OB

and $\angle AOD = \angle AOB$

But these are together equal to $180°$

$\therefore$ $\angle AOD = 90°$

$\therefore$ Diagonals bisect each other at right angles. (H.P.)

Exercise (i)

(1) The diagonals of a rectangle are equal.

(2) If the diagonals of a parallelogram are equal it is a rectangle.

(3) The diagonals of a square are perpendicular to each other.

(4) If the diagonals of a parallelogram are equal and perpendicular it is a square.

(5) AB and CD are 2 parallel lines and a tranversal intersects AB in X and CD in Y. Show that the bisectors of interior angles form a rectangle.

(6) The bisectors of angles of a parallelogram form a rectangles.

(7) If the bisector of the angles of a quad meets at 90°, it is a rhom bus.

Theorom No. 23

If a parallel line is drawn from the middle point of one side of a Δ *to the second side, it bisects the third side.*

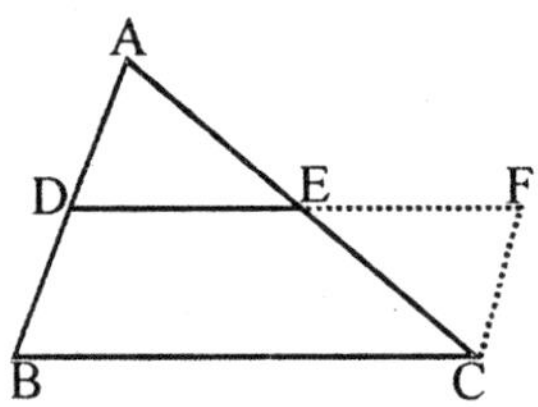

Given:- D is the mid point of AB of a Δ ABC. DE $\parallel$ BC.

To Prove:- E is the middle point of AC.

Construction:- Produce DE beyond E to F.

and draw CF $\parallel$ AB

Proof:- CF $\parallel$ AB (Const.)

DF $\parallel$ BC

DBCF is a parallelogtam.

$\therefore$ DB = FC

by DB = AD (given)

$\therefore$ AD = FC

Also AD $\parallel$ FC and DF is the transversal

$\therefore$ $\angle ADE = \angle CFE$ (Alternate $\angle s$)

Now in Δ s ADE and CFE

because $\begin{cases} AD = FC \\ \angle ADE = \angle CFE \\ \angle AED = \angle FEC \end{cases}$ *(Proved above)* *(Th 3)*

$\therefore$ ΔADE $\equiv$ ΔCFE

AE = EC (H.P.)

Theorem No. 24

The line joining the middle point of 2 sides of a Δ is parallel to the third side and is equal to half of it.

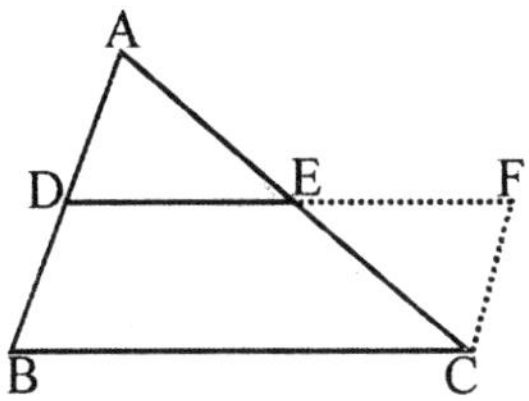

Given:- ΔABC in which AD=DE and AE = EC.

To Prove:- DE $\parallel$ BC and DE = $\frac{1}{2}$BC.

Construction:- Produce DE beyond E so that EF = DE. Join CF.

Proof:- In Δs AED and CEF

because $\begin{cases} AE = EC \\ DE = EF \\ \angle AED = \angle CEF \end{cases}$ *(given)* *(Construction)* *(Th 3)*

$\therefore$ ΔAED $\equiv$ ΔCEF

$\therefore$ $\angle ADE = \angle EFC$

But these are alternate $\angle s$.

$\therefore$ AB $\parallel$ FC (Th 3)

Also AD = FC

AD = DB

$\therefore$ DB = FC

$\therefore$ DFCB is a parallelogram

$\therefore$ DF = BC

But $DE = EF$

$\therefore$ $DE = \frac{1}{2} BC$

DE has been proved to be $\parallel$ to BC. (H.P.)

Exercise (J)

1. In a quad. if the mid points of consecutive sides are joined, it is a parallelogram.
2. In a quad. the mid points of opposite sides are joined, they bisect each other.
3. The quad. formed by joining the consecutive mid points of a rectangle is a rhombus.
4. The quad. formed by joining the mid points of a rhomus is a rectangle.
5. The quad. obtained by joining the mid points of a square is another square.
6. ABC is an isosceles Δ with AB=AC. D, E, F are the mid points of BC, CA and AB. Show that AD $\perp$ EF and is bisected by it.
7. ABCD is a parallelogram. E, F are the mid points. of AB and CD respectively. Prove that the line segments AF and CE trisect the diagonal BD.
8. ABC is a Δ. M is the centre of BC. N is the centre of AM. BN is produced to meet AC in D. Show that $AD = \frac{1}{3} AC$.

Definition

If we draw 2 or more than 2 lines say L_1, L_2, L_3 and a line is drawn that intersects them in points. P, Q, R,.... etc. then PQ QR, ... are called the **Intercepts** made by the line an L_1, L_2, L_3 ... etc and the line PQR.... is called the **Transversal**.

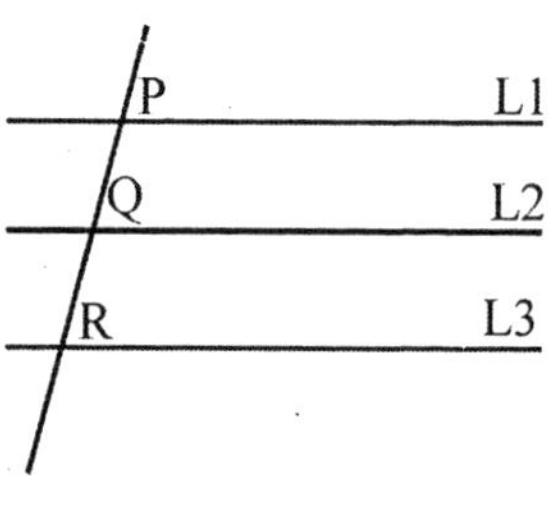

Theorem No. 25

If three or more parallel lines are drawn and if the intercepts made by them on any one transversal are equal, then the intercepts made by them on any other transversal are also equal.

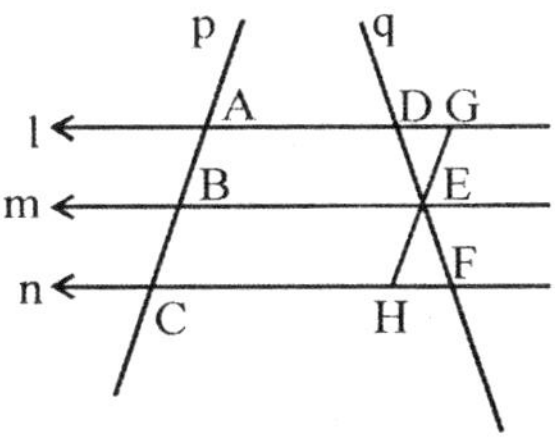

Given:- l, m, n are 3 parallel lines. p is a transversal intersecting them so that AB = BC. Another transversal q is drawn intrersecting l, m, n in D, E and F.

To Prove:- DE = EF

Construction:- Draw a line through E || to p intersecting l in G and n in H.

Proof:- AG || BE

AB || GE

ABEG is a parallelogram

∴ AB = GE ...(1)

Similarly BC || EH

BE || CF

∴ BCHE is a parallelogram

∴ BC = EH ...(2)

From (1) and (2) because AB = BC

GE = EH ...(3)

Because DG || HF

$\angle GDE = \angle EFH$

Because they are alternate $\angle s$

∴ in Δs GDE and HFE

because $\begin{cases} GE = EH & (Proved\ above\ 3) \\ \angle GDE = \angle EFH & (alternate\ \angle s) \\ \angle DEG = \angle HEF & Th\ 3 \end{cases}$

$\therefore$ $\Delta DEG \equiv \Delta HEF$

$\therefore$ DE = EF (H.P.)

Exercise (k)

1. Prove that the middle points of one of the opposite sides of a parallelogram and the middle points of the diagonals form a parallelogram.
2. If the mid pts. of the diagonals of a quad. are joined then it will bisect the mid points of the lines joining both the opposite-sides.
3. A line AB is drawn. From A and B perps. AP and BQ are drawn to another line XY. Then P and Q are equidistant from the midpoints R of AB.
4. In the above example if from R the midpt of AB a line RN is drawn perp. to the line XY then RN= $\frac{1}{2}$ (AP+BQ) or RN = $\frac{1}{2}$(AP–BQ) according as line XY meets AB outside or inside.
5. The line joining the mid pts of inclined sides of a trapezium is parallel to the parallel sides and is equal to $\frac{1}{2}$ sum of parallel sides.
6. If from the four angular points of a parallelogram perpendiculars are drawn to any line outside the paralellogram then the sum of the four perpendiculars is 4 times the perpendicular from the center of the diagonals to the line.
7. If you take a point in the base of any isosecles Δ and draw perpendiculars from it to the sides of the Δ then the sum of these perpendiculars is equal to the perpendicular drawn from one end of the equal sides to the opposite side.
8. If a point is taken by producing the base of an isosceles Δ and perps are drawn to the equal sides then the difference of these perps is equal to the perpendicular from one end of the base to the opposite side.

9. If a point is taken inside an equilateral Δ and perpendiculars are drawn to all the three sides then the sum of these perpendiculars is equal to the perpendiular from the vextex to the opposite side.
10. If a point is taken outside the equilateral Δ, what will be the change in the proof for the above example.

Theorem No. 26

In a right angled Δ the line joining the midpoint of the hypotenuse to the opposite vertex is equal to half the hypotenuse.

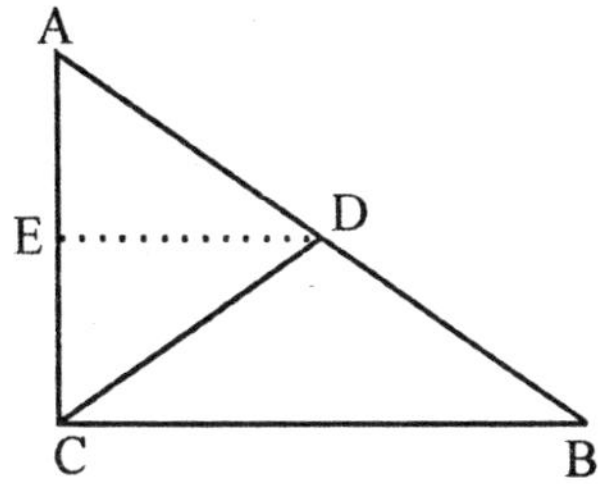

Given:- A right angled Δ right at C. AB is the hypotenuse and D is its middle point.

To prove:- $CD = \frac{AB}{2}$

Construction:- Draw DE || BC

Proof:- D is the mid point of BA.

DE || BC

$\therefore$ E is the mid pt. of AC.

DE || BC

$\therefore$ $\angle DEC = 90° = \angle DEA$

In Δs AED and CED

because $\begin{cases} AE = CE & \text{(Proved above)} \\ ED \text{ is common} \\ \angle AED \equiv \angle DEC = 90° \end{cases}$

$\therefore$ $\Delta AED \equiv \Delta CED$

$\therefore$ $CD = AD = \frac{AB}{2}$ (H.P.)

Practical Geometry - II

(Construction of Δs and quads.)

Construction No. 11

To draw a rectangle having given the hypotenuse and one side.

Given:- Hypotenuse AB and one side AC

To draw:- A right angled Δ ABC

Construction:- Draw AB the hypotenuse with D as its centre and $\frac{1}{2}$AB = distance in the compass draw a semicircle. From A draw AC the side. ABC is the Δ.

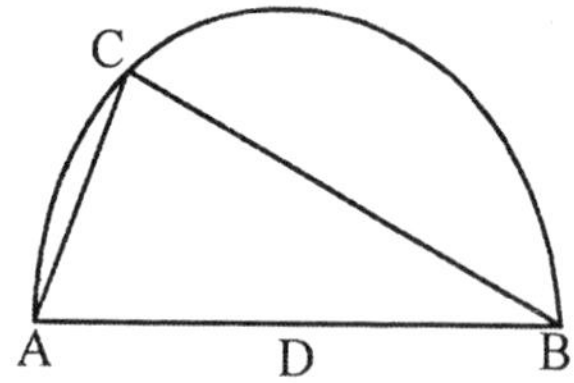

Construction No. 12

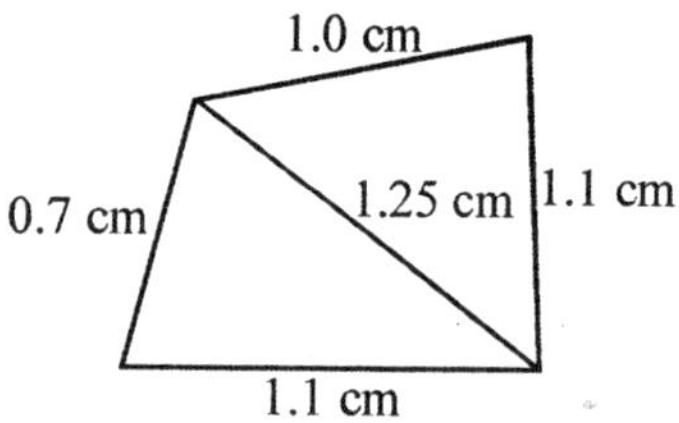

To draw a quadrilateral having given 4 sides and a diagonal.

First draw a rough figure and enter all the dimensions.

Construction:- First draw the Δ of which the 3 sides are known then with remaining 2 sides complete the quad.

Construction No. 13

To draw a parallelogram.

Given:- Two sides and angle between them. Draw the parallelogram

Construction:- Draw the line AB equal to one side. Make $\angle A$ = given angle. Draw AD = 2nd side. Through D and B draw $\parallel$ lines DC and BC. ABCD is the parallelgram.

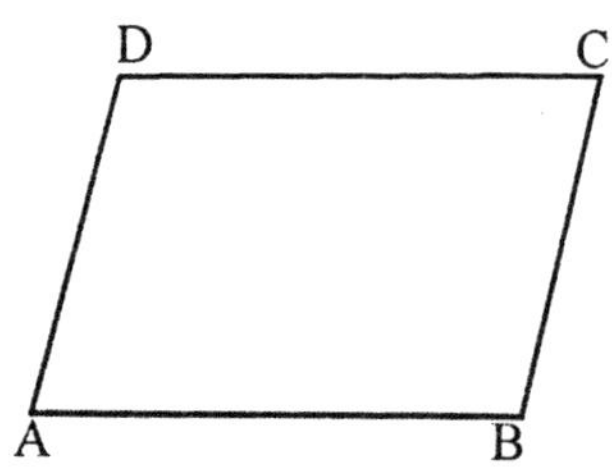

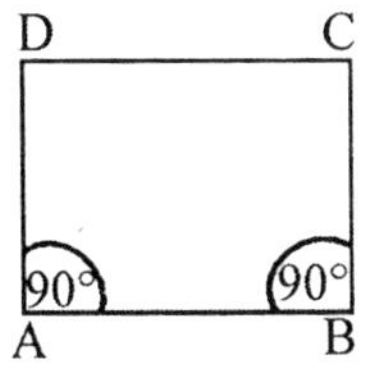

Construction No. 14

To draw a square on a given line.

Draw the side = AB. at A and B make ∠ =90° Make AD and BC=AB.

Construction No. 15

To draw a trapezium having given 4 sides.

DP = DC - AB and AP = BC.

We can draw △ADP And complete the trapezium by drawing parallel through A and C to DP and AD respectivcly.

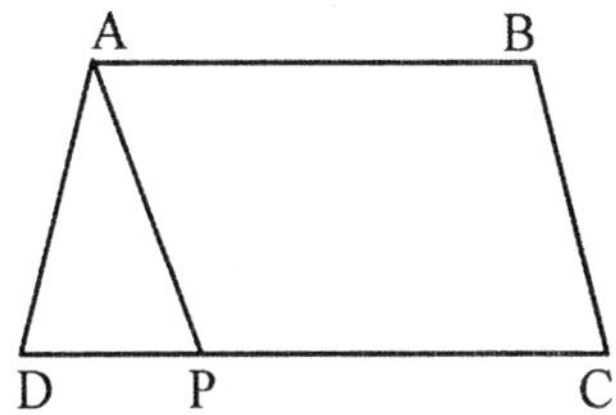

Exercise (I)

Isosceles Triangles

To draw a △ we require 3 parameters. However in the case of isosceles triangles, two sides are equal is already known. Hence only 2 parameters are sufficient.

1. Draw an Isosceles △ with base = 4 cms and ∠ at the vertex = 48°; (Calculate base ∠ s)

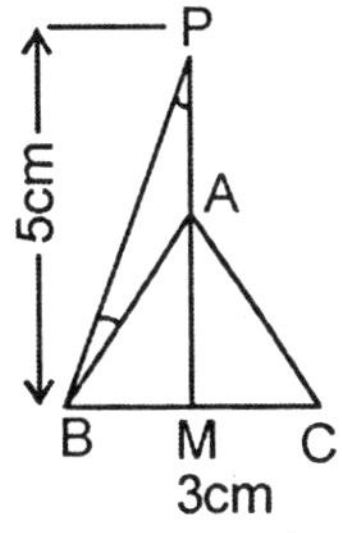

2. Draw an Isosceles △ with base = 3 cm. and height + equal side = 5cm; (Make BC = 3 cms. Find out its mid-pt M. Draw MP ⊥ BC. Make BP = 5 cms. Make ∠BPA = $\angle PBA$. ABC is the △)

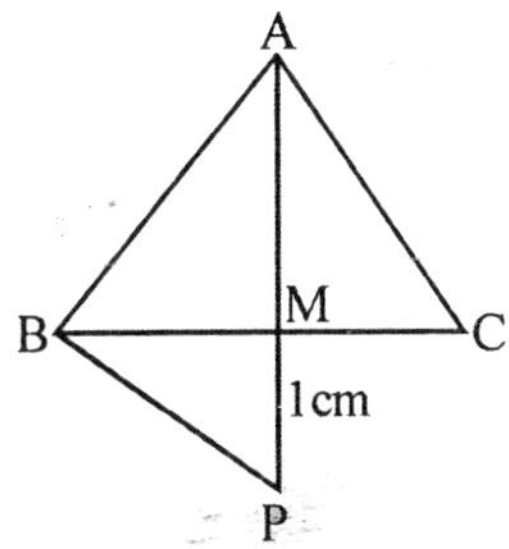

3. Draw an isosceles △ having been given base = 5 cm and the difference between equal side and height = 1cm. Take the base BC = 5 cm. Find its mid-point and 1 cm. below that, make an angle APB= $\angle ABP$ ABC is the △.

4. Draw an isosceles Δ having given base = 5 cm and perimeter =17 cms

(All sides can be calculated)

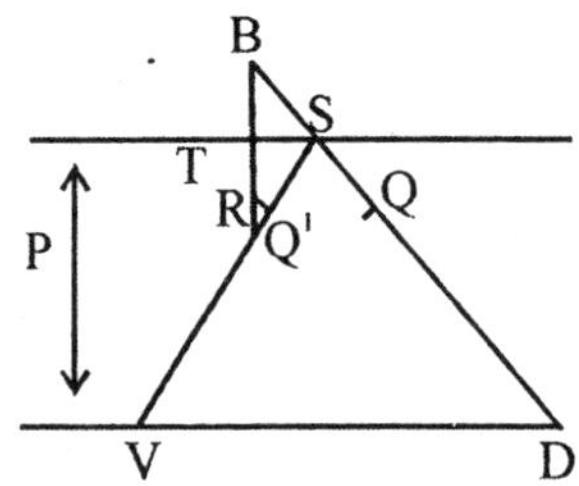

5. Draw an isosceles Δ having given 2 points through which equal sides pass and the the ht of the Δ.

(Draw RT $\perp$ to the parallel lines and produce to B so that RT=TB Join BQ S is the vartax.)

Rt angled Δ

6. One side + hypotenuse = 18 cms, 2nd side = 12 cms

(Make a rt angled Δ with 18 as one side, 12 as the other. Make $\angle ABD = \angle ADB$; ABC is the Δ).

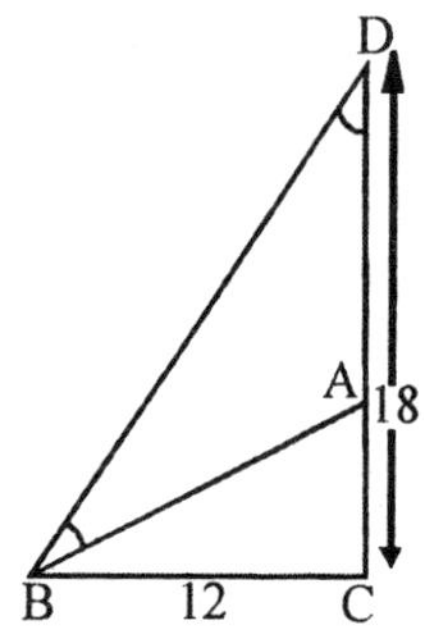

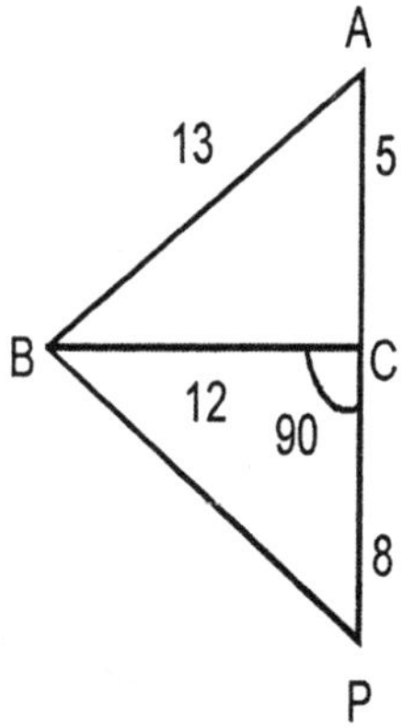

7. In right $\angle d$ Δ one side = 12 cms; Difference between hypotenuse other side = 8 cms. Draw the Δ.

(Make one side BC = 12 cms. From one end below make CP = 8 cms at right angles to BC.

Make $\angle APB = \angle ABP$. ABC is the Δ. AB is the diagonal)

8. In rt angled Δ, hypotenuse = 13 cms.

The sum of other sides = 17 cms.

(At B make an angle = 45° with A as centre and with 13 cm. radius cut off at C and draw CD $\perp$ AB ACD is the Δ.

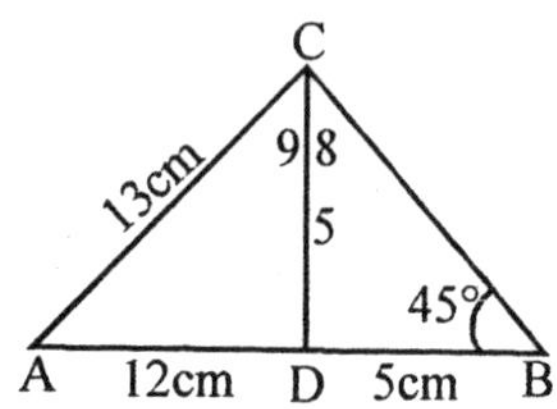

9. One side = 5 cms. The median bisecting the hypotenuse=6.5 cms.
10. Hypotenuse = 13 cms median bisecting one side = 7 cms.

    (Pythagoras theorem which occurs at S.No. 43 is assumed to be known)

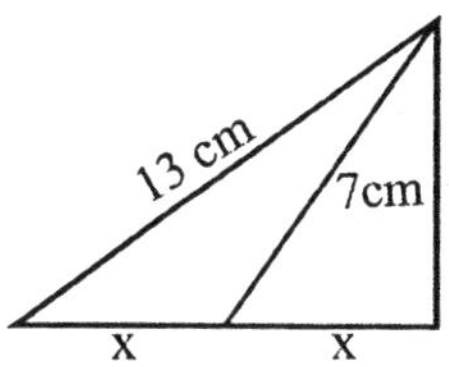

$13^2 - 4x^2 = 7^2 - x^2$

$13^2 - 7^2 = 3x^2$

$169 - 49 = 3x^2$

$120 = 3x^2$

$40 = x^2$

$x = 2\sqrt{10}$

11. AB = hypotenuse 13 cms; Diff. of 2 sides = 7 cms.
12. Hypotenuse = 10 cms, Diff. of 2 $\angle$s = 30°

Triangles in General

13. Height = 3 cms; One side = 5 cms; base = 6 cms. Draw the Δ.
14. Base = 6 cms. The foot of perp from one side = 2 cms Second side = 4 cms. Draw the Δ.
15. Base = 6 cms. $\angle$ at the vertex = 40°, Diff of base $\angle$s = 48°.

    Draw the Δ. (Calculate the base $\angle$s)
16. Given the mid pts of the sides of the Δ. Draw the Δ.
17. Median bisecting the base and the angles it makes with the other sides. Draw the Δ.
18. Two sides and median bisecting the third side. Draw the Δ.
19. Base, Sum of the other sides, one base $\angle$ is given. Draw the Δ.
20. Perimeter and both the base $\angle s$ angles are given, draw the Δ.
21. Base, $\angle$ at the vertex, sum of 2 sides are given. Draw the $\angle s$.
22. Base $\angle$ at the vertex and difference of the other sides. Draw the $\angle s$.

Quadrilaterals

1. To draw a quadrilateral, a 4 sided figure we need five parameters. By drawing a diagonal we divide the quadrilateral into 2 Δs. We

require 3 parameters for the first Δ. To draw the second Δ we need only 2 parameters because already one side as diagonal is already drawn. Hence 5 parameters are enough.

(b) To draw a trapezium we need only 4 parameters because 2 of its sides are parallel is already in its definition.

(c) To draw a parallelograme we require only 3 parameters because both sides are parallel to each other, is known in its definition.

(d) To draw a rectangle we need only 2 parameters because in addition to both its opposite sides are parallel its angle is known to be 90° by its definition.

(e) To draw a rhombus only 2 parameters are necessary because both sides being parallel to each other, all sides are equal is known in its definition.

(f) Finally to draw a square use need only one parameter because both sides are parallel, in addition each $\angle$ being 90°, all sides are equal is known by its definition.

23. Draw a quadrilateral having given AB = 3 cms, BC = 4.5 cms, AC = 5.0 cms, AD = 4.0 cms, CD = 4 cms.

24. Draw a quadrilateral ABCD having given AB = 4.00 cms, BC = 4.5 cms, $\angle B = 70°$, $\angle C = 85°$ and $\angle D = 60°$.

25. Draw a quadrilateral having given AB = 3.5, $\angle B = 70°$, BC = 4.00 cms, CD = 4.5 cms, DA = 5.0 cms.

26. Draw a quad. ABCD where AD = 2 cms; AB = 3 cms, $\angle A = 120°$, $\angle C = 40°$; DC + CB = 7 cms.

27. Draw a quad. in which AB = 4 cms, $\angle A = 115°$, $\angle B = 80°$, $\angle D = 75°$; CD = 5 cms.

28. Draw a quad. in which $\angle A = 75°$, $\angle D = 125°$, AB = 4.00 cms, BC = 6 cm, CD = 5.5cm.

29. Draw a quad. in which AB = 2 CD, AB || CD, AD = 3 cms, BC = 4 cms $\angle BAD = 60°$.

30. Draw a parallelogram having AB = 4.5 cms and AD = 3.6 cms and $\angle A = 75°$.

31. Draw a rectangle with sides 5 cms and 3.5 cms.
32. Draw a rhombus, each side = 4.5 cms and one angle = 45°.
33. Draw a rhombus where diagonals are 16 cms and 30 cms respectively.
34. Draw a parallelogram in which AB = 3 cms, BC = 5 cms and AC = 4.5 cms.
35. Draw a rhombus whose one side measures 6 cms and one diagonal measures 8 cms.
36. Construct a rectangle of one side = 6 cms and diagonal = 10 cms.
37. Construct a trapezium in which AB = 2CD, AB || CD, AD = 3 cms $\angle BAD = 60°$.
38. Construct a trapezium in which AB = 4.5 cms BC = 2.2 cms, CD = 2.8 cms, AD = 1.9 cms and AB || CD.
39. Draw a rectangle whose diagonal is 5 cms and whose parameter is 14 cms.
40. Draw a square of difference of its diagonal and one side=4 ($\sqrt{2}$ -1).

LOCI

LOCI is the plural of locus.

Definition

The locus of a point is the path is describes when it moves according to a given condition.

(1) A point moves so that it is at a fixed distance, called the radius, from another fixed point called the centre.

Obviously it is a circle.

(2) A point moves so that it is at a fixed particular distance from a given line. The locus is composed of 2 parallel lines from a fixed line on both sides of it.

In problems on locus it is necessary that we must prove it on both counts.

1. Every point on the locus satisfies the given condition.
2. Every point that satisfies the given condition is on the locus.

Thus in case I above we will have to prove that every point on the circle satisfies that it is at the same distance from the fixed point. And we also have to prove that every point that is at the same distance is on the circle.

Below are 2 Theorems on locus.

Theorem No. 27

The locus of a point which moves so that its distance from two given fixed points is equal is the penrpedicular bisecter of the line joining these points.

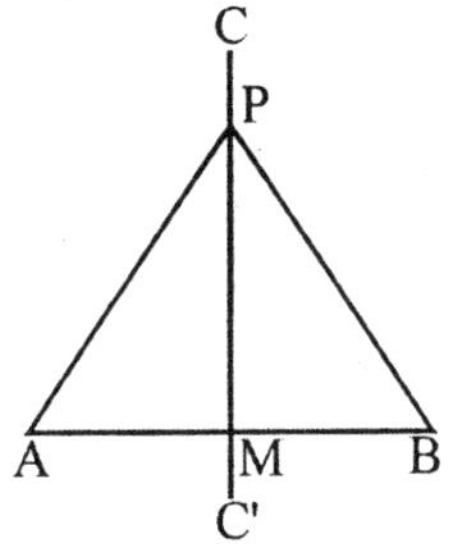

Given: Two fixed points A and B. A point P moves so that PA = PB. M is the mid-point of AB and a perpendicular is drawn to AB at M.

CMC^1 is the perp. bisector of AB.

To Prove: That CMC^1 is the locus of the point P in question or CMC^1 is equidistant from A and B.

That is (1) Every point on CMC^1 is equidistant from A and B.

(2) Every point that is equidistant from A and B is on $\perp$ bisector of AB.

Proof: (1) A point P is taken on the locus and joined to A and B.

because $\begin{cases} AM = MB & (M \text{ is the mid pt.}) \\ \angle AMP = \angle BMP & (\text{Both } 90^\circ) \\ XM \text{ is common} \end{cases}$

$\Delta AMX \equiv \Delta BMX$ (*SAS congruency Axiom*)

$\therefore$ $AP = PB$

(2) Suppose P is such that $AP = PB$. Join MP.

In Δs AMP and BMP.

because $\begin{cases} AP = BP \\ AM = BM \ (M \text{ is the midpt.}) \\ MP \text{ is common} \end{cases}$

$\therefore$ $\Delta AMP \equiv \Delta BMP$ (Th. 13)

$\therefore$ $\angle AMP = \angle BMP$

But these adjacent $\angle s$ are together 180°

Hence each is $= 90^\circ$

Theorem No. 28

The locus of a point which moves, so that its distance from two intersecting lines is equal, IS THE PAIR OF LINES THAT BISECT THE LINES IN QUESTION.

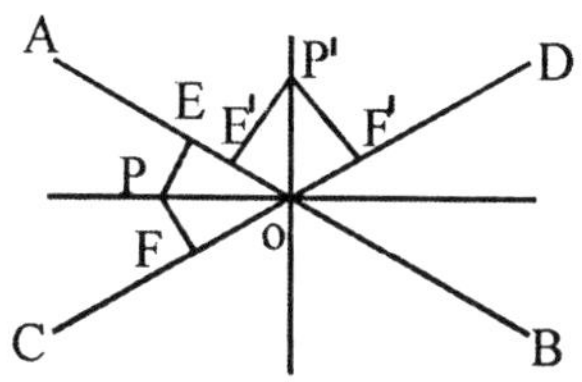

Given: AB and CD are 2 lines intersecting in O. P is a point so that $PE \perp AB$ and $PF \perp CD$ and PE=PF. And P'is a point $P'E' \perp AB$ and $P'F' \perp CD$.

To Prove: (1) POQ and ROS are the bisecting the angle between AB and CD are the required Loci.

That is if PE = PF then $\angle POE = \angle POF$ or if $\angle POE = \angle POF$ then PE=PF. Similarly for P'

Construction : Join OP and OP'

Proof: (A) In Δs OPE and OPF

because $\begin{cases} PE = PF \\ OP \text{ is common} \\ \angle PEO = \angle PFO \end{cases}$ (both 90°)

$\therefore$ These Δs are congruent

$\therefore$ $\angle POE = \angle POF$ (Th. 18)

P is on the bisector of $\angle AOC$

(B) In Δs POE and POF

because $\begin{cases} \angle POE = \angle POF \\ OP \text{ is common} \\ \angle OEP = \angle OFP \end{cases}$ (both 90°)

$\therefore$ $\Delta POE = \Delta POF$

PE = PF

Similarly for P[1] (H.P.)

Theorem No. 29

The perpendicular bisectors of the three sides of a Δ *are concurrent.*

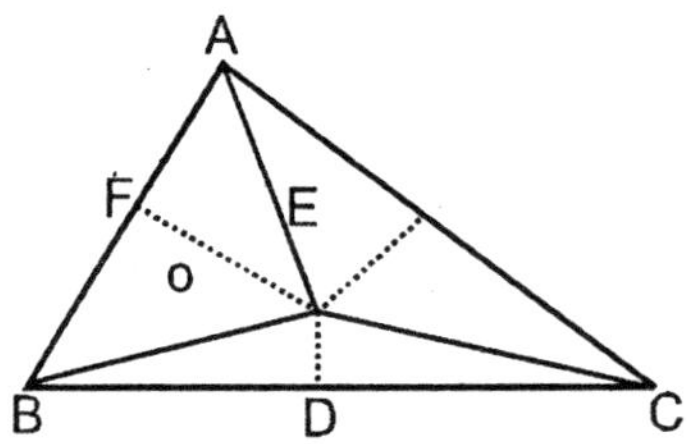

Given: ABC is a Δ. D, E, F are the mid-pts of the three sides. $OD \perp BC$, $OE \perp CA$, OD and OE meet at O. Join OF.

To Prove: $OF \perp AB$

Proof: All pts on OD are equidistant from B and C

$\therefore$ OB = OC ...(1)(Th. 27)

All pts on OE are equidistant from O and E.

$\therefore$ OC = OA ...(2)(Th. 27)

From (1) and (2) we get,

OB = OA

O must lie on the $\perp$ bisector of AB (Th. 27)

$\therefore$ $OF \perp AB$ (H.P.)

Definition: 'O' is known as the **centre of curvature** of the ΔABC i.e. of the circle passing through A, B, C. (OA = OB = OC)

Theorem No. 30

The angle bisectors of the three angles of a Δ are concurrent.

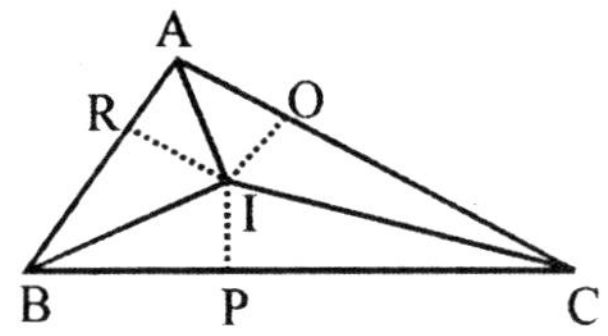

Given: IA is the bisector of $\angle A$

IB is the bisector of $\angle B$

Let IA and IB meet in I. Join IC.

To Prove: IC is the bisector of $\angle C$

Proof: IA is the bisector of $\angle A$

IQ = IR ...(1) (Th 28)

Similarly IB is the bisector of $\angle B$

$\therefore$ IP = IR ...(2) (Th 28)

From (1) and (2) we get

IP = IQ

$\therefore$ I must lie on the bisector of C (Th 28)

$\therefore$ The 3 bisectors of a Δ are concurrent. (H.P.)

Definition

'I' is known as **incentre** of the circle. It is the centre of the circle touching the sides of the Δ. (IP = IQ = IR)

Theorem No. 31

The 3 medians of a Δ are concurrent.

Given: Let BL and CK be the medians of the Δ. They meet in G. AG is joined and produced to meet BC in I.

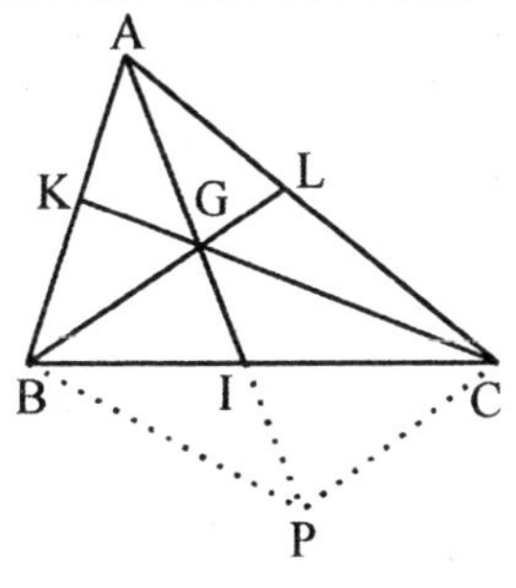

To Prove: AI is the third median of the Δ.

Construction:- Produce AI and draw a $\|$ CP to BL.

Proof: In Δ APC from L the mid pt of AC

LG is drawn parallel to CP.

$\therefore$ AG = GP (Th 23)

$\therefore$ G is the mid point of AP

K is the mid point of AB

$\therefore$ KG $\|$ BP (Th 24)

$\therefore$ BGCP is a parallelogram

$\therefore$ GP and BC bisect each other at I.

$\therefore$ I is the mid point of BC.

$\therefore$ AI is the median (H.P.)

Corollary: The point of intersection of the 3 medians of a Δ is a point of trisection of the medians.

$$AG = GP = 2GI$$

$$AG = 2GI$$

$$\therefore \quad AI = 3GI$$

Similarly $KG = \frac{1}{2}BP = \frac{1}{2}GC$

$$\therefore \quad KG = \frac{1}{3}CK$$

and $GL = \frac{1}{2}PC = \frac{1}{2}BG$

$$\therefore \quad GL = \frac{1}{3}BC$$

Therefore G is the point of trisection of the 3 medians of the Δ.

Definition

G is called the **centroid** of the Δ. A plane lamina can be balanced horizontally by a pin on this point.

Theorem No. 32

The 3 perpendiculars drawn from the 3 vertices of a Δ are concurrent

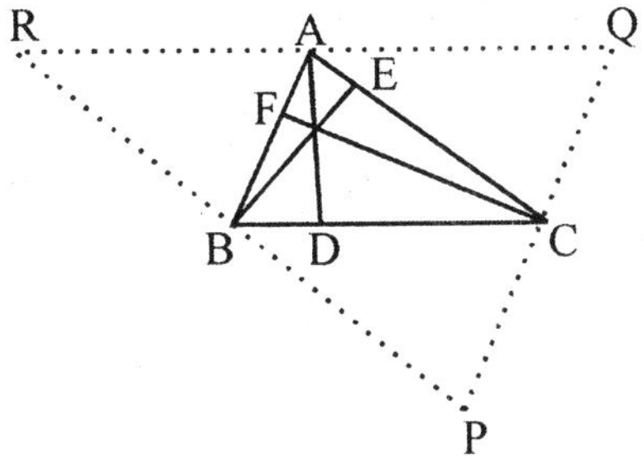

Given: ABC is a Δ in which $AD \perp BC$, $BE \perp AC$, $CF \perp AB$.

To Prove: The 3 perpendiculars are concurrent i.e. AD, BE and CF meet in a point.

Construction:- Through A draw $QR \parallel BC$

Through B draw $RP \parallel CA$

Through C draw $QP \parallel AB$

Proof: $AR \parallel BC$

$BR \parallel AC$

$\therefore$ RACB is a parallelogram.

$\therefore$ $BC = RA$ ...(1)

Similarly $AQ \parallel BC$

$CQ \parallel AB$

$\therefore$ ABCQ is a parallelogram

$\therefore$ $BC = AQ$ ...(2)

From (1) and (2) $AR = AQ$

$\therefore$ A is the mid pt of PQ

Moreover $AD \perp BC$

$\therefore$ $AD \perp RQ$ (Since $RQ \parallel BC$)

$\therefore$ AD is the perpendicular bisector of RQ.

Similarly BE and CF are the perpendicular bisectors of PR and PQ respectively.

∴ AD, BE and CF of a Δ are concurrent.

∴ AD, BE, CF are the perp biscetors of Δ PQR. They are the perpendiculars from the vertics of a Δ ABC on opposite sides.

∴ AD, BE, CF are concurrent.

Exercise (m)

1. A line AB is gvien. Find out a point on it which is equidistant from 2 other given points. When is this not possible?
2. AB, CD and EF are 3 lines in the same plane. Find out a point P so that its perp. distance from AB, CD and EF is the same. When it is not possible?

***3. Two sticks are fixed at right angle to each other. One more stick moves so that one end is on one stick and the other is on the other. What is the locus of the mid point of the third stick.**

4. Line AB is supposed to be the hypotenuse of a right angled Δ ABC. What is locus of C?
5. From one point A, a no. of lines are drawn to another line. What is the locus of mid pts. of such lines?
6. From a point A are drawn a no. of lines to another line and they are further extended by equal distance. What is the locus of ends of these lines?
7. From a point a no. of lines are drawn to another line. And these lines are extended away from AB in the reverse direction by an equal distance. What will be the locus of ends of these lines?
8. From two lines which are mutually perpendicular the sum of the distance of a point is 6 cm. What is the locus such points?
9. From two lines which are mutually perpendicular the difference of distances of a points is 2 cm. What is the locus of the point?
10. One point and one line is given. The point is 5 cms from the line. Find out a point which is 4cms from the point and 3cms from the line. How many such points are possible?
11. (a) Base, height and the median bisecting the base is given. Draw the Δ. How many Δs are possible?

    (b) Base, one side and median bisecting the base the side is given. Draw the Δ.

12. Three mediarns are given. Draw the Δ.
13. Find out the bisector of 2 lines. Which do not meet on paper.
14. From a points lines are drawn to another line. What is the locus of a point which is trisecting these lines?

*15. Base of a Δ is produced on both sides. The exterior $\angle$s formed in this way are bisected. These bisectors are concurrent with the bisector of the angle at the vertex. Prove this.

Exercise (n)

Miscellaneous Examples - I

1. In a Δ, the three medians are together more than $\frac{3}{4}$ of the perimeter of the Δ.
2. AOB and COD are 2 lines intersecting in O. COAE and DOBF are 2 rhombuses. Show that E, O, F are in one straight line and AC $\parallel$ DB.
3. In a ΔABC, AB = 2 AC. The exterior $\angle A$ is bisected and it meets the base BC produced in D. Prove that BC=CD (From C draw CE $\parallel$ AD).
4. In a quad. ABCD, AB = CD and AC = BD. Prove that BC $\parallel$ AD.
5. ABC is a Δ. From the mid point D of AB a line is drawn parallel to BC. The bisector of $\angle$B meets it in E. Prove that $\angle$AEB = 90°.
6. AB and CD are st. lines. There is a point X in AB. Find out Y is AB so that XY = distance of Y from CD.
7. ABC is a Δ. PE is drawn $\parallel$ to the base and such that PE = (BP+CE). How to draw the line. (Bisect $\angle B$ and $\angle C$. Let them meet in O. Draw a line $\parallel$ base through this pt)
8. What is each angle of a regular seven sided figure? A regular seven sided figure has the ends of one side connected to the single vertex opposite. Prove that the vertical angle of the Δ so formed is 3 times of each of the base $\angle$s.

***9. All the four corners of a quadrilateral are turned inside about the mid pt. of the sides. They meet in a point. Show that the diagonals of such a quad. are at right angles to each other.**

10. The side BC of a square ABCD is produced to E. A point G is taken on BC. And GF is drawn ⊥ AG meets EF the bisector of $\angle DCE$ in F. Prove that AG = GF.

    (HINT. Take a point H on BA so that BH = BG. Prove that Δ AHG = Δ GCF, AH = GC, Prove $\angle AHG = \angle GCF = 135°$. Also $\angle HAG = \angle FGC$.

11. ABCD is a square. An equilateral Δ ABX is drawn. Prove that $\angle ACX = \frac{1}{2} \angle ABX$. This is true if X on either side of AB.

    [HINT. BC = AX, $\angle BCX = 15° \therefore \angle ACX = 30°$]

12. In a Δ ABC, $\angle A = 45°$; AD and BE are perpendiculars from A and B to the opposite sides they meet in O. Prove that OE = EC.

    (Apply Theorems 32 and 12)

13. In an isosceles Δ, AB = AC. Draw a line ∥ to BC which meets AB and AC in P and F so that BP = PF = CF (Bisect one of the base $\angle s$ to meet the equal side and draw a parallel to base).

14. In an isosceles Δ ABC $\angle A = 90°$, D is a point on the hypotenuse. AD is joined. BP and CF are perpendiculars to it draw from B and C. Prove that (1) BP=AF and (2) BP-CF=PF.

15. ACB is a st. line ABX and ACY are 2 equilateral Δ s drawn on opposite sides of AB. Prove that CX = BY.

16. If a Δ ABC, M is the centroid. If AM = BC then prove that $\angle BMC = 90°$.

17. ABCD is a rectangle. AP, BF, CQ and DR are equal distances on AB, BC, CD and DA on the four sides. Prove that PFQR is a parallelogram (Prove Δ BPF ≡ Δ DQR)

18. ABC is isosceles right angled Δ of which AC is the hypotenuese. CA is produced and the exterior angle at A is bisected. It meets CB produced in P. Prove that BP=AB + AC.

    (Drop a perp. from P to CA)

19. In a Δ ABC 3 points X, Y, Z are taken inside anywhere. Prove that the perimeter of Δ ABC is greater than the perimeter of Δ XYZ.

(Produce the sides of Δ XYZ to meet the sides of ABC and apply Theorem No. 16).

20. ABC is a Δ in which AB = AC. A point P is taken on AC and BA is produced to F so that AF = AP.

    Prove that FP $\perp$ BC.

21. Draw a line from a given point to two parallel lines, whose intercept on the parallel lines will be of given length.

***22. In a quadrilateral in case the sum of the three verties from each angular point is the same then the quadrilateral is a rectangle.**

23. ABC is a Δ. BXYC is a square drawn on BC. AX and AY are joined and parallel lines are drawn to them through B and C. These parallel lines meet in P. Show that PA when produced is perpendicular to BC.

***24. In a Δ ABC, BE $\perp$ AC and CF $\perp$ AB. OP and OQ are perpendicular bisectors of AB and AC. If D is where BE and CF meet. Prove that 2 OP = DC and 2 OF = DB.**

25. Difference of the $\angle s$ at the base, sum of the remaining 2 sides other than the base and the difference of the parts of the base made by the perpendicular to the base from the vertex are given. Draw the Δ.

26. A and B are points whose position is fixed. A X B is an obtuse angle. The perpendicular bisectors of AX and BX meet AB in points P and F. Show that the perimeter of P X F is fixed.

27. In a Δ ABC, P and F are the centres of AB and AC. From A a line is drawn $\parallel$ to the base BC and a point D is taken on this line. DP and DF is produced till they meet BC in pts E and G.

    Prove that BC = EG.

28. D and E are points on BC, AC respectivcely of a Δ ABC. AD and BE meet in P. Prove that (CD + CE) is greater than (PD + PE).

    (HINT:- From D and E draw parallels to BE and AD).

29. In a Δ ABC, side AB > side AC. AF is the bisector of $\angle A$. BP and CF are perpendicular from B and C to the bisetor AF. Prove that

    DP = DF if D is the mid point of BC and DP = DF = $\left(\frac{AC\text{-}AB}{2}\right)$.

30. In a right angled $\triangle$ AC is the diagonal and AB is shortest side. From BC a distance BD is cut off = AB, BA is produced to E so that BE = AC. Prove that DE is perpendicular to the median of AC.

31. The base $\angle s$ of an equilateral $\triangle$ are bisected and from their intersection parallels are drawn to the other 2 sides. Prove that these lines trisect the base.

32. On a parallelogram ABCD a point P is taken on the diagonal BD so that BP = DP. Show that P is the centre of DP.

33. In a $\triangle$ ABC a point D is taken on the base. From D parallels are drown to AC and AB meeting them in P and F respectinely. As D moves along BC what is the locus of the centre of PE.

34. The base of a $\triangle$ is extended both ways and exterior angles are bisected and parallels are drawn to these lines from the vertex. Prove that these meet the base in points which are = (sum of the other 2 sides—base).

35. In a quadrilateral AB=CD, X and Y are the mid poins AD and BC while S and T are the mid points of BD and AC respetively. Prove that SYTX is a rhombus and XY bisects the angle SXT.

36. ABCD is a st. line so that AB = BC = CD. BPFC is a parallelogram. If BP = 2BC then $PD \perp AF$.

37. In a parallelogram one diagonal is such that each of the other sides is less than it.

***38. In any $\triangle$ the angle between the bisector of the vertical angle and the perpendicular on to base from the vertex is half the difference of the base $\angle s$.**

39. A, B, C, D are 4 points in this order at equal distances from another point 'O'. Prove that

$$\angle ABC + \angle ADC = \angle DAB + \angle BCD = 180^\circ$$

40. A polygon (a closed figure having more than 4 sides) is having 'S' sides then thrice the sum of all the interior angles = interior angles of (3S-4) sides.

41. ABCD is a square. on AB, BC, CD and DA four points E, F, G, H, are taken so that $EG \perp FH$. Show that EG = FH.

***42. In any quadrilateral the 2 lines joining the mid points of opposite sides and the line joining the midpts. of the diagonals are concurrent and the point bisect all the lines.**

43. If opposite sides of a quad. are produced and allowed to meet each other in a point and the angles so formed are bisected the angle between the bisectors is equal to half the sums of the sum of opposite $\angle$s of the quadrilateral.

***44. In a right angled Δ if one of the acute angles is half of the other then the side opposite the smaller angle is half of the hypotenuse.**

*45. The intercept made by one inclined line on 2 opposite sides of a sqaure is equal to the intercept made by a perpendicular inclined line on the other pair of opposite sides.

46. ABC is a right angled Δ with right angle at A. On BC is drawn a square, whose diagonals intersect at P. Prove that AP bisects the right angle at A.

47. ABC is an equilateral Δ. On BC, CA and AB are taken 3 pts D, E, F so that DEF becomes another equilateral Δ. Prove that AE + AF = BC.

48. If two medians of a Δ are equal then it is an isosceles Δ.

49. In a parallelogram one diagonal is such that all its sides are less than it.

*50. Draw the quadrilateral ABCD having given that AB = 4 cms, CD = 6 cms, $\angle A = 110°$, $\angle B = 90°$ and $\angle D = 75°$.

*51. Draw a quadrilateral ABCD having given AB = 3.8 cms, BC = 4.5 cms, CD = 3.0 cms, BC and AD when produced make an angle = 28°, BA and CD when produced make an angle = 35°.

2

Area, Proportionality, and Pythagoras Theorem

Just as to measure the length of a line we required a unit viz. 1 cm or 1 metre, or 1 kilometre, so also to measure an area we require a unit for measurement. The smallest unit for measurement of area is 1 sq. cm. That is the area enclosed by a small rectangle 1 cm long and 1 cm. broad. The bigger unit is 1 square meter and still bigger unit is 1 square kilometer. That is the area enclosed by 1 kilometer long and 1 kilometer broad. Intermediate units and their inter-connection are given below.

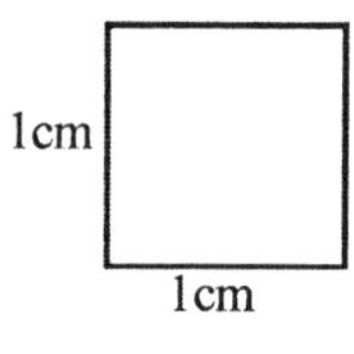

100 sq. mm = 1 sq. cm   100 sq. meter = 1 sq. Dm

100 sq. cms = 1 sq. dm   100 sq. Dekameter = 1 sq. Hm

100 sq. decimeter = 1 sq. m   100 sq. Hectometer = 1sq. Km

The following units are sometimes used

1 are = 100 m^2,   1 hectare = 100 acres = 10000 $meter^2$

1 acre = 4.8 hectares

To measure the area we need to measure the length and breadth in centimeters and multiply the two we get the area in sq.cms.

Theorem No. 33

Two parallelograms standing on the same base and between the same parallel lines are equal in area.

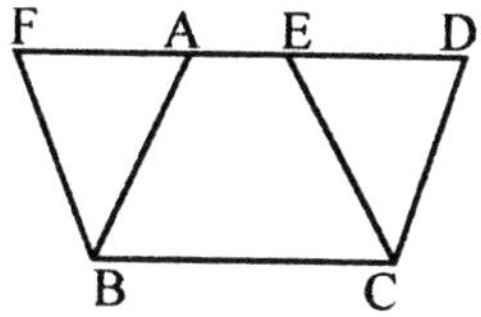

Given: Two parallelograms ABCD and FBCE are standing on the same base BC and are between the same parallel lines BC and FD.

To Prove: Area of □gm ABCD = Area of □gm EBCE

Proof: $AB||CD$ and $FB||EC$

∴ $\angle FBA = \angle ECD$

Now in Δs FBA and ECD

because $\begin{cases} AB = DC & (Th.\ 20) \\ FB = EC & (Th.\ 20) \\ \angle FBA = \angle ECD & (Proved\ above) \end{cases}$

∴ $\Delta FBA \equiv \Delta ECD$

They are equal in area

$\Delta FBA + \square ABCE \equiv \Delta ECD + \square ABCE$

∴ □ gm ABCD = □ gm. FBCE in area. (H.P)

Corollary 1. A parallelogram is equal in area to a rectangle if they are standing on the same base and are between the same parallel lines.

∴ Area of parallelogram = Area of rectangle = length x breadth.

Exercise (o)

1. Draw a parallelograms with adjacent sides, 4 cms and 3 cms and the angle between these sides = 30°.

   What is the area of the parallelogram. Draw a rectangle and a rhombus of equal area on the same base.

2. Draw a parallelogram with base = 6cm and adjacent sides 4cm.

Angle between the sides as 45° and 135°. What is its area. Draw the 2 heights of the parallelogram and find out its area in 2 ways.

3. Draw a parallelogram with base = 6 cms and adjacent sides 4 cm and angle between the sides = 60°. Find out the area.

4. Draw a parallelogram and from the intersection of the diagonals draw a line meeting the opposite sides. Show that the line divides the parallelogram into two.

Theorem No. 34

If a triangle and a rectangle are standing on the same base and are between the same parallel lines the area of the Δ *is* $= \frac{1}{2}$ x *area of rectangle.*

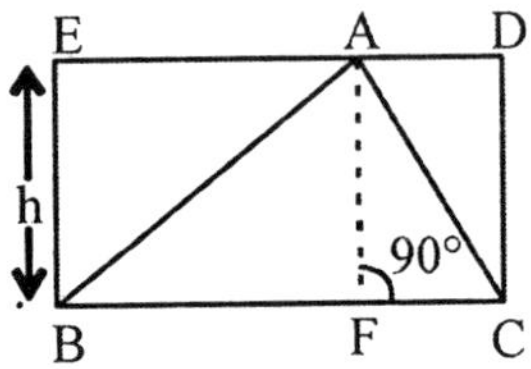

Given: A ΔABC and a rectangle EBCD. They are standing on a common base and are locked between the same parallel lines BC and ED.

To Prove: Area of the $\Delta = \frac{1}{2}$ Area of rectangle $= \frac{1}{2}$ BC x EB

Construction: Draw $AF \perp BC$

Proof: Area of $\Delta ABF = \frac{1}{2}$ Area of rect AEBF

Area of $\Delta AFC = \frac{1}{2}$ Area of rect AFCD

By adding (1) and (2)

Area of ΔABF + Area of ΔAFC

$= \frac{1}{2}$ Area of rect AEBF + $\frac{1}{2}$ Area of rect AFCD

$\therefore$ Area of $\Delta ABC = \frac{1}{2}$ (Area of rect AEBF + $\frac{1}{2}$ Area of rect AFCD)

$\therefore$ Area of $\Delta ABC = \frac{1}{2}$ Area of rect angle BCDE

$$= \frac{1}{2} BC \times BE$$

$$= \frac{1}{2} \times a \times h = \frac{1}{2} ah. \qquad \text{(H.P.)}$$

Theorem No. 35

If from a point on the base of a triangle parallels are drawn to the other sides, the area of parallelogram so formed is the largest, when the point is the mid point of the base of the Δ.

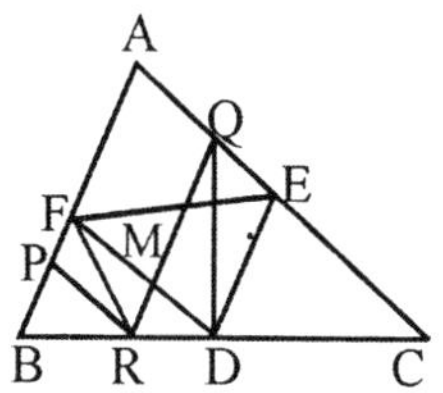

Given: (1) ABC is a Δ. D is the mid-point of the base BC. DE || AB and DF || AC

(2) R is any point $RQ \| AB$ and $RP \| AC$

To Prove: Area of parallelogram DFAE > Area if parallelogram RPAQ

That is Area of parallelogram DFAE is the largest and that is larger than area of similar parallelograms from any pt in the base.

Proof: Area of ΔQRD > Area of ΔERD because they are on a common base RD but the height QRD>height ERD.

$\therefore$ Area of ΔQRD > Area of ΔFRD

Subtracting Area of MRD from both

Area of ΔQMD > Area of ΔFMR

2 Area of ΔQMD > 2Area of ΔFMR

Area □ gm MQEB > Area □ gm PFMR

Adding Area of AQMF to both sides

Area of □ gm DFAE > Area of □ gm PAQR

∴ Area of □ gm DFAE is largest.

Exercise (p)

1. Draw the Δs with the following data

   (a) a = 8.2 cm, b = 6.4 cm, c = 3.8 cm

   (b) a = 6.5 cm, $\angle B = 50^\circ, \angle C = 75^\circ$

   (c) a = 5.0 cm, c = 6.8 cm, $\angle B = 65^\circ$

   Draw a ⊥ from each of the vertices to the opposite sides. Find out the area in each case and compare the result.

2. Draw a right angle Δ with the following data AB = 6 cms, BC = 7 cms, $\angle B$ = 90°. Find out the area of this Δ. Measure the hypotenuse. Find the perp. to the hypotenuse.

3. The median to the base divides the Δ into 2 triangles of equal area.

4. Taking into account that 2 diagonals of a rhombus are at right angle to each other prove that the area of a rhombus is half the product of the diagonals.

5. G is the centriod of the Δ. Prove that the area of Δ = 3 area of ΔGBC.

6. The line joining the midpts of inclined side of a Trapezium is half the sum of parallel sides Prove this.

7. If 2 sides are given and the included angle is increased from 0° to 180° the area of Δ is maximum when the included angle is 90°.

8. If 2 sides of a Δ are equal to 2 sides of another Δ and the included angle is supplementary the areas of the Δ are equal.

9. In an isosceles Δ if a point is taken in the base and perpendiculars are drawn to sides then the sum of these perpendiculars is equal to perp from one end of the base to the opposite side.

10. If 2 Δ s are equal in area and are standing on the same base then the line joining their vertices is parallel to the base.

11. In a trapezium the line joining the mid pts. of the inclined sides is parallel to the base.
12. If each of the two diagonals of the quadrilateral bisect each other then it is a parallelogram.
13. If the diagonals of a quad. divides it into 4 Δ s of equal area then is a parallelogram.
14. If a no. of Δ s of equal area are drawn on the same base then the perimeter of the Δ is minimum when it is isosceles.

Areas of Various Quadrilaterals

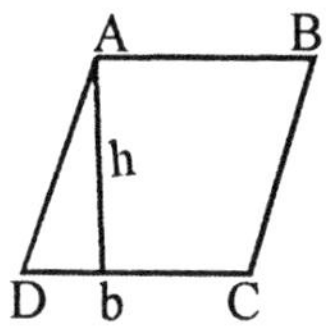

1. Parallelogram

Area = b $\times$ h = side $\times$ corresponding height

2. Trapezuim

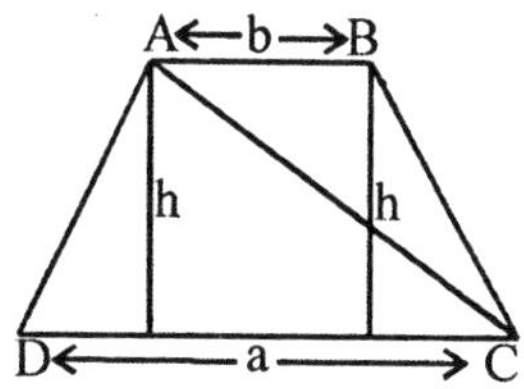

Area of trapezium

= Area of Δ ADC + Area of Δ ABC

$= \frac{1}{2}$ ah $+ \frac{1}{2}$ bh $= \frac{1}{2}$ h(a+b) $= \frac{1}{2}$ x (height) (sum of sides)

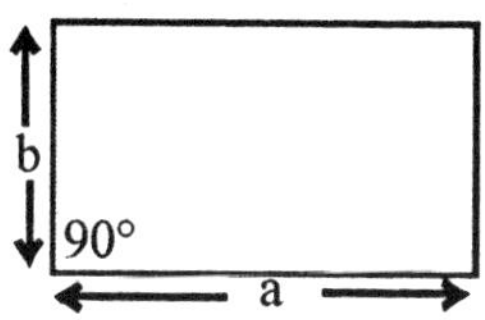

3. Rectangle

Area of rectangle

= a x b

4. Rhombus

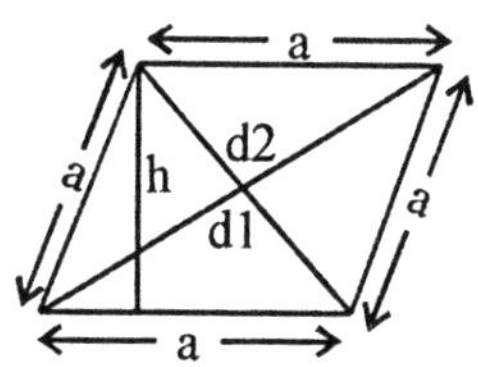

Area of rhombus = ah = $\frac{1}{2} d_1 d_2$

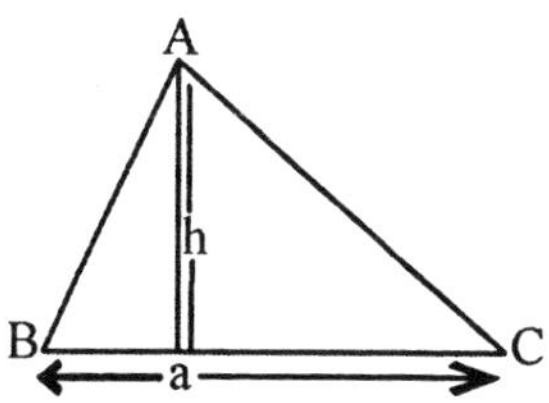

5. Triangle

Area of $\Delta = \frac{1}{2}$ ah.

Theorem No. 36

(Basic Proportionality Theorem)

(Thales Theorem)

A line parallel to the base of a triangle intersecting the other two sides in distinct points divides them in the same ratio.

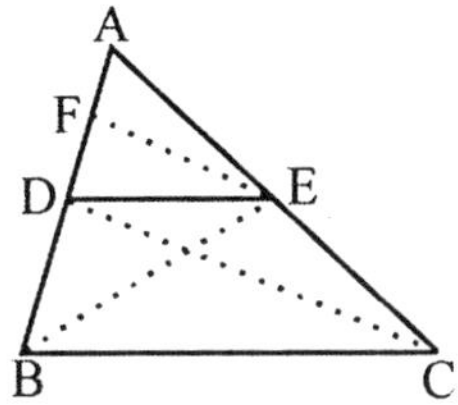

Given:- A Δ ABC in which DE is $\parallel$ BC. It interesects AB and AC in D and E respectively.

To Prove:- $\frac{AD}{DB} = \frac{AE}{EC}$

Construction:- Join BE and CD and drop EF $\perp$ AB.

Proof:- $$\frac{Area\ of\ \Delta ADE}{Area\ of\ \Delta BDE} = \frac{\frac{1}{2}AD.EF}{\frac{1}{2}BD.EF} = \frac{AD}{BD}$$

Similarly $$\frac{Area\ of\ \Delta ADE}{Area\ of\ \Delta CDE} = \frac{AE}{EC}$$

But as Area of Δ BDE= Area of Δ CDE as they are standing on the same, base are between the same parallel lines. These rations are equal

$$\frac{AD}{BD} = \frac{AE}{EC} \quad ...(A)$$

Adding (1) to both sides (H.P.)

$$\frac{AD}{BD} + 1 = \frac{AE}{EC} + 1$$

$$\frac{AD + BD}{BD} = \frac{AE + EC}{EC}$$

$$\frac{AB}{BD} = \frac{AC}{EC} \quad \text{...(B)}$$

By invertendo of (A)

$$\frac{BD}{AD} = \frac{EC}{AE}$$

Adding (1) to both sides

$$\frac{BD + AD}{AD} = \frac{EC + AE}{AE}$$

$$\therefore \quad \frac{AB}{AD} = \frac{AC}{AE} \quad \text{...(C)}$$

$\therefore$ (A), (B), (C) are true i.e.

$$\frac{AD}{BE} = \frac{AE}{EC};\ \frac{AB}{BD} = \frac{AC}{EC};\ \frac{AB}{AD} = \frac{AC}{AE}$$

Theorem No. 37

(Converse of Th 36)

If a line divides two sides of a Δ in the same ratio it must be parallel to the third side.

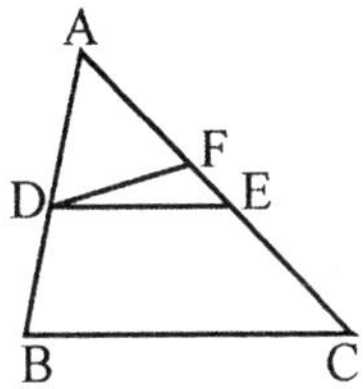

Given:- A Δ ABC and a line l intersecting AB in D and AC in E such that

$$\frac{AB}{DB} = \frac{AE}{EC}$$

To Prove:- $DE \parallel BC$

Proof:- Let us suppose that line l is not parallel to BC. Then through D there must be another line parallel to BC.

Let $DF \parallel BC$

Then by Basic Proportionality Theorem

$$\frac{AD}{DB}=\frac{AF}{FC}$$

But we are given that

$$\frac{AB}{DB}=\frac{AE}{EC}$$

$$\therefore \qquad \frac{AE}{EC}=\frac{AF}{EC}$$

Adding (1) to both sides

$$\frac{AE+EC}{EC}=\frac{AF+FC}{FC}$$

$$\therefore \qquad \frac{AC}{EC}=\frac{AC}{FC} \;\therefore\; \text{EC}=\text{FC}$$

This is possible only when F coincides with E.

$\therefore$ DE is $\parallel$ to BC.

Theorem No. 38

(Bisector Theorem)

The bisector of the angle of a Δ divides the opposite side into 2 parts whose ratio is equal to the ratio of the other 2 sides.

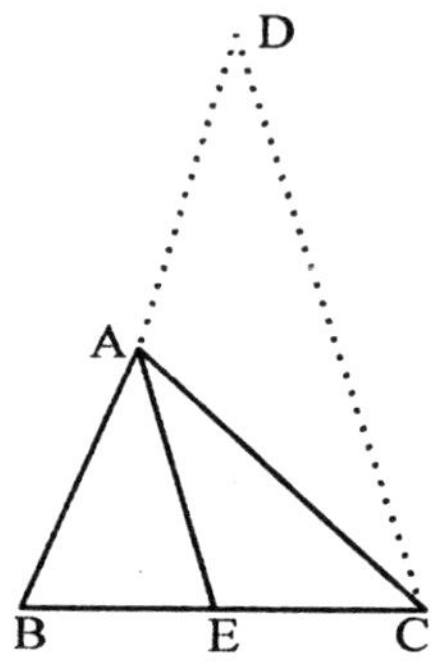

Given:- ABC is a Δ in which AE is the bisector of $\angle A$

To Prove:- $\frac{AB}{AC}=\frac{BE}{EC}$

Construction:- Draw $CD \parallel AE$ and produce BA beyond A to meet it in D.

Proof:- $AE \parallel DC$ (Construction)

AC meets it

$\therefore$ $\angle EAC = \angle ACD$ alternate $\angle S$ ...(1)

Similarly $AE \parallel DC$

AD meets it

$\angle BAE = \angle ADC$ Corresponding $\angle S$ ...(2)

But we know that $\angle EAC = \angle BAE$

From (1) and (2)

We get $\angle ACD = \angle ADC$

$\therefore$ AD = AC

$$\frac{BA}{AD} = \frac{BE}{EC} \quad \text{(Basic Proportionality Theorem)}$$

$$\frac{AB}{AC} = \frac{BE}{EC} \quad \text{(H.P.)}$$

as AD = AC

Definition

A line is said to be divided harmonically in the ratio $\frac{a}{b}$

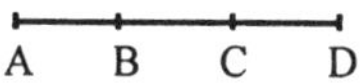

when $\frac{AB}{BC} = \frac{AD}{CD} = \frac{a}{b}$

That meams AC is divided internadly at B and externally at D in the same ratio a : b

Exercise (q)

1. Divide a line segment AB = 5 cms into 2 parts which are 2 : 3.
2. M, N are points on the sides of PQ and QR respectively of a Δ PQR. For each of the following. State whether $MN \parallel QR$.

   (i) PM = 4 cms, QM = 4.5 cms, PN = 8 cms and NR = 9 cms

(ii) PQ = 1.28, QR = 2.56, PM = 0.16 and QM = 0.32

3. The bisector of exterior $\angle A$ of the Δ ABC intersects the side BC produced in D. Prove that $\frac{AB}{AC} = \frac{BD}{CD}$

   (It is customary to say that D divides BC externally in the ratio of AB : BC).

4. Determine the point which divides the given line segment internally in the ratio of
   (i) 1 : 2 (ii) 3 : 4.

5. Determine the point which divides a given line segment externally in the ratio of
   (i) 1 : 2 (ii) 3 : 4.

6. Prove that the diagonals of a parallelogram divides each other proportionally and the ratio is equal to the ratio of parallel sides.

7. If the diagonals of a parallelogram divides each other proportionally then it is a trapezium.

8. State and prove the converse of bisector theorem.

9. The diagonals BD of the quadrilateral bisect $\angle B$ and $\angle D$. Prove $\angle B$ and $\angle D$.

10. O is any point inside the Δ ABC. The bisector of $\angle AOB, \angle BOC, \angle COD$ meet the sides AB, BC and CA in D, E and F respectively. Prove that AD.BE.CF = DB.EC.FA.

11. The internal and external bisector of $\angle A$ of the ABC divide the base harmonically.

Theorem No. 39

If 2 triangles are equiangular, then their corresponding sides are proportional (i.e. they are similar)

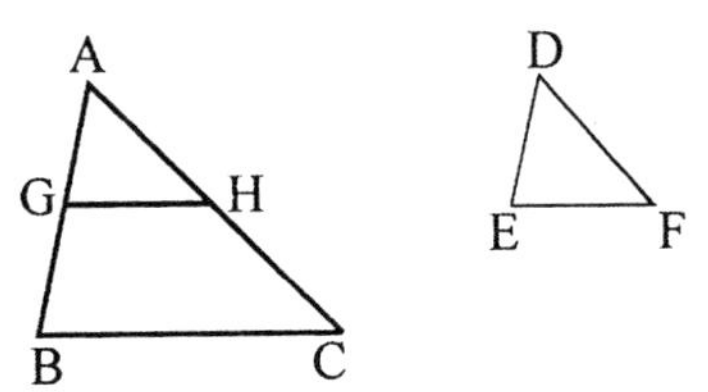

Given:- ABC and DEF are 2 Δs in which

$$\angle A = \angle D, \angle B = \angle E, \ \angle C = \angle F$$

To Prove:- $$\frac{AB}{DE} = \frac{BC}{EF} = \frac{AC}{DF}$$

Construction:- We mark the points G on AB and H on AC such that AG = DE and AH = DF. Join GH.

Proof:- In Δs AGH and DEF

because $$\begin{cases} AG = DE & (Const.) \\ AH = DF & (Const.) \\ \angle A = \angle D & (given) \end{cases}$$

ΔAGH = ΔDEF

$$\angle E = \angle AGH$$

$$\angle E = \angle B$$

$\therefore$ $$\angle B = \angle AGH$$

$\therefore$ $$GH \parallel BC$$

$$\frac{AG}{AB} = \frac{AH}{AC}$$

$\therefore$ $$\frac{DE}{AB} = \frac{DF}{AC}$$

$\therefore$ $$\frac{AB}{DE} = \frac{AC}{DF} = \frac{BC}{EF} \qquad \text{(H.P.)}$$

$$\frac{AG}{GB} + 1 = \frac{AH}{HC} + 1$$

$$\frac{AB}{GB} = \frac{AC}{HC}$$

$$\frac{AB}{AG} = \frac{AC}{AH}$$

$$\frac{AB}{DE} = \frac{AC}{DF}$$

Similarly $\frac{AB}{DE} = \frac{BC}{EF}$

$\therefore$ $\frac{AB}{DE} = \frac{AC}{DF} = \frac{BC}{EF}$

Trigonometry

The previous theorem gives us a departure into another branch of mathematics called Trigonometry.

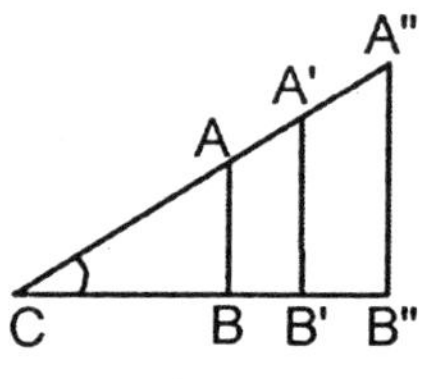

Suppose we draw a right $\angle$d ΔABC, right angled at B. We draw A^1B^1 another perpendicular and we get another right $\angle$d ΔABC. We can draw another perpendicular $A^{11}B^{11}$ and we still get another right angled $\Delta A^{11}B^{11}C$. We can see that the Δs are similar. It is easy to see that

$$\frac{AB}{AC} = \frac{A^{||}B^{||}}{A^{||}C^{||}} = \frac{A^{||}B^{||}}{A^{||}C^{||}}$$

This ratio will be constant throughout. As long as the Δs remain similar The ratio $\frac{AB}{AC}$ will be constant. This ratio is called Sin θ in trignometry which is abbrebiated as Sin θ. Similarly the ratio $\frac{BC}{AC}$ is called cosine θ. It is called cos θ in trigonometry. Similarly $\frac{AB}{BC}$ will be constant. It is called tan θ from tangent θ. The reciprocal of these ratios are called cosecant θ, secant θ and cotangent θ. These are abbrebiated as cosec, sec θ and cot θ

$$\therefore \quad \sin\theta = \frac{Perp}{Hypotenuse}, \cos\theta = \frac{Base}{Hypotenuse}, \text{ and } \tan\theta = \frac{Perp}{Base}$$

The reciprocal of these ratios are

$$\operatorname{cosec}\theta = \frac{Hypotenuse}{perp}, \sec\theta = \frac{Hypotenuse}{base}, \text{ and } \cot\theta = \frac{Base}{Perp}$$

It is not proposed to go deeper into trigonometry but it becomes easier if we utilise some of the formula in geometry. These are given below and very occasionally used

$\sin^2\theta + \cos^2\theta = 1$ | $\sin(A+B) = \sin A\cos B + \cos A\sin B$

$1 + \tan^2\theta = \sec^2\theta$ | $\cos(A+B) = \cos A\cos B - \sin A\sin B$

$1 + \cot^2\theta = \operatorname{cosec}^2\theta$ | $\sin(A-B) = \sin A\cos B - \cos A\sin B$

$\cos(A-B) = \cos A\cos B + \sec A\sin B$

$\sin^2 A = 2\sin A\cos A$

$\cos^2 A = 2\cos^2 A - 1 = 1 - 2\sin^2 A$

Theorem No. 40

If the corresponding sides of 2 Δs are propaortional, the Δs are equliangular.

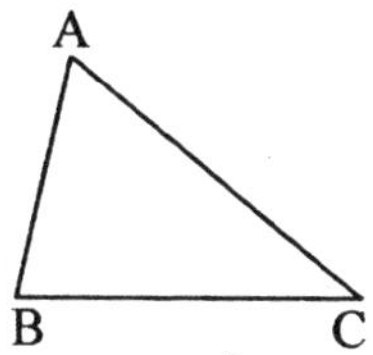

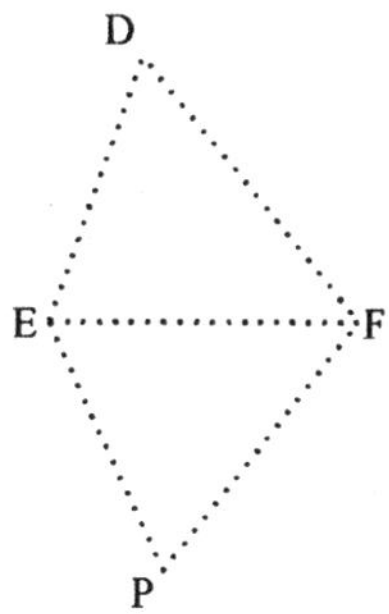

Given:- In Δs ABC and DEF

$$\frac{AB}{DE} = \frac{BC}{EF} = \frac{AC}{DF}$$

To Prove:- $\angle A = \angle D, \angle B = \angle E, \angle C = \angle F$

Construction:- One side of EF opposite to that on which D is located draw a Δ PEF in such a way $\angle PEF = \angle ABC$ and $\angle PFE = \angle ACB$.

The third angle $\angle EPF$ will automatically be $= \angle BAC$

Proof:- Since Δ ABC is equiangular to Δ PEF

$$\frac{AB}{PE} = \frac{BC}{EF}$$

But we are given that

$$\frac{AB}{DE} = \frac{BC}{EF}$$

$\therefore$ DE = PE

Similarly we can prove that DF = PF

Now in Δs DEF and PEF

because $\begin{cases} DE = PE \\ DF = PF \\ EF \text{ is Common to both} \end{cases}$

$\therefore$ Δ DEF $\equiv$ Δ PEF

$\therefore$ Δ DEF is equiangular Δ PEF

But Δ PEF is equiangular to Δ ABC

$\therefore$ Δ ABC is equiangular to Δ DEF (H.P.)

Important Note

When we say that two Δs ABC and DEF are similar we mean A corresponds to D, B corresponds to E and C corresponds to F. ABC and DEF must be written in identical order in which the Δs are correspondingly similar. We should not write. ABC and EDF are similar. Also AB corresponds to DE and BC corresponds to EF and CA corresponds to FE.

We recognise their corresponding angles by locating the sides opposite the corresponding $\angle s$. Similarly if we know corresponding sides, we recognise corresponding angles by means of locating angles opposite to corresponding sides. In the figure given herewth $PQ \parallel RS$.

Since P corresponds to S, Q to R and $\angle POQ = \angle SOR$ Similarly PO Corresponds to SO, RO to QO and PQ to SR.

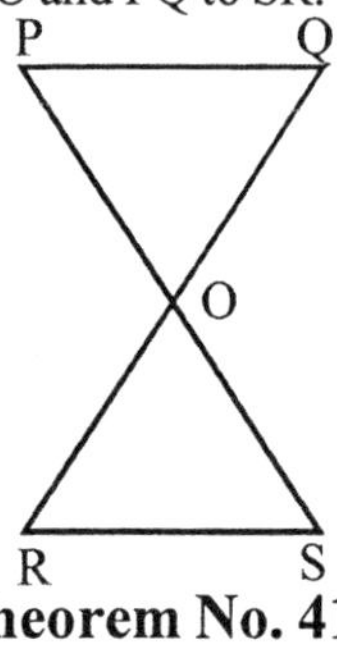

Theorem No. 41

If in two triangles one pair of corresponding sides are proportional and the angle included in this pair are equal then the triangles are similar.

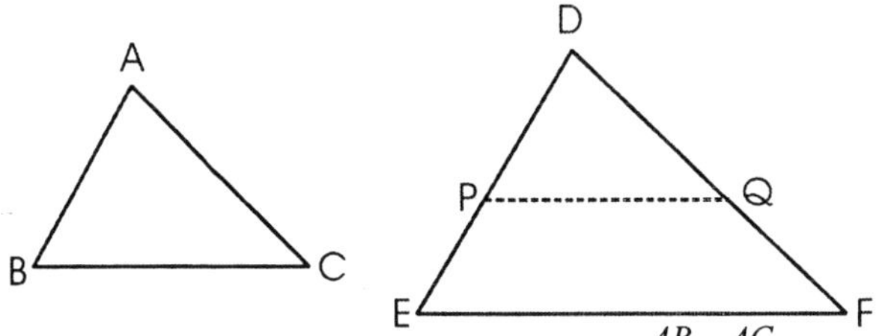

Given: Two Δs ABC and DEF are such that $\frac{AB}{DE}=\frac{AC}{DF}$ and $\angle A=\angle D$.

To Prove: $\Delta ABC \sim \Delta DEF$

Construction : Mark P on DE such that AB = DE. Mark Q on DF so that AC = DQ. Join PQ.

Proof: In Δs ABC and DPQ

Because $\begin{cases} AB=DP & \text{construction} \\ AC=DQ & -do \\ \angle A=\angle Q=\angle D & \text{given} \end{cases}$

$\therefore$ $\Delta ABC \equiv \Delta DPQ$ (SAS condition)

But $\frac{AB}{DE}=\frac{AC}{DF}$ (given)

and AB = DP

AC = DQ

$\therefore$ $\frac{DP}{DE}=\frac{DQ}{DF}$

$\therefore$ $PQ \| EF$

$\angle DPQ=\angle E$

and $\angle DPQ=\angle F$

Because ABC and DPQ are 2 congruent triangles.

Theorem No. 42.

The ratio of the areas of 2 similar triangles is equal to the ratio of squares of any two corresponding sides.

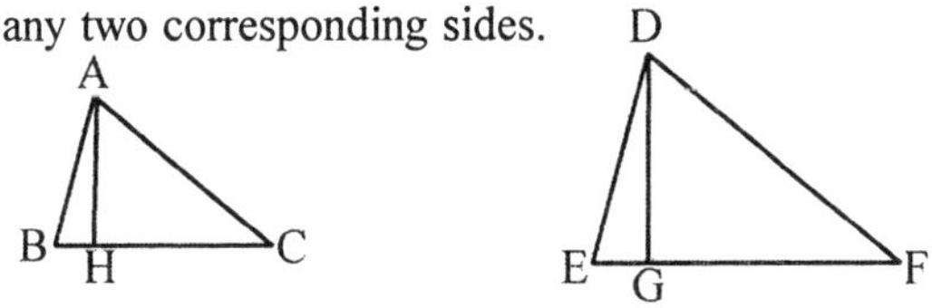

Given: $\Delta ABC \sim \Delta DEF$

To Prove: $\dfrac{Ar\,\Delta ABC}{Ar\,\Delta DEF}=\dfrac{AB^2}{DE^2}=\dfrac{BC^2}{EF^2}=\dfrac{CA^2}{DF^2}$

Construction: Draw $AH \perp BC$ and $DG \perp EF$

Proof: $\dfrac{Ar\,\Delta ABC}{Ar\,\Delta DEF}=\dfrac{\frac{1}{2}BC\times AH}{\frac{1}{2}EF\times DG}=\dfrac{BC\times AH}{EF\times DG}$ ...(1)

Now ABH and DEG are equiangular because $\angle B=\angle E$ and $\angle AHB=\angle DGE=90^\circ$ and the remaining $\angle s\ \angle BAH=\angle EDG.$

$\therefore$ $\dfrac{AB}{DE}=\dfrac{AH}{DG}=\dfrac{BC}{EF}$

$\therefore$ Substituting in (1) above

$$\frac{Area\,of\,\Delta ABC}{Area\,of\,\Delta DEP}=\frac{AH\times AH}{DG\times DG}=\frac{AH^2}{DG^2}=\frac{AB^2}{DE^2}=\frac{BC^2}{EF^2}\ \text{(H.P.)}$$

Exercise (r)

1. D and E are points on the sides AB and AC of a ΔABC. If AD = 3 cms and DB = 6 cms, and AE = 5 cms, and EC = 10 cms. Prove that BC = 3 DE.
2. Prove that the line segments joining the mid pts of a Δ forms four Δs each of which is similar to the original Δ.
3. If the angles of one Δ are respectively equal to the angles of another Δ, the ratio of corresponding sides is proportional to the ratio of corresponding (i) medians (ii) Bisectors of the angles (iii) altitudes of the Δ.
4. Prove that the areas of two similar Δs are in the ratio of the squares of corresponding (i) medians (ii) bisectors of the angles (iii) altitudes of the Δ.
5. In a Δ the median bisecting the base is itself bisected by the line joining the midpts of the other 2 sides.

6. If 2 sides and a median corresponding to one of these sides of a Δ are respectively proportional to two sides and the corresponding median of another Δ then prove that the Δs are similar.

7. If two sides and a median bisecting the third sides of a Δ are respectively proportional to the two sides and the corresponding median of another triangle then prove that the Δs are similar.

8. The bisector of the interior angle A of a ΔABC meets the base in D and the bisector of the exterior angle of the A of this Δ, meets the base produced in E then prove that D and E divide the base internally and externally in the ratio of the sides.

9. Two isosceles Δs have their vertical $\angle s$ equal. Their areas are in the ratio of 25:36. Find the ratio of their corresponding heights.

10. In a parallelogram ABCD, M is the mid point of side CD. BM is joined. It intersects AC in L and AD produced in E. Show that EL=3BL.

11. A vertical stick 18 cms long casts a shadow 12 cms on the ground. A tower casts a shadow 48 cms long. What is height of the tower.

12. A line PQ drawn parallel to the base, BC of the ΔABC. It divides ΔABC into 2 parts of equal area. Find $\frac{BP}{AB}=\left(\frac{\sqrt{2}-1}{\sqrt{2}}\right)$.

13. D is a point on the side BC of the ΔABC such that $\angle ADC=\angle BAC$. Prove that $\frac{CA}{CD}=\frac{CB}{CA}$.

14. If a perpendicular is drawn from the vertex to the hypotenuse of a right angled Δ then prove that the Δs on each side of the perp. are similar to each other and to the original Δ. Also prove that the square on the perp. is equal to the product of the lengths into which the diagonal is divided.

15. Two Δs ABC and ABD are standing on a common base with their vertices C and D on the same side of AB. From a point E in the base parallel lines are drawn to AC and AD respectively there meet BC and BD in F and G respectively. Show that $FG||CD$.

16. A ΔPFL is drawn and a point X is taken in the side FL. A parallel line AB is drawn to FL, metting PX in C. Prove that AC : CB = FX : XL.

17. In a ΔDEF the bisector of $\angle D$ meets the base EF in A and O is the in centre of the Δ then prove that DO : OA = (OE + OF) : EM.

18. In an isosceles Δ the diameter of the circumcircle is as many times the equal side of the Δ as is equal side of the Δ to the perpendicular from the vertex to the base of the Δ .

19. ABCD is a parallelogram. From A line AXY is drawn meeting BC in X and DC produce in Y. Show that BX.BY is always constant.

Theorem No. 43

Pythagoras Theorem

The square on the hypotenuse of a right angled Δ *is equal to the sum of the squares on the other two sides.*

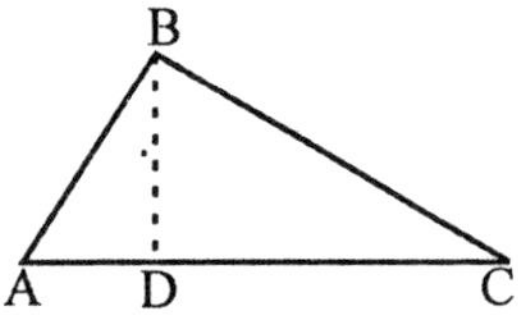

Given: A right angled ΔABC in which $\angle B = 90^\circ$.

To Prove: $AC^2 = AB^2 + BC^2$

Construction: Draw $BD \perp AC$

Proof: In Δs ADB and ABC

$\angle BAD = \angle BAC$ (common angle)

$\angle ADB = \angle ABC$ (each 90°)

$\therefore$ ΔADB is equiangular to ΔABC

$\therefore$ $$\frac{AB}{AC} = \frac{AD}{AB}$$

$\therefore$ $AB^2 = AC.AD$ ...(1)

Similarly in Δs BDC and ABC

$$\frac{BC}{CD} = \frac{AC}{BC}$$

$BC^2 = AC.CD$ ...(2)

(1) and (2) by Addition

$AB^2 + BC^2 = AC.CD + AC.AD$

$= AC(AD + CD)$

$= AC \times AC = AC^2$

$\therefore \quad AC^2 = AB^2 + BC^2$ (H.P.)

Theorem No. 44

(Converse of Pythagoras Theorem)

In a triangle if the sum of the squares on two sides is equal to the square on the third side the angle opposite to the third side is a right angle.

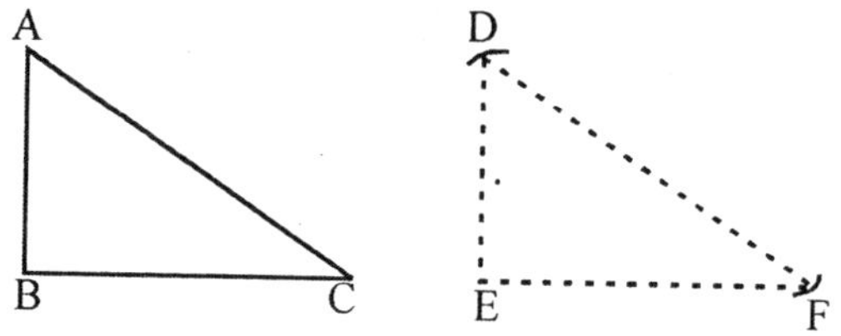

Given: A ΔABC in which

$AB^2 + BC^2 = AC^2$

To Prove: ABC is a right angled Δ with $\angle B = 90^\circ$

Construction: Construct a right angled Δ with side DE = AB and EF = BC, $\angle E = 90^\circ$

Proof: Because $\angle E = 90^\circ$

$DE^2 + EF^2 = DF^2$ (Th. 43)

But $DE = AB$ and $EF = BC$

$\therefore \quad AB^2 + BC^2 = DE^2 + EF^2 = DF^2$

But $AB^2 + BC^2 = AC^2$ (given)

$\therefore \quad AC^2 = DF^2$

$\therefore \quad AC = DF$

In Δs ABC and DEF

because $\begin{cases} AB = DE & \text{given} \\ BC = EF & \text{– do –} \\ AC = DF & \text{proved above} \end{cases}$

$\therefore \quad \Delta ABC \equiv \Delta DEF$

$\therefore \quad \angle ABC = \angle DEF = 90°$ (H.P.)

Exercise (s)

1. A ladder erected 5 m from the house reaches a point 12 m high. What is the length of the ladder? (13 m)
2. A ladder 41 m long reaches a window 40 m high. How far is its end from the house? (9 m)
3. A man goes 11m towards the South and then goes 24 m towards the west. He again goes 4 m. north. How far is he from the starting point? (25 m)
4. A ladder 73 m long is tied to a point and reaches 55 m high. Then it is turned on the other side when it reaches a window 48 m high on the other side. What is the width of the road? (103 m)
5. What is the length of a diagonal of a square field 20 m in length? $(20\sqrt{2}\,m)$
6. Length of the diagonal 41 m and one side = 40 m. What is the length of the other side. (9 m.)
7. Length of the diagonal of a square field = 60 m. What is its area. (1800 sqm)
8. An isosceles Δ has base = 120 m and equal side = 65 m. What is its area? (1500 sqm)
9. Length of perimeter of rectangle = 14 m and diagonal = 5 m. What is its area? (12 sqm)
10. Perimeter of a rhombus = 260 m One diagonal = 32 m. What is the other diagonal and the area. (126 m, 2016 m^2)
11. A right triangle has a diagonal = 16 m and its area = 40sq. meters. What is its perp.? (5 m)
12. An equilateral Δ has one side = 9 cms. Prove that its height $= \frac{9\sqrt{3}}{2}$ m and its area $= \frac{81\sqrt{3}}{4} m^2$.

13. AD is perp. to the base of a ΔABC. Prove that $AB^2 - AC^2 = BD^2 - CD^2$.

14. The sum of the squares on the side of the rhombus is equal to the sum of the squares on both the diagonals.

15. ABC is a Δ in which AB = AC and D is any point on the base BC. Prove $AB^2 - AD^2 = BD.CD$.

16. Point O is taken in the interior of a rectangle ABCD. It is joined to the vertices A, B, C, D. Prove that $OB^2 + OD^2 = OC^2 + OA^2$.

17. In an isosceles ΔABC, AB = AC. BD is the perpendicular from B to the side AC. Prove that $BD^2 - CD^2 = 2CD.AD$.

    $BD^2 - CD^2 = AB^2 - AD^2 - CD^2 = AC^2 - AD^2 - CD^2 = 2CD.AD$

18. In an equilateral ΔABC. The side BC is trisected at D. Prove that $9AD^2 = 7AB^2$.

19. The perpendicular AD on the base BC of a ΔABC intersects it at the point D, so that DB = 3 CD. Prove that $2AB^2 = 2AC^2 + BC^2$.

20. From a point O in the interior of the ΔABC perpendicular OD, OE, OF are drawn to the sides BC, CA and AB respectively. Prove that

    (i) $AF^2 + BD^2 + CE^2 = OA^2 + OB^2 + OC^2 - OD^2 - OE^2 - OF^2$

    (ii) $AF^2 + BD^2 + CE^2 = AE^2 + CD^2 + BF^2$

*21. P and Q are the mid-pts of the sides CA and CB respectively of a ΔABC right angled at C. Prove that

    (i) $4AQ^2 = 4AC^2 + BC^2$

    (ii) $4BP^2 = 4BC^2 + AC^2$

    (iii) $4AQ^2 + 4BP^2 = 5AC^2$

*22. ABC is a right angled Δ with $\angle C = 90^\circ$. If p is the length of the perp. to AB from C. Side BC = a and AC = b then prove that

    (i) PC = ab and (ii) $\frac{1}{p^2} = \frac{1}{a^2} + \frac{1}{b^2}$.

23. Can the above example be solved trignometrically

$\sin\theta=\frac{b}{b}$ $\cos\theta=\frac{a}{b}$ $\sin^2\theta+\cos^2\theta=1$

$\frac{b^2}{a^2}+\frac{b^2}{b^2}=1$ $\therefore \frac{1}{a^2}+\frac{1}{b^2}=\frac{1}{b^2}$

*24. In a ΔABC AC > AB. D is the midpoint of BC and $AE\perp BC$. Prove that

(1) $AC^2=AD^2+BC.DE+\frac{1}{4}BC^2$

(2) $AB^2=AD^2-BC.DE+\frac{1}{4}BC^2$

(3) $AC^2+AB^2=2AD^2+\frac{1}{4}BC^2$

*25. Prove that three times the sum of squares of the sides of a Δ is equal to four times the sum of the squares of the median of the Δ.

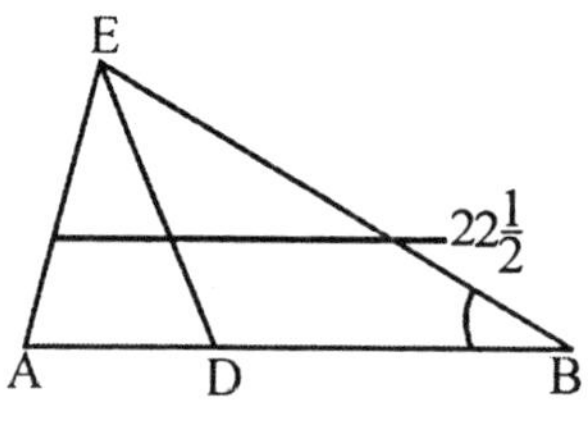

26. Divide a line in 2 parts so that the square one part is twice the square on the other.

(**Hint:-** Suppose D is the point on the given line AB so that $2AD^2=DB^2$ so that $ED^2 = DB^2$ $\therefore$ $\angle EBD=22\frac{1}{2}^\circ$)

27. Divide a line in 2 parts so that the square on one part = thrice the square on the other.

(**Hint:-** Suppose AB is the line. Make $\angle BAE=45^\circ$ and make $\angle DBE=30^\circ$. Drop, E is required point)

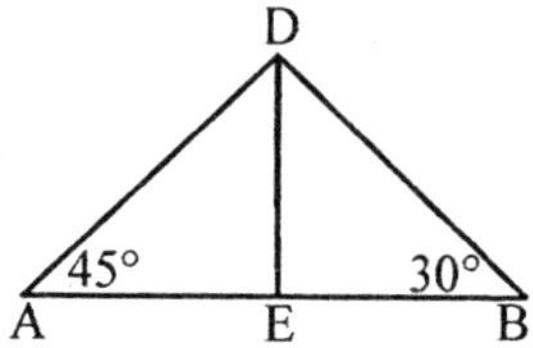

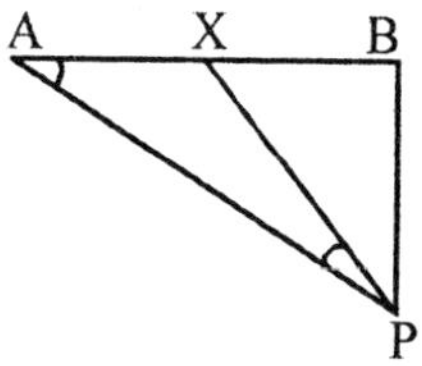

28. Divide a line in 2 parts so that the difference in square on the two parts is equal to a square on a given side.

(**Hints:-** At B make at ⊥ to the line so that BP = a give square side. At P make an angle XPA = XAP. X is the requidred point.)

*30. The hypotenuse + one side is given. Draw the isosceles right Δ. (Draw AD = hypotenuse + side.

At D make an angle = $22\frac{1}{2}^{\circ}$ meeting the perp. from A in point B. At B make equal angle CBD = $22\frac{1}{2}^{\circ}$ at B. ABC is the Δ)

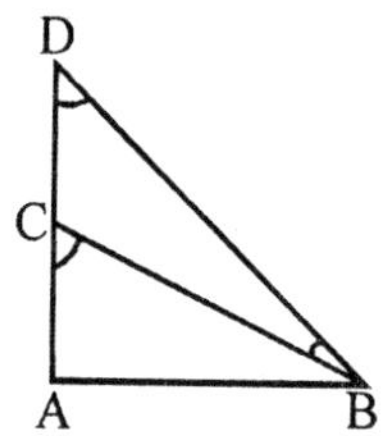

*31. A peacock is sitting quietly on tree 12 meter high. There is a snake in its hideout at a distance of 18m from the pole. The snake crawls towards the pole from the hideout. At the same time the peacock flies towards the snake in a st. line and catches it enroute. The speed at the peacock and the snake are the same. When and where will it catch the snake. At what angle with the vertical will the peacock fly?

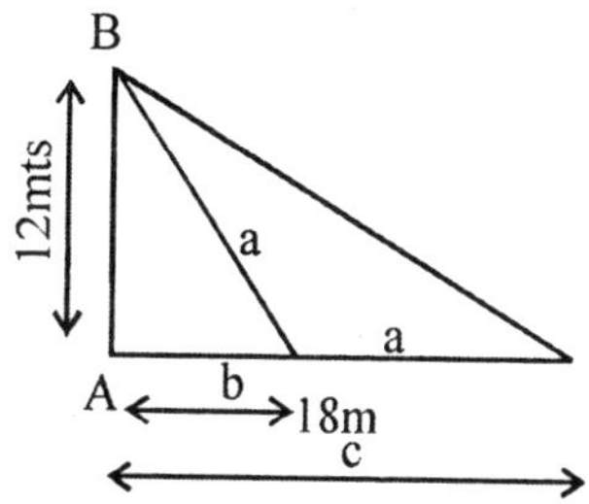

$a^2 = b^2 + 144$ $\quad 2a = 26$ (LEELAVATHI)

$a^2 - b^2 = 144$ $\quad a = 13$

$a + b = 18$ $\quad b = 5$

$a - b = 8$

*32. A lotus with the vertical stem stands erect in a pond. Its head out of water is $\frac{1}{2}$ cm. When the wind comes the lotus falls sideways a distance of 2cm and is completely submerged. What is dept of pond. (LEELAVATHI)

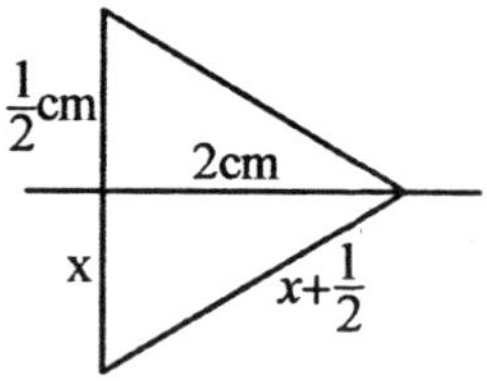

$$\left(x+\frac{1}{2}\right)^2 = x^2+4$$

$$x^2+x+\frac{1}{4}=x^2+4$$

$$x=4-\frac{1}{4}=\frac{15}{16}m.$$

*33. A monkey is sitting on a tree 50m. high and jumps straight a distance of 100m. from it. Another monkey is sitting at the foot of the tree first jumps vertically upwards and then jumps to the same spot as the first mokey. If the distance travelled by both monkeys be the same how high the second monkey jumps. (LEELAVATHI)

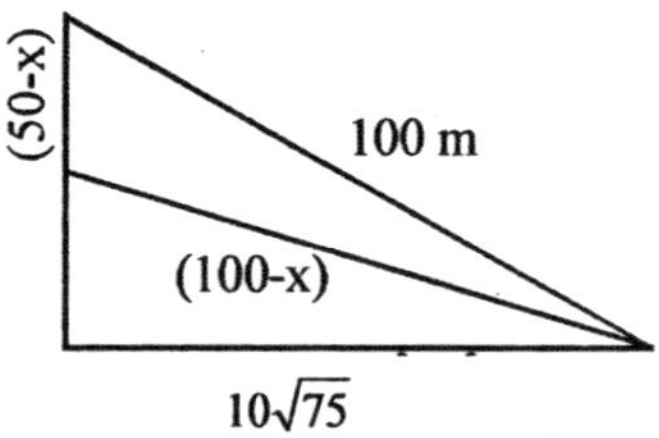

$$(100-x)^2 = x^2+7500$$
$$10000-200x+x^2=x^2+7500$$
$$2500=200x$$
$$12.5m=x$$

34. ABCDEF is a regular hexagon. Each end point has 3 diagonals. The one in the centre is twice each side of the diagonal while the other are $\sqrt{3}\times$ one side.

Practical Geometry - III

(Construction of quadrilaterals of areas equal to other figures)

Construction No. 16

To draw a square = twice a given square

= thrice a given square

= four times a given square

= five times a given square

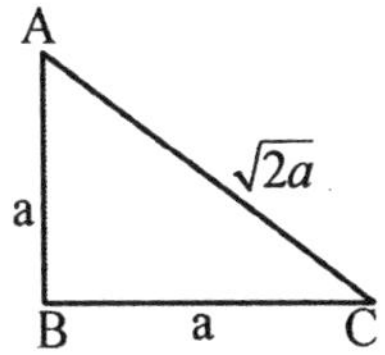

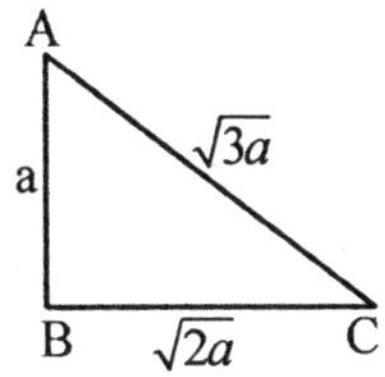

AB side of a square

$$AC = \sqrt{2}a$$

$\therefore$ $AC^2 = 2a^2$

$\therefore$ AC is a side twice a square

like that $AD = \sqrt{3}a$

$AE = 2a$

and $AF = \sqrt{5}a$

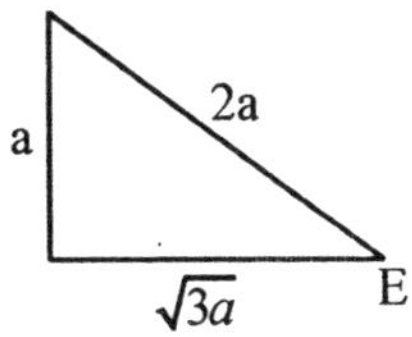

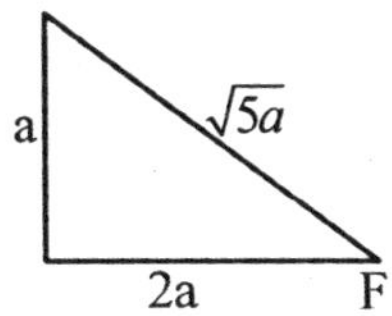

Construction No. 17

To find the area of Triangle (Hero's formula)

Construction: Draw $AD \perp BC$.

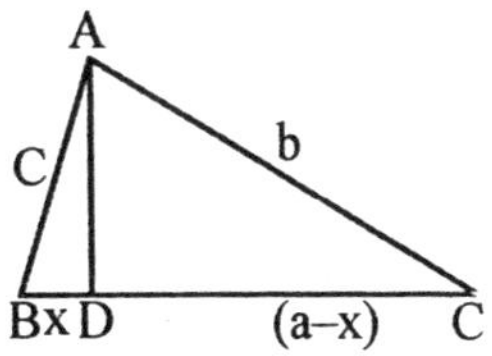

$$AD^2 = c^2 - x^2$$

$$AD^2 = b^2 - (a - x)^2$$

$$\therefore \quad c^2 - x^2 = b^2 - (a-x)^2$$

$$c^2 - x^2 = b^2 - a^2 - x^2 + 2ax$$

$$a^2 + c^2 - b^2 = 2ax$$

$$x = \frac{c^2 + a^2 - b^2}{2a}$$

$$\therefore \quad AD^2 = c^2 - \left\{\frac{a^2 + c^2 - b^2}{2a}\right\}^2 = \frac{4a^2c^2 - \left(a^2 + c^2 - b^2\right)^2}{4a^2}$$

$$= \frac{\left(2ac + a^2 + c^2 - b^2\right)\left(2ac - a - c^2 + b^2\right)}{4a^2}$$

$$= \frac{\left[(a+c)^2 - b^2\right]\left[b^2 - \left(a^2 + c^2 - 2ac\right)\right]}{4a^2}$$

$$= \frac{(a+b+c)(a-b+c)(b+a-c)(b-a+c)}{4a^2}$$

$\therefore$ $2a.AD = \sqrt{(a+b+c)(-a+b+c)(a-b+c)(a+b-c)}$

$\therefore$ $4\Delta = \sqrt{2s \times (2s-2a)(2s-2b)(2s-2c)}$ If $(a+b+c) = 2s$

$\therefore$ $\Delta = \sqrt{s(s-a)(s-b)(s-c)}$

Construction No. 18

To draw a parallelogram equal in area to a given Δ but whose angle is equal to a given angle.

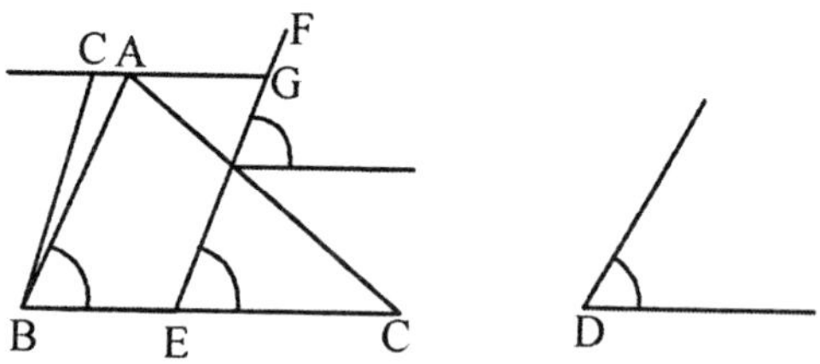

Given: Let ABC be the given Δ and D be the given angle.

To Draw: A parallelogram equal in area to ΔABC and one its angles equal to $\angle D$.

Construction: Bisect BC at the point E and at E make an angle = $\angle D$ so that $\angle FEC = \angle D$. Through A draw AEG parallel to BC. Through B draw a parallel EF to EG, EGCB is the parallelogram.

Proof: Area of $\Delta ABC = \frac{1}{2} BC \times h$

Where h is the distance between AG and BC.

Area of parallelogram EGCF

$$= EC \times h = \frac{1}{2} BC \times h$$

$$= \text{Area of } \Delta ABC$$

Construction No. 19

To draw a triangle equal in area to a given quadrilateral.

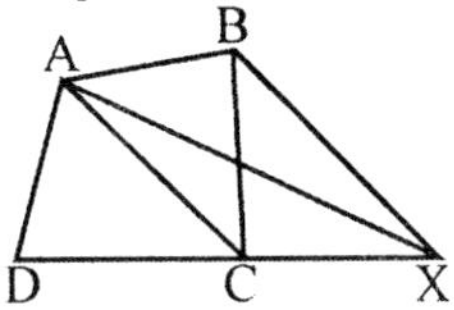

Given: ABCD is a given quadrilateral.

To Draw: A Δ equal in area to quadrilateral ABCD.

Construction: Draw the diagonal AC.

Through B draw BX parallel to AC.
Let it meet DC in X. Join AX. AXD is the Δ equal in area to quad. ABCD.

Proof: ΔABC and ΔAXC stand on the same base AC and between the same parallel lines AC and BX

$\therefore \quad \Delta ABC = \Delta ACX$

$\therefore \quad \Delta ADC + \Delta ABC = \Delta ADC + \Delta ACX$

quad. ABCD = ΔACX

Construction No. 20

To draw a square equal in area to a given rectangle.

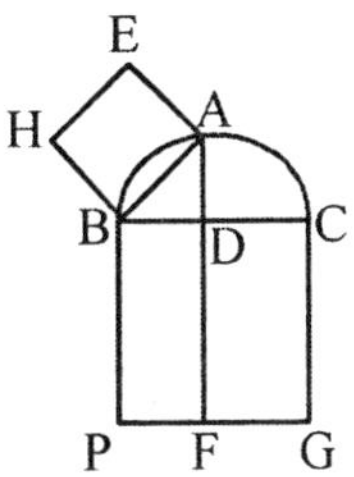

Given: A rectangle BPFD

To Draw: We are required to draw a Square equal in area to this rectangle BPFD

Construction: (i) Complete the square BPGC by extending BE upto BC

BC = BP.

(ii) Draw the semi circle on BC.

(iii) Extend FD to A so that A is on the circle.

(iv) Draw square on BA.

(v) Square on BA = Rectangle BPFE.

Miscellaneous Exercise (U)

(1) In any quadrilateral the middle point of the four sides are joined. The parallelogram so formed is half the area of the original quadrilateral.

(2) Prove that if the diagonals of a quadrilateral are at right angles then the sum of the squares on opposite sides are equal to each other.

(3) **If the diagonals of a quad. are at right angles then its area is equal to half the area contained by the diagonals.**

(4) If the length of the diagonals and the angle between them is given, the area of the quad. is the same howsoever you draw the quad.

(5) In a right angled Δ the sum of the squares on the three medians is equal $1\frac{1}{2}$ times the square on the diagonal.

(6) ABCD is a parallelogram. E is a point outside or inside. The distance of AE from C is either the sum or difference of its distance from B and D according as D is inside or outside the parallelogram.

(7) AB and CD are 2 straight lines intersecting at X. If the area of ΔAXC=area of ΔBXD, then $AD \parallel CB$.

(8) ABC and DBC are 2Δs standing on a common base BC and are between the same $\parallel$ lines BC and AD. Parallelogram ABCE is completed. AC and BD meet in F. Prove that

Area of ΔDCE = (Area of ΔFAD) ~ (Area of ΔFBC)

(9) From the vertices of a ΔABC three lines are drawn which are parallel to each other. They meet the opposite sides in D, E, and F. Show that the area of ΔDEF = twice the area of ΔABC.

(10) In a rectangle ABCD, the centre of BC is E and F is a point on CD such that CF = 3FD. Prove that = $\frac{7}{16}$ rectangle ABCD in area.

(11) In ΔABC AB is produced to D, DC is produced to E, and CA is produced to F is such away that BD, CE and AF are each equal to twice AB, BC and CA respectively.

Prove that $\Delta DEF = 19\ \Delta ABC$.

(12) A point is taken in a regular figure of any no. of sides then the sum of its perpendicular distances from all the sides is the same irrespective of where you take the point.

(13) In a ΔABC AB and AC are bisected at D and E. While the side

BC is trisected in F and G. DF and EG meet at H outside the Δ.

Prove that $\Delta HFG = \frac{1}{3}\Delta ABC$.

(14) G is the centroid of the ΔABC. Δ is formed with the sides AG, BG and CG. Prove that its area $= \frac{1}{3}\Delta ABC$.

(15) ABCD is a parallelogram. E and F are the mid pts of CD and DA. BE and BF are joined and they meet AC in M and F. Show that pentagon FMNED $= \frac{1}{3}$ parallelogram ABCD.

(16) In a ΔABC, D and E are the midpts of AB and AC. From D and E two parallel lines DP and EF are drawn meeting BC in P and F. Prove that the parallelogram DPFE $= \frac{1}{2}$ area of ΔABC.

(17) ABCD is a parallelogram. $\overline{CF}$ is drawn meeting CB produced in F and AB in P. Prove that ΔABF.

(18) In a quad. if the sum of the squares of one pair of opposite sides is equal to the sum of the squares of the other opposite sides then the diagonals of the quadrilateral are at right angles.

(19) In a Δ squares are drawn on the sides outside the Δ. Six points are formed outside the Δ. The free points of the adjacent squares are joined by 3 straight lines so that figures are formed of 3 Δs + 3 squares in addition to the main Δ. Prove that the four Δs are equal in area.

(20) The angles of a parallelogram are bisected and they form a rectangle. Prove that the diagonals of the rectangle is equal to the difference in the two adjacent side of the main parallelo-gram and the diagonals of the rectangle coincide with the intersection main parallelogram.

(21) ABC is a Δ with A as a right angle. On side AB an equilateral Δ ABP is drawn on the opposite side of the Δ ABC. Also on BC is drawn an equilateral Δ BCF also outside the Δ ABC. Prove that

$\Delta AFB = \Delta ABP + \frac{1}{2}\Delta ABC.$

(22) A, B, C, D are fixed points. E is such a point that Δ EAB+ Δ ECD is fixed. What is the locus of E?

(23) In the figure on the side

AB = BC = 4 cms

BP = CD = 3 cms

AP = BD = 5 cms

Find the area of shaded figure.

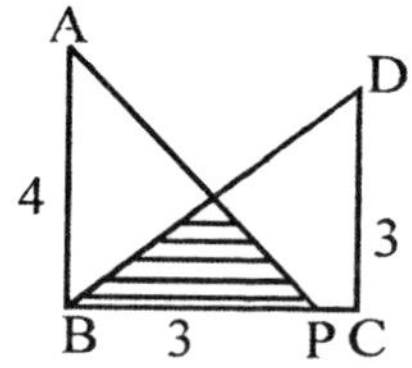

(24) A right angled Δ with perimeter = 2s and diagonal = b. Prove that its areas = s(s-b)

(25) ABC is a right angled Δ with B = 90°. The bisector of $\angle B$ meets the diagonal in P. From P perpendicular are dropped on AB and BC. They meet AB and BC in X and Y respectively. Prove that

$$\frac{1}{PX} = \frac{1}{AB} + \frac{1}{AC}$$

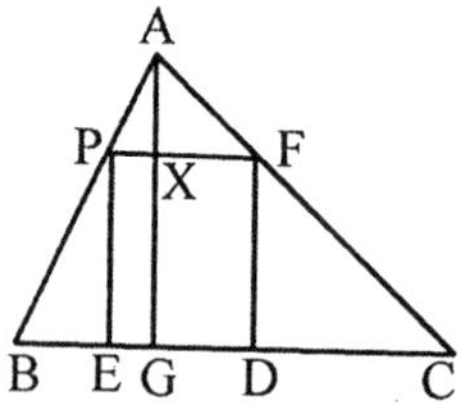

(26) In a figure drawn on R.H.S. AG $\perp$ BC and PFDE is a square drawn in Δ ABC If PF = x, AG = y and BC = z. Prove that $\frac{1}{x} = \frac{1}{y} + \frac{1}{z}$

(27) In Δ ABC, AD is the median. It meets a line || BC and bounded by AB and AC in P and F. PF and AD intersect in G. Show that PG = GF.

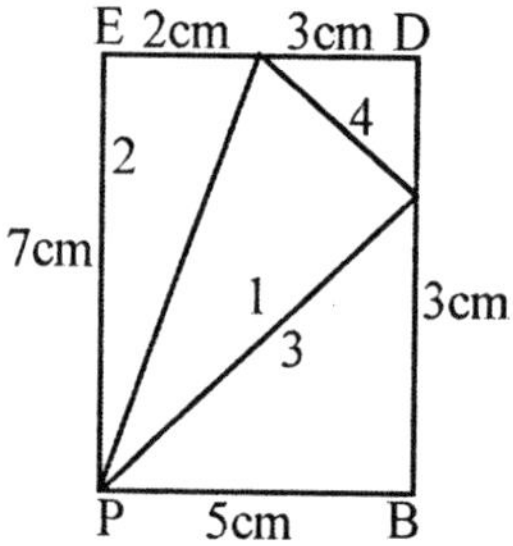

(28) In the figure PEDB is a rectangle 7 x 5 cm. Find out the area of the rectangle. Find all the heights of the Δ s.

(29) Draw a rectangle whose area is twice a given quadrilateral and whose one side = one diagonal of the quadrilateral.

3

Circles and Tangents

Definitions

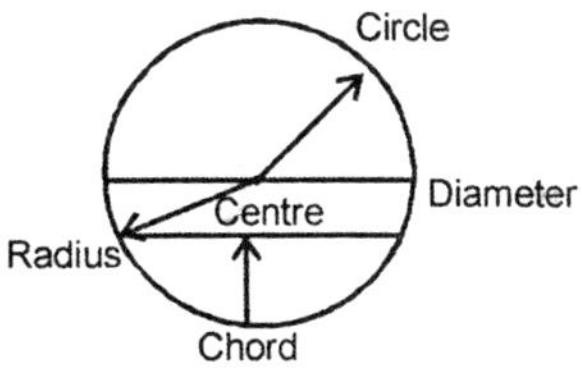

A *circle* is the locus of a point when it moves so that its distance from a fixed point called the *CENTRE* of the circle is always constant. The constant distance is called that *Radius* of the circle.

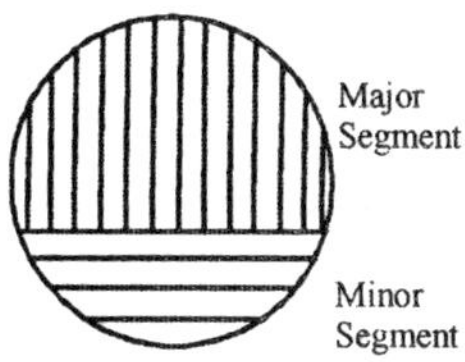

The boundary of the circle is known as the circumference of the circle. Actually circumference of circle is known as the circle. When 2 points on the circumference are joined by a line segment it is known as the *chord* of the circle. *A chord is* not a part of the circle. Only its end points are on the circle. When a chord passes through the centre of the circle it is known as the *Diameter*. There are many diameters of the circle, in fact infinite. A diameter is twice the radius of the circle.

A chord divides the circle into two segments. They are called *MAJOR SEGMENT* and *MIROR SEGMENT* respecetively depending upon it is bigger or smaller. A minor arc is the arc of a circle that subtends an angle less than 180° with the centre while a major arc subtends an angle more than 180° with the centre. Major segment is shown by vertical lines and minor segment is shown by horizontal line.

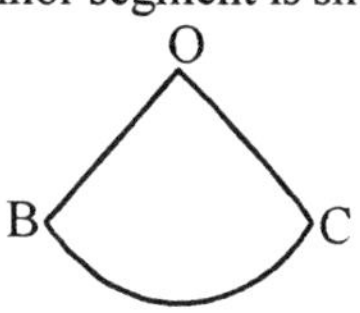

Minor Arc BC
A Minor Are BC Subtends < 180° at the centre

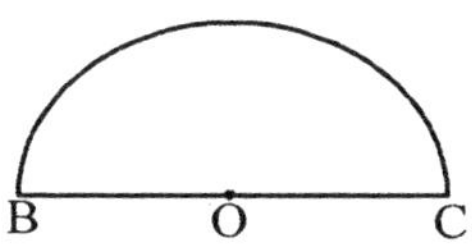

Semi Circle sebtends BOC = 180° at the centre

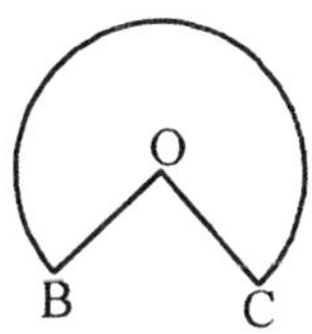

Major Arc BC
Major are BC subtends > 180° at the centre

From the above definition it is quite clear that

(i) The circle is a closed curve.

(ii) When a st. line intersects circle it has to intersect it again before coming out.

(iii) When a point is taken at a distance > = < than the radius depending upon whether is more than, on or inside the circle.

(iv) Two circles of equal radius are known as congruent circles.

(v) Two circles with a comnon centre do not intersect each other and are called concentric circles.

There are the properties of the circle.

Theorem No. 45

A circle is symmetric about any diameter.

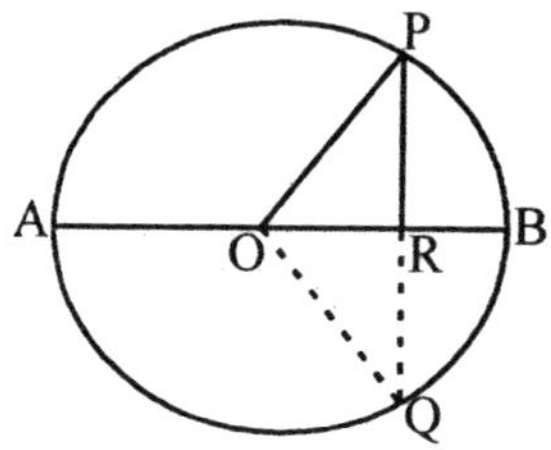

Given:- A circle is given with O as centre and any diameter AB.

To Prove:- The circle is symmetric about this or any diameter.

Construction:- Take a point P on the circumference and draw PR $\perp$ AB. Produce PR beyond R so that RQ = PR. Join O to Q.

Proof:- In Δs OPR and ORQ

because $\begin{cases} OR \text{ is Common} & \\ \angle ORP = \angle ORQ & \text{(both } 90^\circ) \\ PR = QR & \text{construction} \end{cases}$

$\therefore$ Δ OPR $\equiv$ OQR

$\therefore$ OP = OQ

$\therefore$ Q is on the circle

This shows that for every point P on the circle there is a point Q on the other side.

But AB is any diameter.

$\therefore$ The circle is summetric about any diameter. (H.P.)

Theorem No. 46

In a circle the ratio of the circumference to diameter is always the same.

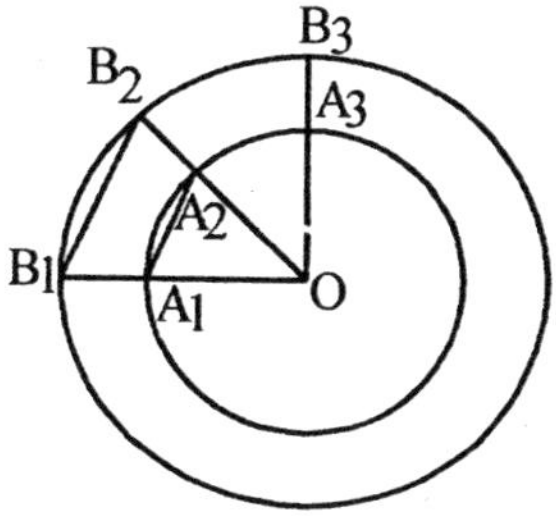

Given:- Two concentric circles with centre O. Let the radio be $OA_1 = r_1$ and $OB_1 = r_2$.

To Prove:- $\frac{\textit{Circumference of the circle}}{\textit{Diameter of the Circle}} = \text{Constant.}$

Proof:- Draw common radii OA_1B_1, OA_2B_2, OA_3B_3 --- $OA_{n-1}B_{n-1} + OA_n B_n$ and back to OA_1B_1

Taking the first 2 Δ s OA_1, A_2 and OB_1B_2. They are similar.

Hence their Corresponding sides are proportional

$$\therefore \quad \frac{OB_1}{OA_1} = \frac{OB_2}{OA_2} = \frac{B_1B_2}{A_1A_2}$$

Extending this to all the similar Δ s round the circle we get

$$\frac{OB_1}{OA_1} = \frac{OB_2}{OA_2} = \frac{OB_3}{OA_3} = \frac{OB_4}{OA_4} \cdots\cdots \frac{B_1B_2}{A_1A_2} = \frac{B_2B_3}{A_2A_3} = \frac{B_3B_4}{A_3A_4} = \cdots\cdots \frac{B_nB_1}{A_nA_1}$$

$$\therefore \quad \frac{r_2}{r_1} = \frac{B_1B_2}{A_1A_2} = \frac{B_2B_3}{A_2A_3} = \frac{B_3B_4}{A_3A_4} = \frac{B_{n-1}B_n}{A_{n-1}A_n} = \frac{B_nB_1}{A_nA_1}$$

$$\therefore \quad \frac{r_2}{r_1} = \frac{B_1B_2 + B_2B_3 + \ldots B_{n-1}B_n + B_nB_1}{A_1A_2 + A_2A_3 + \ldots A_{n-1}A_n + A_nA_1}$$

The no. of Δ s are increased indefinitely so that the right hand of the equation becomes.

$$= \frac{\text{Circumference of outer circle}}{\text{Circumference of inner circle}}$$

$$\therefore \quad \frac{\text{Circumference of outer}\odot}{r_2} = \frac{\text{Circumference of inner}\odot}{r_1}$$

$$\therefore \quad \frac{\text{Circumference of any circle}}{2 \times \text{radius}} = \text{Constant}$$

This ratio is known as π

The value of π cannot be found accurately as it does not terminate to any constant number. It has been found to be equal to 3.14159---. It is approximately taken 3.1416.

$$\therefore \quad \frac{\text{Circumference}}{\text{diameter}} = \pi$$

$$\therefore \quad \text{Circumference} = 2\pi r$$

Corollary 1

To find the area of a circle whose radius is r.

In a circle of radius r draw radii at 15° interval. We will get 6 radii in each quadrant. i.e. 24 radii in 4 quadrants. Now take quadrant No. 1. Cut off sector No. 1 and Sector No. 2 from this quadrant and place them against each other so that the curved portions remain opposite each other and straight portions being equal overlap each other as shown below.

Then after this quadrants No. 3. and No. 4. are similarly joined and placed against quadrant No. 1 and No. 2 as shown below.

Further we also add quadrant No. 5 and No. 6 and the whole area of 6 quadrants from 1 to 6 takes a shope as below.

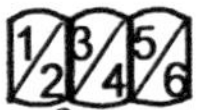

Then we do the same thing for sector No. 7 to sector No. 12 and get one more area as shown below.

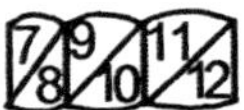

In this way we divide for sector No. 13 to 18 in the third quadrant and sector No. 19 to 24 in the fourth quadrant and we get the entire area of 24 quadrant covered as shown below.

Thus all the 24 quadrants joined together we get the whole area of circles covered in a shape as below.

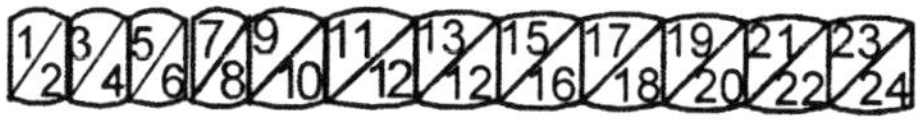

This has become the case of 24 quadrants but we can increase the no. of quadrants indefinitely in which case the area is covered in a rectangle whose one side becomes = radius of the circle and second side

$$= \frac{Circumference}{2} = \frac{2\pi r}{2} = \pi r$$

$\therefore$ The area of the circle = Area of rectangle whose sides are r $\times$ πr

$$= \pi r^2$$

Corollary 2

To find the area of sector of a circle of angle = D°

(a) If 2 radii of the circle make an angle of 1° it will contain an area $\frac{1}{360}$ of the whole area. i.e it will cover an area = $\frac{\pi r^2}{360}$ D

$\therefore$ The area covnered by a sector of an angle = D° will be = $\frac{\pi r^2}{360}$ D

Corollary 3

To find the area of a segment of circle subtending an angle of D° at the centre.

The area of sector of the sector of circle = $\frac{\pi r^2 D}{360°}$

From this we subtract the area of $\triangle$AOB to get the area of the segment.

The area of the segment

$$= \frac{\pi r^2 D}{360} - \frac{1}{2} \times 2r \sin\alpha \times r\cos\alpha$$

$$= \frac{\pi r^2 D}{360} - \frac{r^2}{2} \times 2\sin\alpha \,.\, \cos\alpha$$

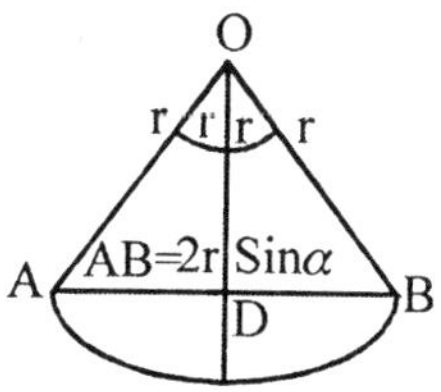

$$= \frac{\pi r^2 D}{360} - \frac{r^2}{2} \sin 2\alpha$$

$$= \frac{\pi r^2 D}{360} - \frac{r^2}{2} \sin D$$

$$= \frac{r^2}{2}\left(\frac{\pi D}{180} - \sin D\right) \text{ square units}$$

$D = 2\alpha$

$\text{Sin } D = \text{Sin } 2\alpha$

$= 2 \sin\alpha \cos\alpha$

For the present we have to assume this formula.

Theorem No. 47

If 2 arcs of a circle or congruent circles are congruent, the corresponding chords are equal.

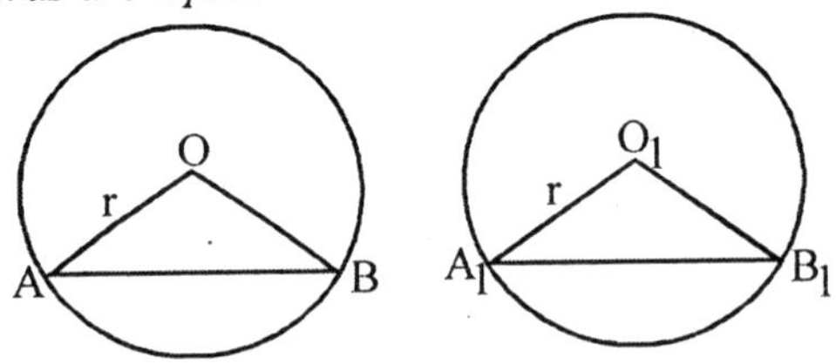

Conversely if two chords of a circle or congruent circles are equal, the corresponding arcs are congruent.

Given:- Two circles with centre at O and O_1 and their radii

$OA = O_1A_1 = r.$

Arc AB = Arc $A_1 B_1$

To Prove:- Chord AB = Chord A_1B_1

Proof (1):- Arc AB = Arc A_1B_1

$\therefore$ $\angle AOB = \angle A_1O_1B$

because the two circles have the same radius.

In Δ s AOB and $A_1 O_1 B_1$

because $\begin{cases} AO = A_1O_1 & (\text{both radii}) \\ BO = B_1O_1 & (\text{both radii}) \\ \angle AOB = \angle AOB & (\text{just proved}) \end{cases}$

$\therefore$ $\Delta AOB \equiv \Delta A_1O_1B_1$ (SAS Axiom)

$\therefore$ $AB = A_1B_1$

Conversely

Given:- Two circles with centre O and their radii

$OA = O_1A_1 = R$ and chord AB = chord A_1B_1

To Prove:- $\angle AOB = \angle A_1O_1B_1$

Proof :- In Δ s AOB and $A_1O_1B_1$

because $\begin{cases} AO = A_1O_1 & (both\ radii) \\ BO = B_1O_1 & (both\ radii) \\ AB = A_1B_1 & (given) \end{cases}$

$\therefore$ There Δs are congruent

$\therefore$ $\angle AOB = \angle A_1O_1B_1$ (H.P.)

Theorem No. 48

One and one circle can be drawn through three points not in the same stranght line.

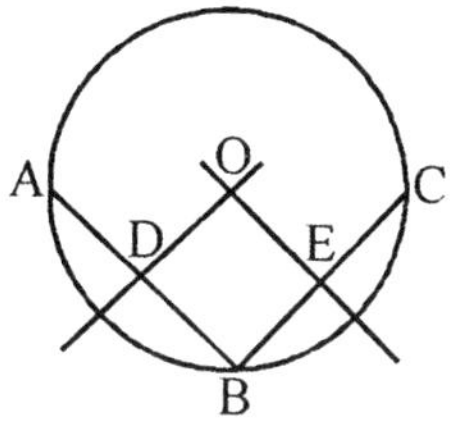

Given:- Three points A B C are given.

They are not in the same straight line.

To Prove:- Only one circle can be made to pass through A, B and C.

Proof :- Let DO be the perpendicular bisector of AB.

All pts on DO are equidistant from A and B (Th 27)

Let EO be the perpendicular bisector of BC.

All pts on EO are equidistant from B and C. (Th 27)

EO and DO intersect at one point only. (Th 1)

That means only one point is equidistant from A, B, and C

Hence one circle can be made to pass through A, B and C.

Corollary:- No circle can be made to pass through 3 points in the same st. line.

Theorem No. 49

The perpendicular from the centre of the circle to a chord bisects the chord.

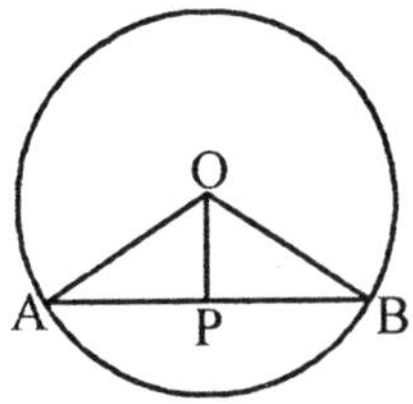

Conversely, if the centre of the circle is joined to the centre of the chord, it is at right angles to the chord.

Given:- AB is a chord of the circle with center at O.

$OP \perp AB$.

To Prove:- AP = PB

Proof :- In Δs AOP and BOP

because $\begin{cases} AO = OB & \textit{(both radii)} \\ OP \text{ is Common} & \\ \angle OPA = \angle OPB = 90° & \textit{(given)} \end{cases}$

As per Th 18 $\Delta AOP \equiv \Delta BOP$

$\therefore$ AP = PB

Converely

Given :- AP = PB and O is the Centre of the circle

To Prove:- $OP \perp AB$

Proof:- In Δs OPA and OPB

because $\begin{cases} AO = OB & \textit{(both radii)} \\ OP \text{ is Common} & \\ \angle OPA = \angle OPB = 90° & \textit{(given)} \end{cases}$

By theorem 13 the Δs are congruent

$\therefore$ $\angle APO = \angle BPO$

But these are adjacent $\angle s$

They add up to 180° (LPA)

$\therefore$ each is 90°

$\therefore$ $OP \perp AB$

Theorem No. 50

Equal chords of a circle are equidistant from the centre.

Conversely, if two chords of a circle are equidistant from the centre they are equal.

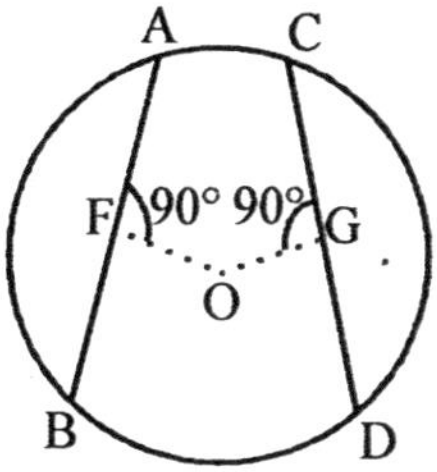

Given:- AB and CD are 2 equal chords of a circle with centre O.

$OF \perp AB$ and $OG \perp CD$.

To Prove:- OF = OG

Proof:- $OF \perp AB$

$\therefore$ $BF = AF = \frac{1}{2}AB.$ (Th 49)

$OG \perp CD$

$\therefore$ $CG = GD = \frac{1}{2}CD.$ (Th 49)

But AB = CD

$\therefore$ BF = AF = DG = CG

In Δs BFO and DGO

$$\begin{cases} BO = DO & (\text{both radii}) \\ BF = DG & (\frac{1}{2}\text{qual chords}) \\ \angle OFB = \angle OGD = 90° & \end{cases}$$

$\therefore$ These Δs are Congruent as per (Th 18)

$\therefore$ OF = OG

Conversely

Given:- OF = OG

To Prove:- AB = CD

Proof :- In two right $\angle d$. Δs BFO and DGO

Because $\begin{cases} BO = CO & (Both radii) \\ OF = OG & (given) \end{cases}$

$\therefore$ The Δs are congruent

Hence BF = DG (Th 18)

$\therefore$ AB = CD

Exercise (A)

1. If AB and CD are 2 diameters of a circle. Prove that ACBD is a rectangle.
2. If AB and AC are two equal chords of a circle, show that the bisector of $\angle BAC$ passes through the centre.
3. Two circles intersect each other in points A and B. A straight line CDEF parallel to AB meets the circles in C and F and D and E respectilely. Prove that CD = EF.
4. Two circles intersect each other in 2 points. Prove that the line joining their centres is perpendicular bisector of their common chord.

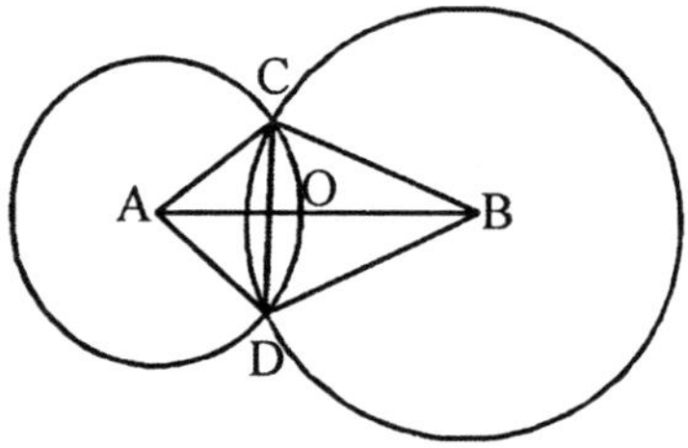

[**Hint:-** Let A and B be their centres. First prove ΔABC $\equiv$ ΔAOD. We get $\angle CAO = \angle DAO$. O being the point of intersection of AB and CD. Then Prove that ΔAOC= ΔAOD $\therefore$ $\angle AOC = \angle AOD$ but their sum is 180° $\therefore$ $\angle AOC = 90°$]

5. A line segment AB is length 5 cms. Draw a circle of 4 cms radius passing through A and B. Can you draw a circle of 2 cms. radius to pass through A and B. Give reasons in support of your answer.

6. Prove that two circles cannot intersect each other in more than 2 points.

7. AB and CD are 2 parallel chords of a circle. AB = 10 cms and CD = 24 cms

   Distance between AB and CD is 17 cms. Find the radius of the circle. **(Ans.** 13)

8. AB and CD are 2 parallel chords of a circle each measuring 6 cms. If the radius of the circle is 5 cms. Find the length of the chord BC. **(Ans.** 9.6 cms)

9. Two parallel chords of a circle are 5 cms and 12 cms respectively. If the distance between them is 13 cms. Find the radius of the circle. **(Ans.** 7.64 cms)

10. Two circles have radii 10 cms. and 17 cms. Their common chord is 16 cms. What is the distance between their centres.

    (Ans. 21 cms)

11. In a circle two chords measure 12 cms and 5 cms. They are parallel to each other. The distance between their centres is 3.5 cms. What is the radius of the circle if they are on the same side of the centre. **(Ans.** r = 6.5 cms)

12. Three circles of the same radius pass through a common point A. The other points in which they meet are B, C, D. Show that AB, AC and AD are at right $\angle s$ to CD, BD, and BC respectively.

13. In a circles a 24 cms chord is at a distance of 5 cms. What is distance of 20.8 cms chord.

    (Ans. 7.8 cms)

14. What is the locus of mid points of equal chords of a circle.

15. In an equilateral Δ, prove that the circum-centre and the centriod coincide.

16. In any circle a chord which is longer is at a shorter distances, than a smaller chord.

17. If through one of the intersections of 2 circles a line is drawn

parallel to line of centres and is produced to meet the circles then it is twice the distance of the line of centres.

18. M is the mid point of the chord AB of a circle. From M two straight lines MP and MR are drawn making equal angles to AB. The points P and Q are on the circle on the same side of AB. Show that MP = MR.

19. AB and CD are parallel chords drawn in a circle. Prove that the line joining their centres is perpendicular to both chords and pass through the centre of the circle.

20. O is the centre of the circle. OP and OQ are two radii at right angles to each other. AB is any diameter. QR and PT are $\perp$ s drawn to the diameter AB. Prove that QR = OT and OR = PT.

21. Distance between the centres of 2 circles is 4 cms. Their common chord is 2.4 cms and one circle has a radius of 3.7 cms find the radius of the 2nd circle.

(**Ans.** 13 cms)

22. ABCD is a rectangle. With A as centre and AB as radius a circle is drawn. This circle meets BA produced at P. Prove PD || AC and CD = AP.

23. Two non-intersecting circles are given and a line is drawn with one end on one circle and the other end on the other. When is this line of minimum length and when it is of maximum length? What difference in the length of the line will it make if one circle happens to be inside the other.

24. CD is a chord of the circle. M is its mid-point. From M, two lines MA and MB are drawn making the same angle with MC and MD. Show that MA = MB.

25. CD is a chord of a circle. A is a point on the circumference. From this point a perpendicular chord is drawn to CD. The difference of the 2 segments of this chord is the same wherever you take the point A. What is the difference equal to.

Theorem No. 51

The degree measure of minor arc of a circle is twice the measure of the angle subtended by it at any point of the remaining circle.

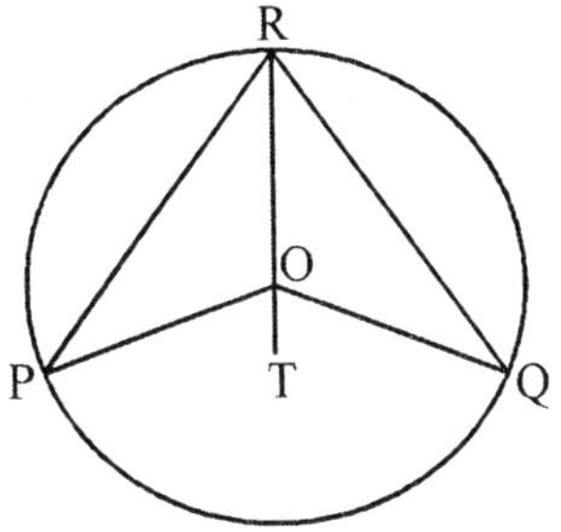

Given: PQ is minor arc of a circle. Its degree measure is $\angle PQQ$. Angle subtended at any point R of the remaining circle = $\angle PRQ$.

To Prove: $\angle POQ = 2\angle PRQ$

Construction: Join R to O. Produce RO to any point T.

Proof: OP = OR = OQ = r = the radius of the ⊙ .

$\therefore$ $\angle ORP = \angle OPR$ and $\angle ORQ = \angle OQR$

$\angle POT = \angle OPR + \angle ORP = \angle ORP + \angle ORP = 2\angle ORP$ ...(1)

Similarly $\angle QOT = \angle OQR + \angle ORQ$

$= \angle ORQ + \angle ORQ = 2\angle ORQ$ ...(2)

$\angle POQ = \angle POT + \angle QOT = 2\angle ORP + 2\angle ORQ$

$= 2(\angle ORP + \angle ORQ) = 2$PRQ (H.P.)

Corollary 1

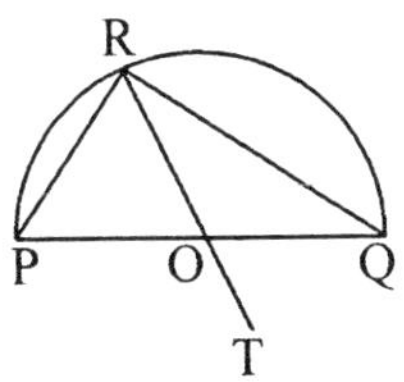

The degree measure of an angle in a semi-circle is 90°.

Given: A semi-circle.

To Prove : Angle in a semi-circle i.e. angle

submitted by the diameter at any point on the remaining part of the circle is 90°.

Proof: Angle submitted at any point R by the diameter PQ is PRQ. It is composed of 2 $\angle s$ viz $\angle ORP$ and $\angle ORQ$.

$\therefore$ $\angle PRQ = \angle ORP + \angle ORQ$

Now $\angle POT = \angle OPR + \angle ORP$

But $\angle OPR = \angle ORP$ as OR = OP

$\therefore$ $\angle POT = 2\angle ORP$

Similarly $\angle QOT = \angle OQR + \angle ORQ = 2\angle ORQ$

$\therefore$ $\angle POT + \angle QOT = 2\angle ORP + 2\angle ORQ$

But $\angle POT + \angle QOT = 180^\circ$

$\therefore$ $2\angle ORP + 2\angle ORQ = 180^\circ$

$\therefore$ $\angle ORP + \angle ORP = 90^\circ$

$\therefore$ $\angle POQ = 90^\circ$

Corollary 2

The degree measure of a major arc is more than 180° but is less than 180°.

$\therefore$ The angle subtended by a major arc at the circumference is more than 90° but less than 180° i.e. it will be obtuse angle.

Theorem No. 52

Angle in the same segment are equal.

Given: $\angle APB$ is one $\angle$ in major segment and $\angle AQB$ is another $\angle$ in the same major segment.

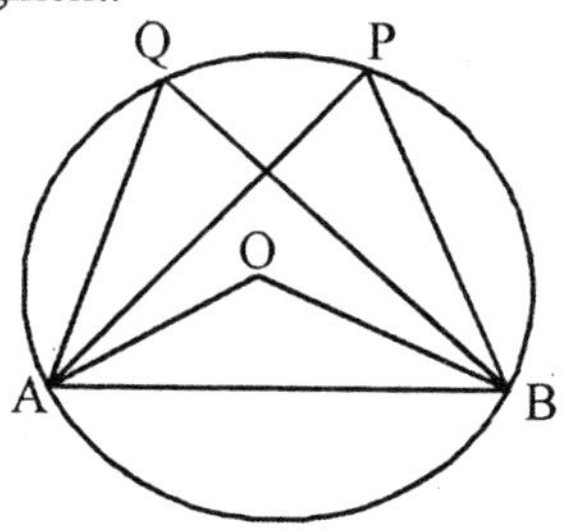

To Prove: $\angle APB = \angle AQB$

Proof: O is the centre of the circle.

$\therefore \quad 2\angle AQB = \angle AOB \quad ...(1)$

Also $\quad 2\angle APB = \angle AOB \quad ...(2)$

Form (1) and (2) we get

$$\angle AQB = \angle APB \quad \text{(H.P.)}$$

Definition

If a circle passes through all the four points of a quadrilateral then it is called a **cyclic quadrilateral**.

Theorem No. 53

Opposite angles of a cyclic quadrilateral are supplementary.

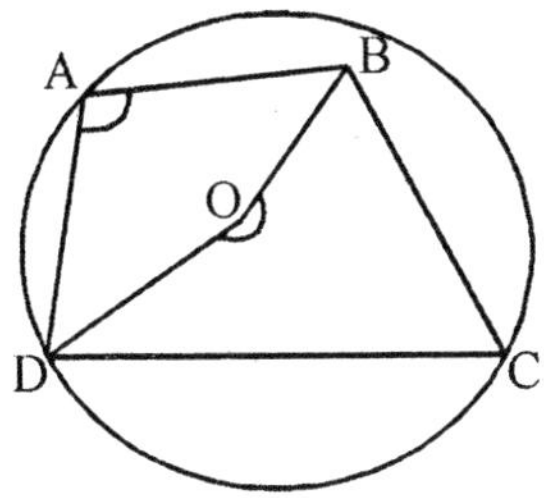

Given: A cyclic quadrilateral ABCD drawn in a circle whose center is 'O'.

To Prove: $\angle A + \angle C = 180^\circ$

$\angle B + \angle D = 180^\circ$

Proof: $\angle DAB = \frac{1}{2}\angle BOD$ (Reflex)

$\angle BCD = \frac{1}{2}\angle BOD$ (abtuse)

$\therefore \quad \angle DAB + \angle BCD = \frac{1}{2}(\angle BOD\,Reflex + \angle BOD\,obtuse)$

$$=\frac{1}{2}\times 360^\circ = 180^\circ$$

$\therefore \quad \angle A + \angle C = 180^\circ$

Similarly $\quad \angle B + \angle D = 180^\circ$

Theorem No. 54

If equal angles are standing on a common base and on the same circle of it, then the 4 points viz. 2 end points of the common side and two angular points are on a common circle.

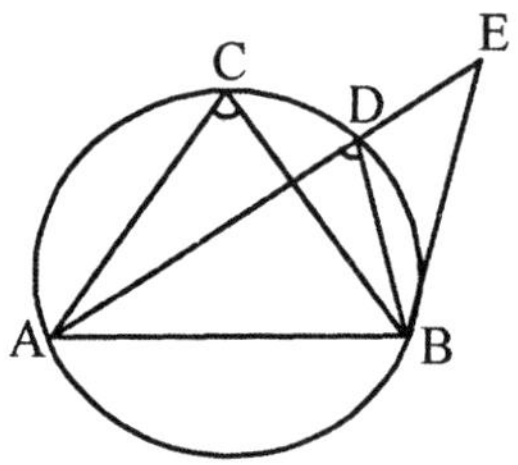

Given: ACB and ADB are two equal angles standing on the same base AB and are on the same side of AB.

To Prove: The four points viz. A and B at the ends of the common base AB and points C and D at which the equal angles are formed are on a common circle.

Construction: Draw a circle through A, B, C.

Proof: A circle can pass through A, B, C but we are not sure if it passes through the fourth point viz. D.

If D is not on the circle then it is either outside the circle or inside the circle. Let it be outside the circle. Let the circle through E instead of D.

Then $\quad \angle ACB = \angle AEB$

But we are given that $\angle ACB = \angle ADB$

$\therefore \quad \angle ADB = \angle AEB$

But this does not seem possible as $\angle ADB$ is exterior $\angle$ of the ΔAEB of which $\angle AEB$ is the interior angle.

$\angle ADB > \angle AEB$

$\therefore \quad \angle ADB \ngtr \angle AEB$

Similarly we can prove that $\angle ADB \nless \angle AEB$

$\therefore$ D coincides with E.

Theorem No. 55

If sum of the opposite $\angle s$ of a quadrilateral is 180° then it is cyclic.

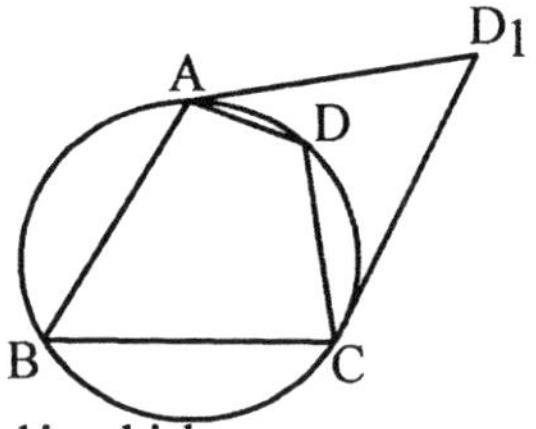

Given: A quadrilateral in which

$\angle A + \angle C = 180^\circ$ and $\angle B + \angle D = 180^\circ$

To Prove: A circle can be made to pass through A, B, C and D.

Construction: Draw a $\odot$ through A, B, C and let it pass D_1 instead of D.

Proof: A, B, C, D_1 are on a circle.

$\therefore \quad \angle B + \angle D_1 = 180^\circ$

But we are given that

$\angle B + \angle D = 180^\circ$

$\therefore \quad \angle D = \angle D_1$

This is possible only if $\angle D$ and $\angle D_1$ coincide

Similarly $\angle D_1$ cannot be inside D.

$\therefore$ D_1 must coincide with $\angle D$.

$\therefore$ A circle must pass through A, B, C, D.

Exercise (B)

1. In cyclic quad. AC and BD are joined. Then on the basis of theorem No. 52 give another proof of theorem No. 53.

2. One side of a cyclic quad. ABCD say BC is produced. Prove that the exterior angle so formed at C is equal to the diametrically opposite angle A of the quadrilateral.

3. If one line makes equal angles with a pair of opposite sides of a cyclic quadrilateral then it makes equal angles with the second pair of opposite sides.

4. If the angles of a cyclic quadrilateral are in the following proportion then find out the fourth $\angle$ of the quadrilateral

   (a) $3:4:9:x$

   (b) $6:4:4:x$

   (c) $8:9:7:x$

5. A trapezium is constructed in a circle. Prove that its diagonals are equal. On the other hand a trapezium has its diagonals equal. Prove that it is cyclic.

6. If a trapezium has its inclined sides equal it is cyclic. On the other hand if the inclined sides are not equal it is not cyclic.

7. ABC is a Δ drawn in a circle. Prove that the bisector of $\angle BAC$ passes through the middle point of the arc BC.

8. ABC is any angle. A circle makes intercepts on the arms AB and BC. The sum or difference of the angles which these intercepts make with the centre of the circle is twice the angle ABC.

9. Two circles intersect in A and B. AC and AD are the two diameters of the two circles. Show that C, B, D are in one straight line.

10. Two circles intersect each other in A and B. Through A a line is drawn cutting the circles in C and D. Show that howsoever you draw the line, angle CBD has the same value.

11. AB is a diameter of the circle with centre O. AC is chord of the circle. A perpendicular is drawn to AB at O. The perp. meets AC in T. Show that T, O, B, C is cyclic and the circle has radius equal to the circum-circle of ATO.

12. ABC is a Δ drawn in a circle. The bisector of 3 angles A, B, C meet the circum-circles in D, E, F respectively. Prove that AD is at right angles to EF and that the angles of DEF are respectively compliments of $\frac{1}{2}\angle s$ of ABC.

13. PQ PR are two equal chords of a circle. M and N are the mid points of the arcs PQ and PR. MN is joined. It meets the chords PQ and PR in C and D. Prove that PC = PD.

14. A and B are points of intersection of two circles. A point P is taken on one circle. PA and PB are joined and are produced to meet the second circle. Prove that the arc intercepted on the second circle is always the same wherever the point P is taken.

***15. A point P is taken inside a circle. From this point a no. of lines are drawn to the circumference. What is the locus of mid-pts of these lines.**

***16. A point is taken inside a circle. From this point of a no. of chords are drawn to the circle. What is the locus of mid-points of the chords.**

***17. A point is taken inside a circle. From this point a no. of lines are drawn to the circle and the lines are trisected. What is locus of the point of trisection.**

***18. In case the point in the above 3 examples is taken outside the circle what would be the difference in the locus? Give a general answer to all the four questions from ex 15 to ex 18.**

19. A point is taken outside the circle and equal lines are drawn to the near circumference of the circle. Show that when extended to the furthermost part of the circumference the extended intercepts are equal.

20. A point is taken outside the circle. From this point draw a line to the distant part of the circle which will be bisected at its intersection of the nearer part of the circle. When is this not possible?

21. ABC is an isosceles Δ drawn a circle. The bisector of $\angle B$ and $\angle C$ meet the circle in C and D. (5 points of a pentagon with 4 sides equal. Show that the points B, E, A, D, C are 5 points of a pentagon with 4 sides equal. What should be the relation between the angles of the triangle if all the five sides are equal? ($\angle s$ of the Δ = $36°, 72°, 72°$)

22. ABCD is a cyclic quad. AB and DC mcet in P while DA and CB meet in F. The circles of Δ of BCP and ABF meet in O. Prove that P, O, F are in one st. line.

23. In case a line makes equal angles with the opposite sides of a cyclic quad. then it makes equal angles with the diagonals also.

24. A is an intersection of two circles. Through A two lines PQ and RS are drawn meeting the circles in P and Q and R and S respectively. Prove that RP and QS make the same angle with each other how so ever you draw the lines PQ and QS.

25. From both the intersections of 2 circles one line each is drawn parallel to the other and they are terminated at the circumference. Prove that the lines joining the ends of these lines are parallel and equal to each other.

26. From one of the intersection of two circles, lines are drawn making equal angles with the common chord and they are allowed to meet the 2 circles. Prove that they are equal.

27. From one of the intersection of two circles draw a chord to one of the circles so that it will be bisected by the other.

28. ABC is a fixed Δ. On the base BC a segment of a circle is drawn which does not pass through A but it meets the lines BA and CA in P and F. Prove that PF is always $\parallel$ to a fixed line.

29. Two equal circles intersect in A and B. Through A a st. line is drawn meeting the two circles in P and F. What is the locus of midpts of PF.

*30. A line is drawn through C, one of the intersection of 2 circles meeting the circumference in P and F. A and B are the centres of these circles PA and FB meet in O. If E be the other point of intersection, prove that

$$\angle POF = \angle ACB = \angle PEF.$$

31. OA and OB are two fixed radii of a circle. P is a point moving on the circle. PM and PT are drawn perp. to the two radii. Show that the length of MT is always fixed.

***32. In any Δ, that part of the base between its centre and the foot of the perpendicular from the vertex, subtends an angle equal to difference of the base angles at any of the other feet of perps.**

*33. AB and CD are two perpendicular chords of a circle meeting in E. Prove that the difference segments of CD viz. CE and DE is always the same wherever CD is drawn.

34. XY is a chord perpendicular to the diameter AB of the circle with the centre M. YCD is another chord through Y and it cuts the

diameter in C. Prove that XA and XB are the internal and external bisectors of angle CXD.

35. A line drawn through one of the two intersections of 2 circles and terminated by them is of maximum length when it is at right angles to the common chord.

36. A line is drawn parallel to the common chord of two circles and meets both circumferences. The intercepts made by the line between the circles are equal.

37. Two circles are intersecting in A and B. A point P is taken on one of them. PA and PB are joined and further extended to meet the second circle in C and D. Prove that the arc CD is always of the same langth wherever you take the point P.

38. Triangle are drawn a common base and the same vertical angle. Prove that of all the Δ s, perimeter is maximum in the case of isosceles Δ.

39. (1) Of all Δ s drawn in circle the perimeter of equilateral Δ is maximum.

    (2) Of all quadrilaterals drawn in a circle the perimeter of a square is maximum.

40. A point is given outside a circle. Draw a line to the circle from that point so that it is bisected by the near point of intersection with the circle.

Tangency

Definitions

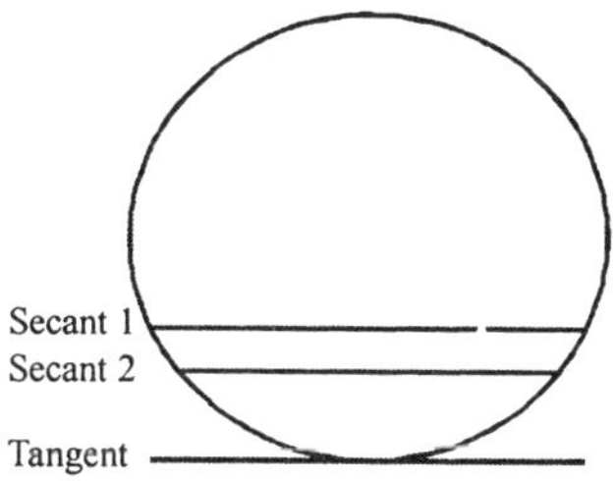

A straight line usually meets the circle in two points. It is then called a *secant*. There are many secants that can be drawn to a circle. If we draw a secant and further move it down parallel to itself, we shall get another secant for it meets the circle on two points but the only difference is that the points are nearer now. we still it move down and get another secant until at last the points coincide. It is then called a tangent. Thus the tangent and the circle is one point in common. That point is called the *point of contact.*

Similarly one circle meets another in 2 points. We can get another circle slipping over the one drawn earlier we get another circle cutting in 2 points. Until at last the 2 points coincide into 1 point only. We then call them as *externally touching circles*. We can as well have 2 circles *touching internally* and have *intenal point of contact.*

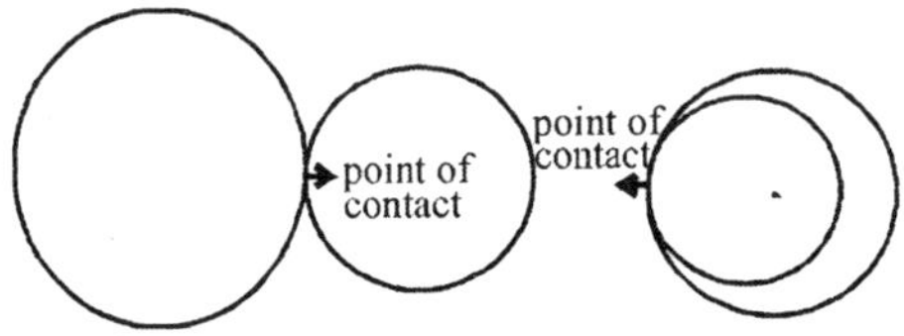

1. Extemally iouching circles 2. Intenally touching circles

Theorem No. 56

A tangent line to a circle is perpendicular to the radius at the point of contact.

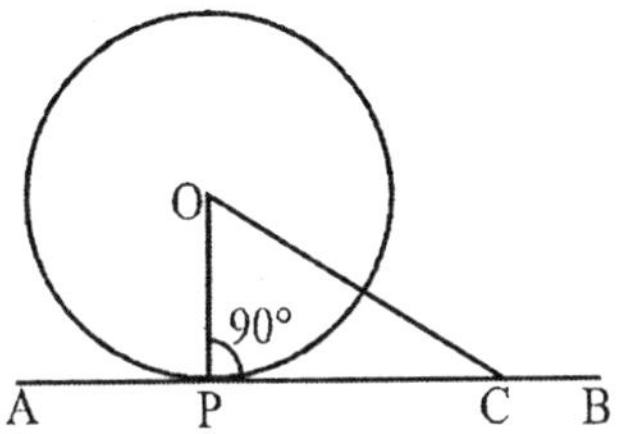

Given:- A circle with centre O. AB is a tangent to the circle at P.

To Prove:- OP is at right angles to AB.

Proof:- All points on the tangent AB except P are outside the circle.

$\therefore$ OC > OP

P is on the circle

and P is on tangent

$\therefore$ OP = r

and P is the only point on AB

that is common to the circle and tangent and is the nearest point.

$\therefore$ $OP \perp AB$ (Th. 17) (H.P.)

Theorem No. 57

Only 2 tangents can be drawn to a circle from a point outside the circle and they are equal in length and the line joining the centre to the point from which the tangent are drawn bisects the angle between the tangents.

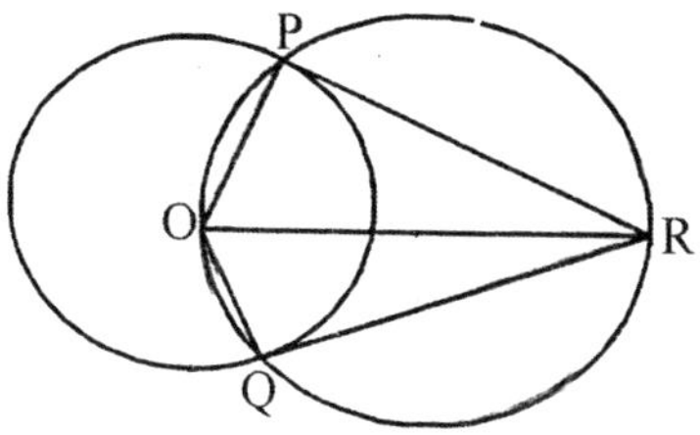

Given:- A Point R outside the circle with centre O.

To Prove:- Only 2 tangents viz RP and RQ can be drawn to the circle and OR bisects the angle PRQ.

Construction:- Join OR. With OR as diameter draw a circle. The two circles meet at P and Q. Join RP and RQ.

Proof:- Because OR is the diameter

$$\angle OPR = \angle OQR = 90°$$

RP and RQ are the 2 tangents to the circle and only 2 are possible because the two circles meet at two points only.

In right angled Δs OPR and OQR

$$\text{because} \begin{cases} OP = OQ & (\text{both radii}) \\ \angle OPR = \angle OQR & (\text{both } 90°) \\ OR \text{ is common} \end{cases}$$

Hence $\Delta OPR \equiv \Delta OQR$

$\therefore$ RP = RQ

$\angle PRO = \angle QRO$

$\angle POR = \angle QOR$ (H.P.)

Theorem No. 58

When 2 circles touch each other internally or externally, their point of contact is in line with the centre.

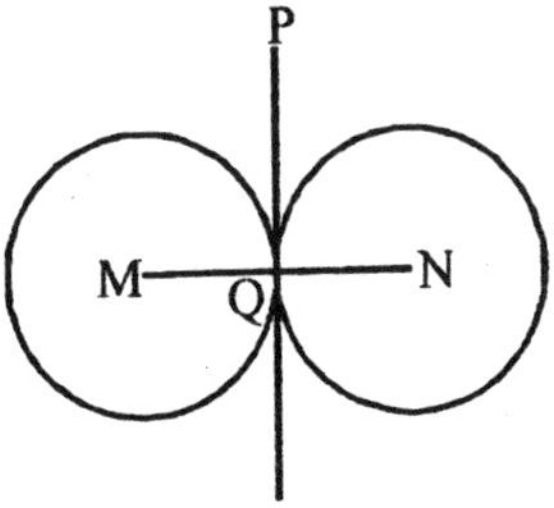

Fig. I

Fig. II

Given:- M and N are centres of 2 circles touching each other externally as in Fig. I and internally as Fig. II.

To Prove:- Q the point of contact is in line with MN.

Construction:- Draw PQ the common tangent to these circles.

Proof:- (a) In fig I

PQ ⊥ QM and PQ ⊥ QN

∴ $\angle MQP = 90°$ and $\angle NQP = 90°$

$\angle MQP + \angle NQP = 180°$

∴ MQN is a st. line

M and N are in line with Q

(b) In fig II

PQ ⊥ MQ

PQ ⊥ NQ

At Q only one line can be drawn ⊥ MN

∴ QNM is one line. (H.P.)

Theorem No. 59

If a chord is drawn to the circle at its point of contact with the tangent, then the angle which this chord makes with the tangent is equal to angle in the alternate segment of the circle.

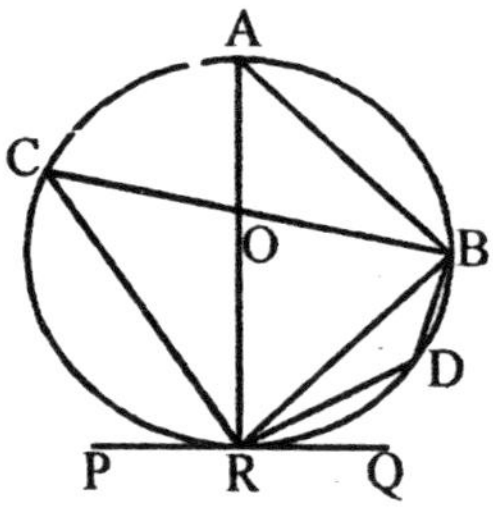

Given:- A circle with centre O. PQ is a tangent to circle at R the point of contact.

RB is any chord making an angle BRQ with the tangent.

To Prove:- $\angle BRQ = \angle BAR$

$\angle BRP = \angle BDR$

Construction:- Draw RA the diameter through R.

Proof:- AR $\perp$ RQ

$\angle ARB + \angle BRQ = 90°$ ...(1)

But $\angle ARB + \angle BAR = 90°$ ($\therefore$ AR is dia) ...(2)

From (1) and (2)

we get $\angle BRQ = \angle BAR = \angle BCR$ or any $\angle$ in the segment

$\angle BRQ + \angle BRP = 180°$ (L.P.A.) ...(3)

$\angle BCR + \angle BDR = 180°$ ...(4)

From (3) and (4)

because $\angle BRQ = \angle BAR$

$\angle BRP = \angle BDR$ (H.P.)

Theorem No. 60

(Converse of Th 59)

If a line drawn through an end point of a chord of a circle, makes with it an angle equal to the angle subtended by the chord at any point in the alternate segment then the line is tangent to the circle at the point of contact.

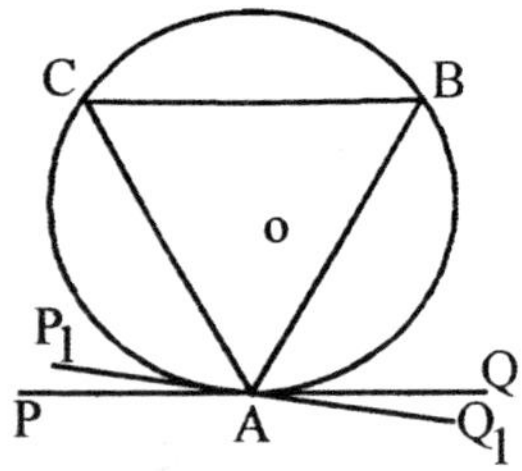

Given:- AB is a chord of the circle (o, r). A line PAQ is drawn through the point A on the circle such that $\angle BAQ = \angle BCA$ where C is a point taken in the alternate arc BA excepting the points B or A.

To Prove:- PAQ is a tangent to the circle at A.

Construction:- If PAQ is not a tangent to the circle another line P_1AQ_1 is drawn which is a tangent to the circle at A.

Proof:- Since P_1AQ_1 is a tangent to the circle at A,

$$\angle BAQ_1 = \angle BCA \qquad \text{...(Th 59)}$$

But $\angle BAQ = \angle BCA$

$\therefore$ $\angle BAQ = \angle BAQ$

This is possible only if AQ, coincides with line AQ.

That means $\angle BAQ$ coincides with $\angle BAQ_1$, that means PAQ is tangent to the circle at A. (H.P.)

Exercise C_1

1. A and B are centres of two circles touching each other. From the point of contact a line is drawn meeting the two circles in P and Q. Prove that AP || BQ.
2. Two circles are touching each other externally at point A. Two lines are drawn through A meeting the two circles in B and C and D and E respectively. Prove that BD || CE.
3. A circle touches the quad. ABCD. Prove that AB + CD = AD + BC.
4. Two circles are touching each other externally. From their point of contact is drawn a line a meeting the two circles in A and B. Show that the tangents to these 2 circles at A and B are parallel.

5. Two concentric circles are drawn. Prove that a tangent to the smaller circle is bisected at the point of contact.
6. Prove that the intercepts made on a tangent to the circle by two parallel tangents subtends a right angle at the centre.
7. ABC and ABD are equal circles. AB is the common chord. If AB=BC, AC touches ABD.
8. Two circles intersect at A and B. From a point P on one of these circles two line segments are drawn intersecting the second circle in C and D. Prove that CD is $||$ to tangent at P.
9. In a right triangle ABC a circle with AB as diameter is drawn to intersect the hypotenuse AC in P. Prove that the tangent to the circle at P bisects the side BC.
10. ABC is an arc of a circle. Arc AB = 2 Arc BC. Through B a tangent is drawn meeting the chord in point D. From A draw AE $\perp$ AC and meets the tangent in E. Prove that EB = BD.
11. ABC is a Δ in which $\angle B = 90°$. The $\perp$ bisector of AC meets AB and BC in P and F respectively. Prove that AC is a tangent to the circle PFC.
12. A line is drawn through one of the points of intersection of 2 circles. It meets them again in points B and C. At B and C tangents are drawn to the circles which meet in point A. Prove that the angle formed at A is equal to the angle between the tangents to the circle at the point of intersection.
13. From a fixed point two tangents are drawn to the circle. A third tangent to the circle meets these 2 tangents in points A and B. Show that howsoever you draw the third tangent, the angles made by AB with the centre is equal to either of the fixed angles.
14. The angle between the tangents to the circles from a point is fixed. What is the locus of such points?
15. Two circles are touching each other externally at point C. They have a common tangent touching them at A and B. Show that the circle with AB as diameter passes through C and the line joining the centre is a tangent to it.
16. Two circles touch each other internally. A line is drawn intersecting two circles. Prove that the intercepts made between the circles makes equal angles with the circumference at the point where the two circle touch each other.

17. AB is a diameter of the circle. AC and AD are two chords of the circle. Through B a tangent is drawn to the circle meeting AC and AD in points P and F. Prove that C, P, F, D are cyclic.

18. XYZ is a Δ drawn in a circle. A line PQ is drawn parallel to XZ from a point in YZ and meets the tangent to the circle XYZ drawn at X in points Q. Show that XPYQ is cyclic.

19. AB is a diameter of the circle and AC is a chord. From C a tangent is drawn to the circle. AD is $\perp$ dropped from A to it. Prove that AC bisects $\angle DAB$.

20. AB and CD are 2 parallel chords of a circle. From D a tangent is drawn to the circle. It meets AB in point M. From M a second line MP is drawn tangential to the circle in P. Show that PC bisects AB.

21. ABC is a Δ drawn in a circle. A tangent to the circle is drawn at A. It meets BC produced in point D. Prove that the bisectors of $\angle BAC$ and $\angle ADC$ are at right $\angle s$ to each other.

22. In a $\Delta ABC, AD \perp BC$. From A a tangent is drawn to the circle ABC. From D a parallel is drawn to AB. It meets the tangent at A in E. Prove that $\angle AEC = 90^\circ$.

23. Two circles intersect each other in P and F. A line A, C, B, D is drawn so as to cut the first circle in A and B and the second circle in C and D show that $\angle APC = \angle BFD$.

24. The in-circle of the ΔABC touches BC in D. Prove that the in circles of ΔABD and ΔACD touch each other.

25. PQRS is a square. From P a line is drawn cutting RS, QR and QS in points X, Y, Z respectively. Show that RZ is a tangent to ΔXRY

26. ABCD is a cyclic quadrilateral. At the angular points A, B, C, D are drawn four tangents which meet in P, Q, R, S. If P, Q, R, S is cyclic then the chord AC makes a right angle with BD.

27. ABC is a Δ drawn in a circle. AD and BE are 2 perps drawn from A and B to the opposite sides. Prove that DE is parallel to the tangent to the circle from C.

28. ABCD is a parallelogram whose diagonals intersect in O. Show that circle ABO and circle ADO touch each other.

29. A, B, C, D are 4 points in that order in a st. line. P is a point so that $\angle APB = \angle CPD$. Prove that the circum circles of APD and BDC touch each other.

30. ABC is a Δ. The bisectors of interior and exterior $\angle s$ of A meet BC in P and F. The tangent to the circle of ABC meets the base BC at M. Show that M is the midpoint of PF.

31. Two circles are drawn touching each other internally. Their centres are A and B. In the space between the circles, a circle is drawn touching one internally and the other externally. If the centre of this circle is P. Then AP + BP is always the same.

32. AB is a chord of the circle and P is any point on this circle. $PM \perp AB$ and AN is $\perp$ to the tangent from P. Prove that $MN \| BP$.

33. One st. line is meeting one circle in points P and F and another in R and S. If the tangent at P to one circle is parallel to tangent at R to the second circle, then the tangent to the 1st circle at F is parallel to the second circle at S.

34. Two circles are touching each other externally at G. A and B are their centres. A line through G is drawn meeting them in P and F. Prove that area of ΔAGF = Area of ΔBGP.

35. A, B, C, D are 4 points in this order in a st. line. P is such a point that the circles ABP and PCD touch each other at P. Show that $\angle APC$ is supplementary to $\angle BPD$.

36. PQ and PR are 2 tangents to a circle from P. O is the centre of the arc QR. Prove that O is either in-centre or the centre of described circle of ΔPQR.

37. The bisector AD of the vertical angle of the ΔABC meets the base in D. From A a line AE is drawn so that $\angle DAE = \angle ADC$. Show that AE is either the tangent to circum circle of ABC or otherwise parallel to BC.

38. Two circles intersect in C and D. A line is drawn through C meeting the two circles in P and F. At P and F straight lines are drawn touching the two circles. They meet in A. Prove that APDF is a cyclic quadrilateral.

39. P, Q, R, S, T are 5 points on a circle in this order. Arc PQ = Arc QR.

SR is joined and is further produced and it meets the tangent to this circle at Q in point B. Prove that $\angle PTS + \angle QBR = 180^\circ$

Prove also that PR || QB.

40. ABCDE is a regular pentagon. O is the centre of the circle about which it is drawn. G is the intersection of AC and BE. Prove that the circum circle of AGO touches AB.

Angle Between 2 Circles

Definition

When 2 circles do not intersect each other there cannot be any angle between them. When they touch each other the angle between them is zero. When they touch each other prominently the angle between them is prominent.

It is the angle between the tangents to the circles at the common point of intersection and the circles are called intersecting circles.

Exercise (C2)

1. Two circles are intersecting at right angles at points A and B, then the angles made by AB with the circumferences of each of the circles are compliment of each other.

   On the other hand if these angles are complimentary the circles intersect at right angles.

2. Two circles intersect each other in A and B. Through A, a line is drawn meeting the circles in P and F. If PBF is right angle then the circles are complimentary.

3. Two complimentary circles intersect each other in A and B. Then in one circle, second one has got arc of the other. One point P is taken on this arc and PA and PB are joined and produced to meet the first circle in C and D. Prove that CD is the diameter of the circle.

Prtical Geometry IV (Construction Problems of Circles and Tangents to Circles)

Construction No. 21

To divide an arc of a circle in 2 equal parts.

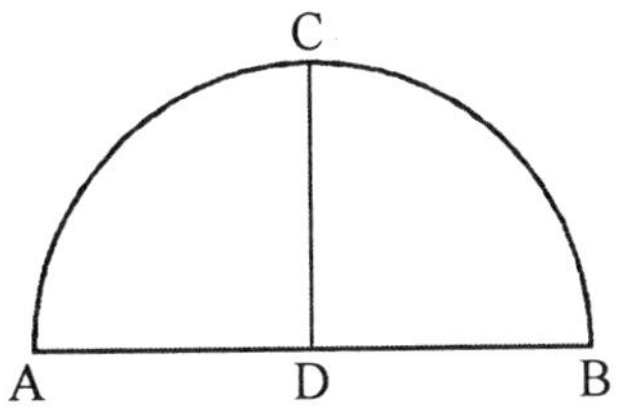

An arc of a circle AB is given to you. It is to be bisected in 2 equal parts.

Join AB. Divide in 2 equal parts at D. Draw ⊥ bisector of AB. Then Arc AC= Arc BC.

Construction No. 22

To draw a tangent to a circle at a point O in it.

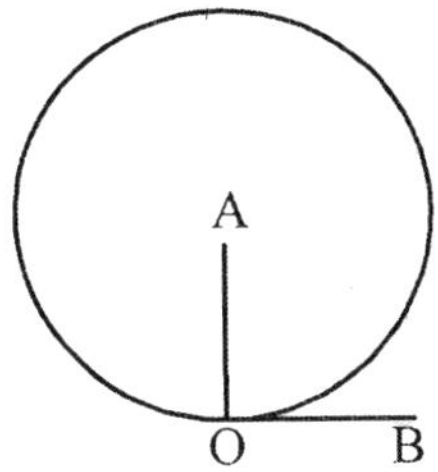

Join O to the centre of the circle. Draw a perp to AO at the point O OB is the tangent.

Construction No. 23

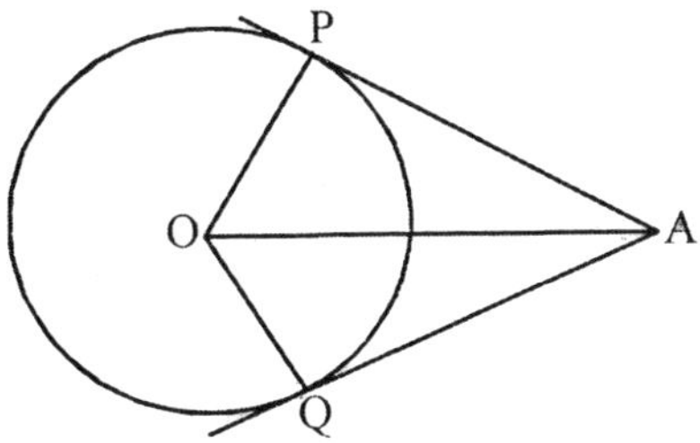

To draw tangents to a circle from a pt. A outside it.

A circle with centre O is given. A is a point outside. You are required to draw tangents from it. Join AO. Draw a semi circle on AO. It intersects the circle in 2 pts P and Q. Join AP and AQ. They are the two tangents.

Direct Common Tangent and Transverse Common Tangents

There are two types of common tangents to two circles. They are Direct common tangents and transverse common tangents. Direct common tangents occur in circles when their centres are on one side of the tangents while transverse common tangents occur when their centres are on opposite side of the tangents. Direct common tangents occur when the centres of the circles can be intersecting or non-intersecting while transverse common tangents occur in non-intersecting circles only.

Construction No. 24

To draw a direct common tangent to two circles.

Given: Two circles with centre at O_1 and O_2 and their radii r_1 and r_2 respectively.

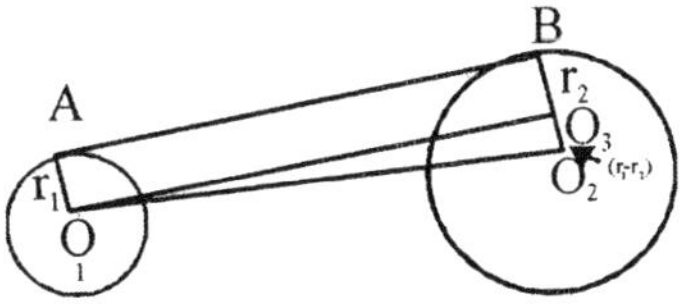

It is immaterial if they intersect or not.

To Draw: A common tangent AB to the two circles.

Analysis: Join O_1 to O_2 and measure the distance O_1 to O_2. Let AB be direct common tangent. $O_1A \perp AB$ and $O_2A \perp AB$.

Let O_1O_3 be parallel to AB. $O_1O_2O_3$ is rt angled Δ.

$$O_1O_2^2 = (O_2O_3)^2 + O_1O_3^2$$

$$\therefore \quad O_1O_3 = \sqrt{O_1O_2^2 - (r_2 - r_1)^2}$$

Method: Join O_1O_2 and draw a semi circle. On O_1O_2. With distance = $O_2O_3 = (r_2 - r_1)$ in the compass cut off the semi circle. Join O_1O_3. Then join O_2O_3 to meet the circle in B from B draw a parallel BA. AB is the tangent

Construction No. 25

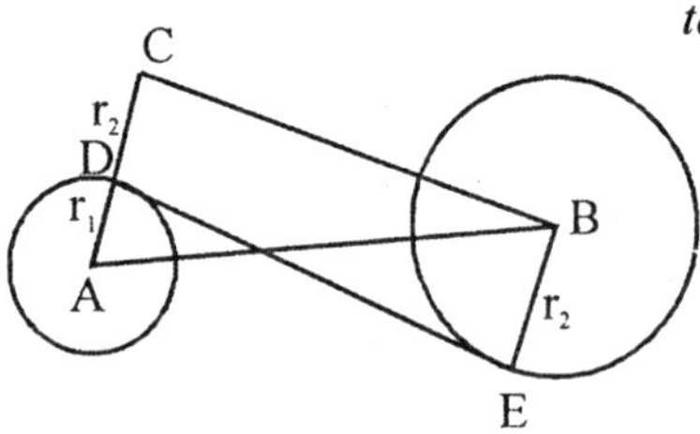

To draw a transverse common tangent to two circles.

Given: Two circles with centres at A and B and raddi = r_1 and r_2 in non-intersecting circles only.

To Draw: A transverse common tangent to these two circles.

Analysis: Let DE be the required tangent. Draw $AD \perp DE$ and draw $BE \perp DE$. Produce AD upto C i.e. DC = r_2 = BE. ACB is rt angled Δ.

$$\therefore \quad AB^2 = AC^2 + CB^2$$

$$\therefore \quad AB^2 = (r_1 + r_2)^2 + CB^2$$

$$\therefore \quad AB^2 = (\text{Required tangent})^2 + (r_1 + r_2)^2$$

$$= \sqrt{AB^2 - (r_1 + r_2)^2}$$

Method: Draw AD. Draw a semi-circle on AD. In the compass take distance = $(r_1 + r_2)$ and cut off at C. Draw the rt angled ΔACB. Draw $BE \| AE$. Join BD. DE is there required distance of the tangent

Exercise D_1

(Theoretical as well as practical)

1. Distance between the centres = 3 cms. Radii = 2.4 cms and 1.0 cm. Draw direct common tangents and measure them.

   (**Ans.** 2.65 cm apperox)

2. Distance between the centres = 4.8 cms, Radii = 2.4 cms and 1.0 cm. Are these circles touching each other or are they intersecting? How many common tangents do they have? What is the length of each common tangent? (4.58 cm 3.39 cm appr)

3. A tangent is drawn to two externally touching circles. Show that it bisects both the direct common tangent to these two circles.

*4. **Two direct common tangents are drawn to two externally touching circles. Prove that they meet on the line joining the centres and bisect the angle between them.**

*5. **The points of contact of the two direct common tangents to two externally touching circles are joined to the common point of contact. Prove that these are at right $\angle s$ to one another.**

6. The area between the circumferences of two concentrie circles is equal to the area of the circle with radius equal to tangent from any point of the outer circle to the inner circle.

7. The distance between the centres of 2 circles is 6.00 cms and their radii are 3.2 cm and 1.6 cm. What is the length of the four their common tangents to these circles.

   (5.78 cms; 5.78 cms, 3.6 cms; 3.6 cms)

8. The distance between 2 centres is 4.2 cm and radii are 3.2 cm and 2.0 cm. What is the length of the common chord and direct common tangents respectively.

9. A point is given outside a circle. Draw a line from this point to the circle so that its intercept on the circle will be equal to a line less than the diameter of the circle.

*10. Draw an equilateral Δ inside the circle so that one side of the Δ will pass through a given point. How many such Δs can be drawn. When is this not possible?

*11. Draw a square inside a circle so that one of its sides will pass through a given point. How many such squares can be drawn? When is this not possible?

12. Draw a Δ equiangular to a given Δ in identical order so that one side will pass through a given point. Total 6 Δs can be drawn. How the position of the point depends upon the no. of triangle.

13. Draw a circle of given radius touching a given circle. When is this not possible?

14. The distance between the centres of the circle is 'a' and is more than x and y the radii of the circle. Find out the length of common tangents to these circles in terms of a, x and y.

(**Ans.** $100+\frac{3\pi}{2}\times 50$)

15. A belt is wound round 2 wheels of diameters 30 cm and 20 cm in such a way that they move in opposite direction and the belt at right angles to each other. What is the total length of the belt.

Construction No. 26

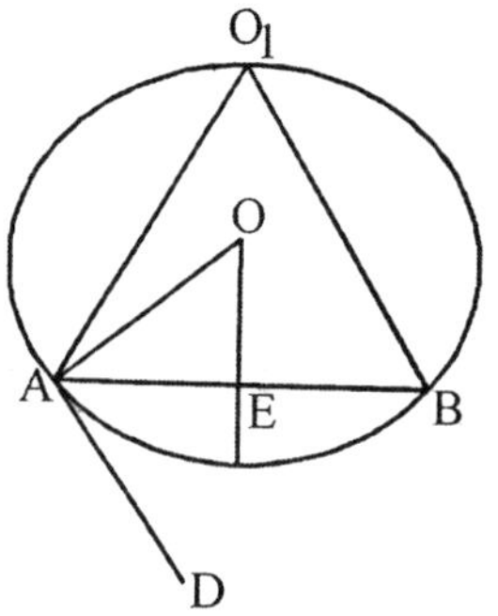

To draw a segment of a circle on a given line, that contains a given angle.

Analysis and Construction: Draw AB of given length. Make BAD equal to a given angle C. Draw $AO\perp AD$. Draw $\perp$ bisector of AB. Let it meet AO in O. With O as centre and OA as radius draw the segment of the circle.

Proof: $OA=OB$ ...(both radii)

$\angle AOE+\angle OAE=90^\circ$ ...(1)

$\angle EOD+\angle OAE=90^\circ$ ...(2)

From (1) and (2) by comparison

$$\angle EAD=\angle AOE=\frac{1}{2}\angle AOB$$

$= \angle AO'B =$ any angle in the segment.

Exercise (D_2)

1. The base and the angle at the vertex is given and the third parameter is one of the following. Draw the Δ.

(a) The foot of the perpendicular from the vertex.

(b) The length of the perpendicular from the vertex to the base.

(c) The area of the Δ.

(d) The length of the perp. from one end of the base to the opposite side.

(e) The median bisecting the base.

(f) The point where the bisector meets the base.

(g) The median of one of the other sides.

(h) The difference of the other sides.

(i) The difference of the square of the other sides.

(j) The difference of the base $\angle s$.

***2. If angle at the vertex is given and the mediam bisecting the 2 sides enclosing the vertical angle is given. Draw the Δ.**

Construction No. 27

To draw the circumcircle of a Δ.

Given :- A Δ ABC

To draw:- The Circum-circle of the Δ.

Construction:- (1) Draw a $\perp$ bisector of the base BC of the Δ at D.

(2) Draw the $\perp$ bisector of any of the other sides say AB at E.

(3) The two $\perp$ bisectors OD and OE meet at O. Draw a $\odot$ with O as centre and OB as radius.

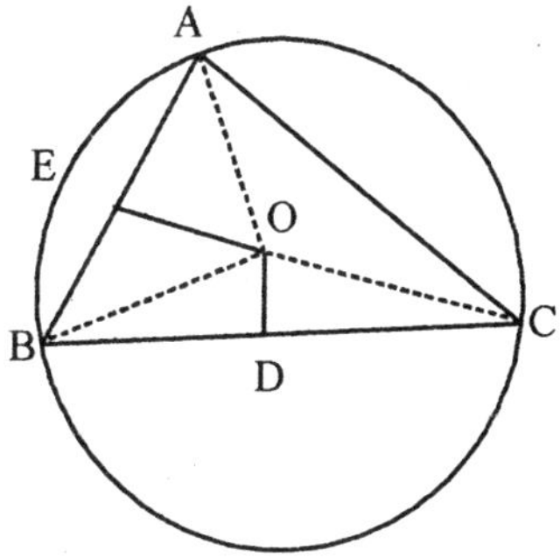

Construction No. 28

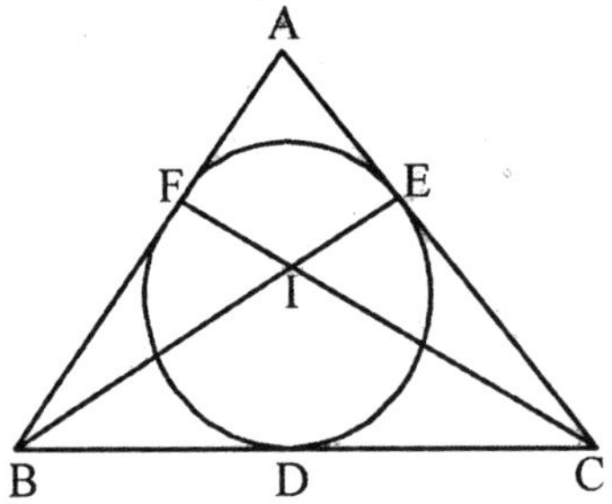

To draw the in-circle of the Δ.

Given :- A Δ ABC

To draw:- A circle touching the 3 sides of the Δ.

Method:- (1) Draw BI the bisector of $\angle B$ of the Δ.

(2) Draw CI the bisector of $\angle C$ of the Δ.

(3) Let them meet in I.

(4) With I as centre and ID $\perp$ from I to Base BC as radius draw a circle touching the 3 sides of the Δ.

Proof:- Obvious.

I is equidistant from the sides BC and CE

$\therefore$ ID = IE

I is equidistant from the sides AB and BC.

$\therefore$ I is the centre and ID the radius tauches the 3 sides.

Construction No. 29

To draw the escribed circle of the Δ *with respect to one side say base.*

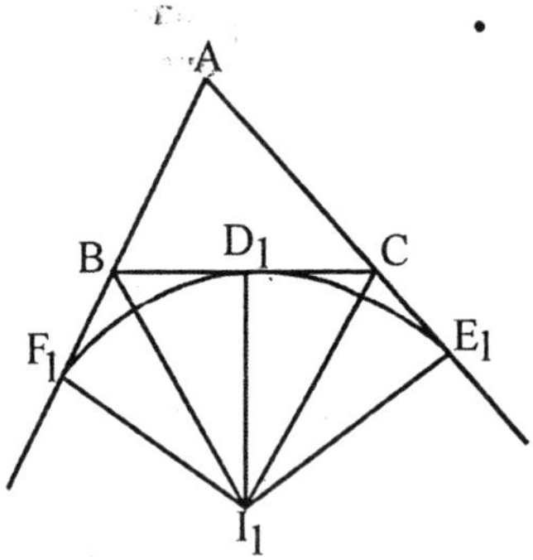

Given :- A Δ ABC

To draw:- The circle which touches AB and AC praduced the side BC from outside is known as escribed circle of the Δ.

Construetion:- (1) Bisect external $\angle B$ of the Δ ABC by the line B_1I_1.

(2) Bisect external $\angle C$ of the Δ ABC. by the line CI_1.

(3) Let them meet in I_1

(4) I_1 is the centre of Δ which meet AB and AC produced and BC from outside

$\therefore$ I_1 is the centre of the escribed of the Δ.

Proof:- $I_1D_1 = I_1F_1$

because $\Delta I_1D_1B \equiv \Delta I_1F_1B$

$I_1D_1 = I_1E_1$

because $\Delta I_1D_1C \equiv \Delta I_1E_1C$

$\therefore$ $I_1D_1 = I_1E_1 = I_1F_1$

Exercise (E)

(a) In the diagram given below. I is tha In-centre of the ΔABC. D, E, F are the points where In circle touches the sides BC, AC and AB respectively.

(b) Similary I_1 is the centre of the escribed circle corresponding to the base BC of the Δ, and D_1, E_1, F_1 are the point where it touches the sides BC, AC produceds, AB produced respectively.

(c) Similarly I_2 is the centre of the escribed circle corresponding to the side AC and D_1, E_1, F_1 are the points where it touches the sides BC produced, AC and AB produces respectively.

(d) Likewise I_3 is the circle of the escribed circle corresponding to the side AD and D_3, E_3, F_3 are the points touching the sides BC produced, AC produced and AB respectively.

(e) Futher let a, b, c denote the sides r_1, r_2, r_3 the radii of excribed cricles and r, R denote the radii of the incircle, circmscribed circles.

(f) Further let S denote the semi perimeter of the Δ.

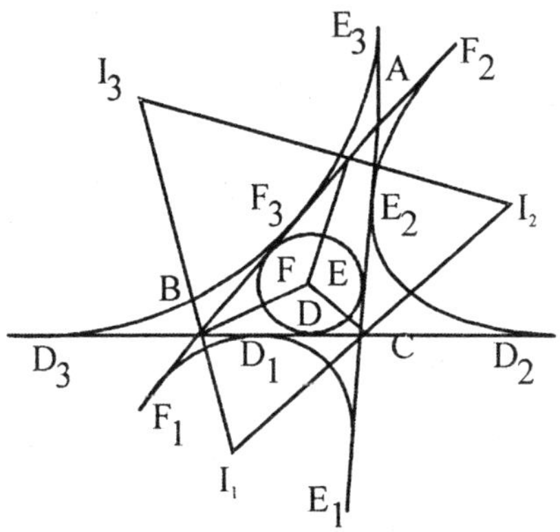

Then prove the following:

(a) (i) AE = AF = (s–a)

(ii) BD = BF = (s–b)

(iii) CD = CE = (s–c)

(b) $AE_1 = AF_1 = BF_2 = BD_2 = CD_3 = CE_3 = s$;

(c) $CD_1 = CE_1 = (s-b)$;

$AE_2 = AF_2 = (s-c)$;

$BF_3 = BD_3 = (s-a)$

(d) $EE_1 = FF_1 = a$

$DD_2 = FF_2 = b$

$EE_3 = DD_3 = c$.

(e) Area of Δ ABC $= r.\ s = r_1\ (s-a) = r_2\ (s-b) = r_3 (s-c)$

(f) $AI\ I_1$ are collinear so are BII_2 and CII_3.

(g) The Δs BI_1C, CI_2A and AI_3B are equiangular to one another.

(h) The four circles which pass through 3 of the points I, I_1, I_2, I_3 are all equal.

(i) $\Delta\ I_1, I_2, I_3$ is equiangular to Δ DEF.

(j) The points $I_2\ A\ I_3$ are collinear so are $I_3\ BI_1$ and $I_1\ CI_2$

(k) Of the four pts I, I_1, I_2, I_3 each is the orthocentre of the Δ whose vertices are other three.

2. With figure on prepage show that the circles whose centres are I, I_1, I_2, I_3 touch BC at D, D_1, D_2, D_3 then

(i) $DD_2 = D_1\ D_3 = b$,

(ii) $DD_3 = D_1D_2 = c$,

(iii) $D_2D_3 = b + c$,

(iv) $DD_1 = b \sim c$.

3. Show that the orthocentre and the vertices of the Δ are the centres of the inscribed and escribed circles of the pedal Δ (sec for defenition of the pedal Δ at the end of the next Theorem).

4. Given the base and the vertical angle A of the Δ find the locus of the escribed circle touching the base.

5. Given the base and the vertical angles of the Δ ABC, find the locus of centre of the escribed circle which touches AC.

6. Given the base and the vertical angle and the point of contact with the base of the incircle, construct the Δ.

7. Given the base and the vertical angle and the point of contact with the base produced of an inscribed circle, construct the Δ.

8. I is the centre of the inscribal circle and I_1, I_2, I_3 are the centres of the three escribed circles show that I I_1, I I_2, $I_1 I_3$ are bisected by the circumference of the circum circle.

9. ABC is a Δ and I_2, I_3 are the centres of the escribed circles which touch AC and AB. Show that the points B, C, I_2, I_3 lie on a circle whose centre is on circmference of the Δ.

10. With 3 given points as centres, describe three circles touching one another two by two. How many solutions will there be?

11. Given the contres of the three escribed circle construct the Δ.

12. Given the centre of the incribed circle and centre of two of the 3 escribed crical construct the Δ.

13. Given the vertical and the radic of the inscribe circle and the perimeter of a constrict the Δ.

14. Given the vertical angle, and the radius of the inscribed and the length of the perpendicular from the vertex to the base. Construct the Δ.

15. In a Δ ABC, I is the centre of the inscribed circle. Show that the centres of circumscribed circles of the Δ s BIC, CIB, AIC lie on the circumference of the main Δ.

Theorem No. 61

The perpendiculars drawn from the three vertices of a Δ to the opposite sides are concurrent.

(This is another proof of Theorem No. 32.)

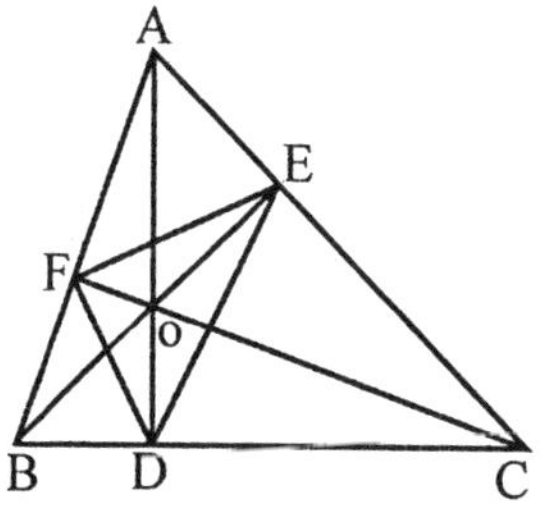

Given:- ABC is a Δ. AD ⊥ BC and BE ⊥ AC. They meet in O, CO is joined and is produced to meet AB in F.

To Prove:- CF ⊥ AB

Construction:- Join DE

Proof:- $\angle AEB = \angle ADB = 90°$ (given)

$\therefore$ A, E, D, B are concyclic

$\therefore$ $\angle ABE = \angle ADE$ ...(1)

because they are angles in the same segment of the circle ABDE.

Similarly $\angle OEC = 90°$ and

$\angle ODE = 90°$

$\therefore$ O, E, C, D, are cyclic

$\angle ODE = \angle OCE$ ...(2)

($\angle s$ in the same segment)

From (1) and (2) we conclude that

$$\angle ABE = \angle OCE$$

$$\angle ABE + \angle A = \angle OCE + \angle A$$

But $\angle ABE + \angle A$ are 2 $\angle s$ of the Δ ABE

$\angle OCE + \angle A$ are 2 $\angle s$ of the Δ BFC

$\therefore$ $\angle CFA$ must be equal to $\angle BEA$

But $\angle BEA = 90°$

$\therefore$ $\angle CFA = 90°$ (H.P.)

Definition: The triangle formed by joining the feet of perpendiculars from the 3 vertices to the opposite sides is known as the PEDAL Δ.

Thus in the figure of this theorem DEF is pedal Δ of ΔABC.

Exercise (F)

1. Theorem No. 61 is proved when the main Δ is acute angled. Prove the theorem if the main Δ is obtuse angled.
2. In an acute angled Δ the sides of the pedal Δ are equally inclined to the sides of the main Δ.
3. In an acute angled Δ the orthocentre of the main Δ is the incentre of pedal Δ.

4. In an acute angled Δ, the vertices of the main Δ are the centres of 3 escribed circle of the pedal Δ.

5. In a ΔABC if O is the orthocentre then $\angle BOC$ is supplementary to $\angle BAC$.

6. In the figure of last theorem the perpendicular bisector of DE passes through the midpts of AB and OC. (These midpts are in the same st. line).

7. The angles of pedal Δ are supplementary to twice each of the main Δ.

8. The base and angle at the vertex is given. What is the locus of the orthocentre.

9. ABC is a Δ and O is the orthocentre then the Δ formed by any three of the four pts A, B, C, O, the fourth pt. is the orthocentre.

10. ABC is a Δ and O is the orthocentre. AO meets the base in D and is further produced to meet the circumference in P. Prove that OD = DP.

11. In a ΔABC, O is the orthocentre. If M is the centre of BC then ME is the tangent to the circle AFOE.

12. In a ΔABC, O is the ortho centre. It is joined to the mid point M of the base BC. OM is produced to meet the circle ABC in point X. Then prove that OM= MX and that OB X C is a parallelogram.

    Prove also that AX is the diameter of the circumcircle of the Δ.

13. ABCD is a parallelogram. Perpendiculars are dropped BM on BC on MD on CD. They meet in M. Show that $MA \perp BD$.

***14. Prove that the distance between the orthocentre and the vertex is twice the distance of the circum centre from the side of the Δ opposite the vertex.**

15. Given the vertex, the orthocentre and the circumcentre draw the Δ.

16. Two points are given on circle. From these points two lines are drawn then the arc intercepted by these 2 is the same, what is the locus of the point of intersection of these 2 lines.

17. AB is a st. line and a chord of the circle CD is moving diameter then AC and BD intersect in X. What is the locus of X?

Exercise (G)

Comparatively Simple Examples on Th.1 to Th. 61)

1. ABC is an equilateral Δ. A piont D is taken on BC and point E on CA Such that BD=EC. AD & CE ment in O. Prove that $\angle AOB = 120^0$.
2. PQRS is a square. The angle PRS is bisected. It meets QS in point D. Prove that QD = RS.
3. MK is the radius of the circle with centre M. TP is any chord of the circle parallel to MK. TK is joined. It meets MP in point X. Prove that $3\angle KTP = \angle KXP$.
4. In a ΔABC X, Y, Z are taken on BC, CA and AB respectively Prove that (AX + BY + C Z) > $\frac{1}{2}$ (AB + BC + CA).
5. ABC is a Δ in which $\angle B = 90^\circ$. BP $\perp$ AC. Angle C is bisected. The bisector meets AB and BP in N and M respectively. Show that BN = BM.
6. ABC is an equilateral Δ. P is a point on BC. AC is produced and a point F is taken on it so that CF = BP. Prove that AP=FP.
7. PQRS is a square. Four points A, B, C, D are taken on the sides PQ, QR, RS and SP respectively. If $AC \perp BD$ then prove that AC = BD.
8. On the sides BC, CA and AB of a Δ ABC, 3 points X, Y, Z are taken in that order so that $XY \| AB$ and $ZY \| BC$. Prove that Area of ΔABX = Area of ΔACZ.
9. PQRS is a rhombus. Two points M and N are taken on the sides QR and RS such that QM = RN. PM and QN are joined. They meet in O.

   Prove that Area of ΔPOQ = Area of quad. NOMR.
10. ABC is Δ drawn in a circle AB < BC. The smaller arc BC is folded about the chord BC. It cuts AC in point D. Prove that BD = AB.

11. In a quad. ABCD, $\angle ABC = \angle DAB = 60^\circ$ and $\angle ADC = 90^\circ$, show that $AB + BC = 2AD$.

12. In a cyclic quad. the bisector of any angle and the bisector of the exterior angle obtained by extending the side of opposite angle meet in a point on the same circle.

13. A triangle is drawn in a circle. The angles in the 3 segments opposite the $\angle s$ of the Δ = 4 right angles. If a quadrilateral is drawn in a circle then the angles in the 4 opposite segments = 6 Rt angle.

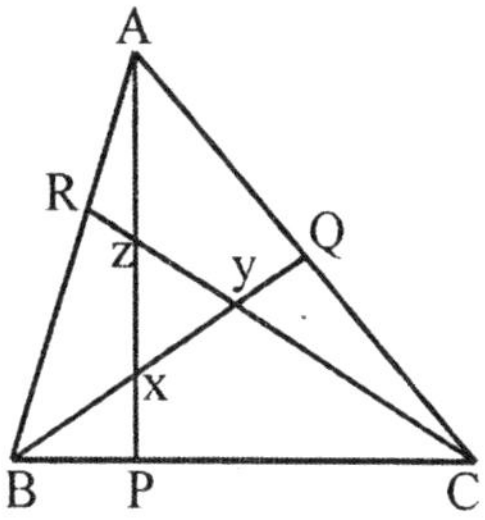

14. In the diagram given on the side $\angle APC = \angle BQA = \angle CRB$.
Prove that ΔABC is equiangular to ΔXYZ.

15. On the face of a clock prove that the line joining 4 and 7 is perpendicular to the line joining 5 and 12.

16. Two tangents are drawn to a circle and are making the same angle with one another. What is the locus of the point of intersection of such tangents.

17. Three points are given. With these as centres draw 3 circles each touching the other two.

18. When the 3 points are 13 cms, 14 cms and 15 cms from each other find the radii of the 3 circles mentioned in the last example.

***19. If three circles are such that each touches the other two show that the common tangents are concurrent.**

***20. One square and an equilateral Δ is constructed in the same circle. If the side of the square x and the side of the equilateral Δ is y then prove that $3x^2 = 2y^2$.**

***21. In any Δ prove the following**

(a) Radius of the in-circle $= \dfrac{\textbf{Area of } \Delta}{\textbf{semi perimeter of the } \Delta}$

(b) Radius of the escribed circle with respect to side b $= \dfrac{\textbf{Area of } \Delta}{(s-b)}$

22. AB is a diameter of a circle. From B a tangent BC is drawn to the circle. A chord r is drawn from A meeting the circle in P and the tangent in C. The mid pt of BC is T. Prove that TP is a tangent to the circle

23. ABC is an isosceles Δ. On line AC, a point D is taken and isosceles Δ PBD is constructed which is equiangular to ΔABC. Show that $AD \| BC$.

24. The in-centre of the Δ ABC is O. AO is joined and is produced to meet the circle in point D. Prove that D is the circum centre of the Δ ABC.

25. ABC is a Δ in which $\angle A = 90^\circ$. From A is drawn a perpendicular AD on BC. From D parallels are drawn to AC and AB meeting them at P and F. Show that P, F, B, C are cyclic.

26. AB and CD are two intersecting chords of a circle. AP and CF are drawn perpendicular to CD and AB respectively. Show that $PF \| BD$.

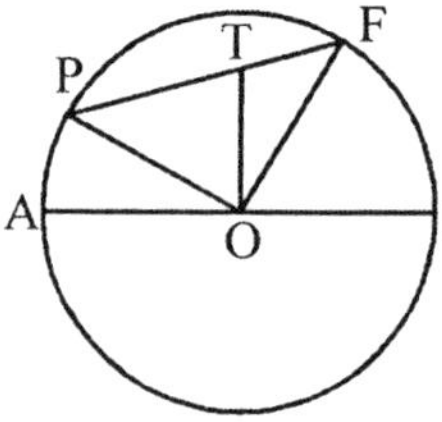

27. In the adjoining diagram, AB is diameter of the circle. O is the centre PF is a chord such that OP=AO=OB=PT. $\angle AOT = 90^\circ$. Thus AO = PT = OB. Prove that Arc AP = $\frac{1}{3}$ Arc FB.

*28. ABC is an equilateral Δ drawn in a circle AX is chord which cuts BC in D. Prove that AX = BX+ CX.

29. ABCD is a square drawn inside a circle. In the same circle an equilateral Δ APQ is drawn. Prove that PQ $\|$ BD. Find all the angles of Δ PCD.

30. AB is a diameter of the circle. CD is a chord of $\frac{1}{4}$ of the circle. AD and BC intersect in P. Prove that AC = CP.

31 O is the circum centre of ΔPQR. A point B is taken on QR. Prove that the circum circles of BOQ and BOR are congruent.

32. ABCD is a quad. and O is the intersection of the diagonals. Prove that the centres of circumcircles of AOB, BOC, COD and DOA from a parallelogram.
33. Two circle intersect in A and B. Draw a line PQ through B so that area of PAQ is the largest Δ possible.
34. AB is a chord of fixed length of a circle. P is a point moving on the circle. AD and BE are the perpendiculars drawn from A and B to BP and AP respectively. What is the locus of midpoint of DE.
35. If base is given and the sum of other sides is given then howsoever the Δ s are drawn, the feet of the perpendiculars drawn from the ends of the base on the bisectors of at the angle the vertex are on a common circle.
36. If one line makes equal $\angle s$ with the opposite pairs of sides of cyclic quad. Then it makes equal angles with the diagonals.
37. Enclose a square in another square so that the angular point is at a point on the side of the given square.
38. A Δ has base = 5 cm, height = 6 cms. A square is drawn in this Δ.

    Prove that its side $= \frac{30}{11}$ cms. Inscribe a square in this Δ How many squares can be drawn.
39. Inscribe a circle in the sector of a circle.
40. Inscribe a square in the sector of a circle.
41. In a Δ the vertices are joined to the circum-centre. Then these sides are $\perp$ to the respective sides of the pedal Δ.
42. If a polygon (n = 4 or) more but even no is constructed about a circle touching it an all sides then (a) the sum of the odd sides is equal to the sum of the even sides (b) The bisectors of the angles are concurrent at the centre of the Δ.
43. If a polygon of n sides (n = 4 or more but even) is constructed inside a circle then (a) the total sum of even angles = total sum of odd angles (b) the perpendicular bisectors are concurrent.
44. Circles are drawn on the sides of a quad. as diameters then the points where the circles meet again on the diagonals of quad. only and the common chords of alternate circles are parallel.
45. Two equal circles are intersecting at right angles to each other.

    The area common to both the circles + Square on the radius $= \frac{\pi r^2}{2}$.

46. Two concentric circles are drawn. Find out a chord of bigger circle which is double of the chord of smaller circle.

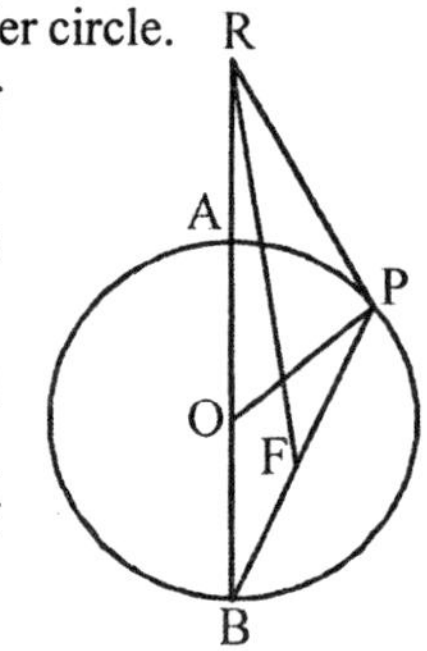

47. In the adjoining diagram AB is a diameter of the circle. RP is a tangent. RF bisects $\angle PRB$. O is the centre of the circle. Prove that $\angle RFP = 45°$.

48. In a Δ ABC, M is the incentre AM is joined and produced to meet the circum circle in point P. Prove that P is the circum center of the Δ BMC.

49. ABC is a triangle drawn in a circle. AD, BE, CF are the three perpendiculars drawn from A, B and C to opposite sides. They meet the circumference in P, Q, R. Prove that C, A, B are the centres of PQ, QR and RP respectively.

50. OP and OQ are two perpendicular radii of circle. From P and Q two perpendicular chords PS and QR are drawn. Prove that PR || QS on the other hand if PS || QR then PR $\perp$ QS.

51. AB is a diameter of the circle. A chord CD makes an angle of 45° with this diameter and meets it in P. Prove that $2CP^2 + 2DP^2 = AB^2$.

52. In an isosceles Δ ABC, $\angle B = 90°$. AC is produced to E so that CE= BC. Prove that $BE^2 = AE.AC$.

53. In an isosceles Δ ABC, D is the mid-point of BC and O is the orthocenter. Prove that DA x DO = OB^2.

54. Two circles are intersecting at right angles. Prove that the common chord makes with the circumference of first circle an angle which is the complement of the angle it makes with the 2nd circle.

55. Enclose a square in a given rhombus or a given regular octgon.

56. If a square is enclosed in a triangle the reciprocal of its side is equal to the sum of reciprocals of the base and the height of the triangle.

57. Two circles are touching each other externally at O. One of their common tangents touch them in P and F. Prove that the circle drawn with PF as diameter passes through O and the line of centres is a tangent to it at O.

58. Two segments of a circle are constructed over a line. A point on the outer segment is connected to the end point of the line. Prove

that the intercept made by them on the inner segment is the same wherever you take the point on the outer segment.

59. **In a right angled Δ, the difference between the radius of the escribed circle corresponding to the hypotenuse and the radius of the incircle of the triangle is equal to the hypotenuse.**

60. In a cyclic quad. ABCD $\angle A = \angle D = 45°$. A perpendicular is dropped from B to AD and is produced further to meet the circle E. Prove that CE is a diameter of the circle and is parallel to the tangeut to the circle from A.

61. In a Δ PQR O is the ortho centre. PO is joined. It meets the base QR in point B and the circm-circle of Δ PQR in C. From C a perpendicular CMA is dropped on PQ meeting PQ in M and QR in A. Prove that AB=BR.

62. (a) The opposite sides of cyclic quadrilateral are produced to meet each other to form 2 acute angles. The sum of these 2 acute angles is supplementary to the angle of the quadrilateral which is farthest from it.

    (b) If the 2 angles obtained by producing the opposite sides of a cyclic quadrilateral are 20° and 40°. Find all the angles of the quadrilateral.

63. BC is fixed chord of a circle. A is any point on the circumference. With A as centre a circle is drawn touching BC. Then two other tangents are drawn from B and C to this circle. Find the locus of the meeting point of these two tangents as A moves on the segment of the circle.

64. From one of the inter-sections of 2 circles a straight line is drawn meeting the 2 circles in P and F. C and D are 2 fixed points on the 2 circles. One on each one of them. What is locus of PC and DF as PF moves on the two circles.

65. On the circum circle of the Δ ABC, a point P is taken and perpendiculars are PD and PE are drawn on AB and AC respectively. What is locus of circum-circle of Δ PDE.

66. ABC is Δ. Points P, Q, R are taken on BC, CA and AB respectively such that BP=$\frac{1}{3}$BC, CQ=$\frac{1}{3}$CA, and AR=$\frac{1}{3}$AB. AP, BQ and CR meet each other in X, Y, Z. Prove that Area of Δ PQR=$\frac{1}{3}$ Area of Δ ABC.

Exercise (H)

(Moderately Difficult) (Ex on Th. 1 to Th. 61)

1. In a ΔABC $\angle BAC = 120°$ AB=AC. The perpendicular bisector of AB meets BC in X. Show that BC = 3BX.

2. ABC is drawn in a circle. AB =AC. On BC produced a point E is taken. AE is joined and it cuts the circle in point D. Show that $\angle DCA = \angle DEC$. Hence prove that AC is tangent to the circum circle of ΔDCE.

3. In the side lined figure. ABCD is a cyclic quad. DA and CB both produced meet at F forming an angle Z. Similarly BA and DC produced meet in E and form an angle y and x is the angle BCD. Prove that
   (i) If y = 32° and z = 40° then x = 54°
   (ii) If y + z = 90° then x = 45°
   (iii) 2x + y + z = 180°

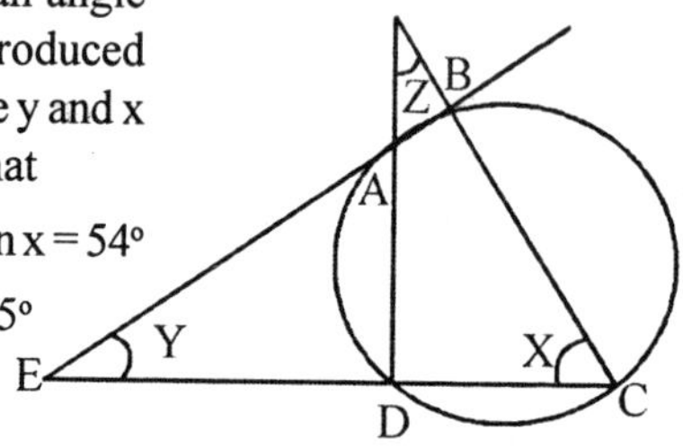

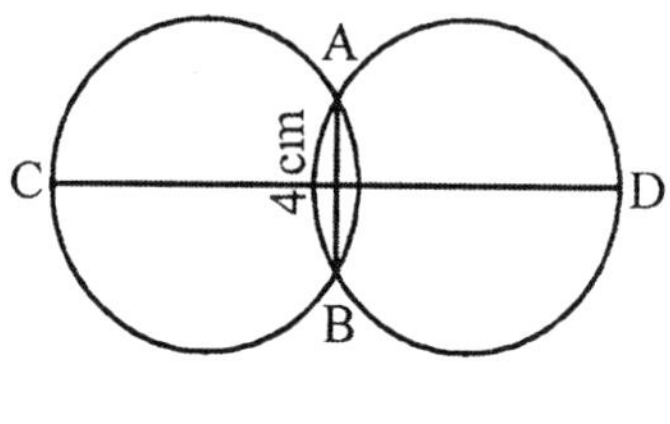

4. In the figure given on the side circles of the same radii are drawn. Common Diameter = 10 cms. Common chord AB = 4 cms. Prove that the radius of the circle = 2.9 cms.

5. In a quadrilateral ABCD the angles at B and C are each equal to 90°. A line is drawn perp. to AD at P meeting BC in F. Show that $\angle BPC = \angle AFD$.

6. AB is a diameter of a circle AMC and BMD are 2 chords of a circle with $\angle CAB = \angle DBA = 22\cdot5°$. Prove that $AM^2 = 2CM^2$.

7. ABCD is a square metal box resting against a railway compartment wall. AB = 13 cms. The edge A is resting on the wall of the

carriage. Point P which is the centre of edge BC is resting on a bar which is attached to the wall of the carriage and is at a distance of 11cms from the same wall. Show that the edges B, C, D are at distances of 5 cms, 17 cms and 12 cms from the wall.

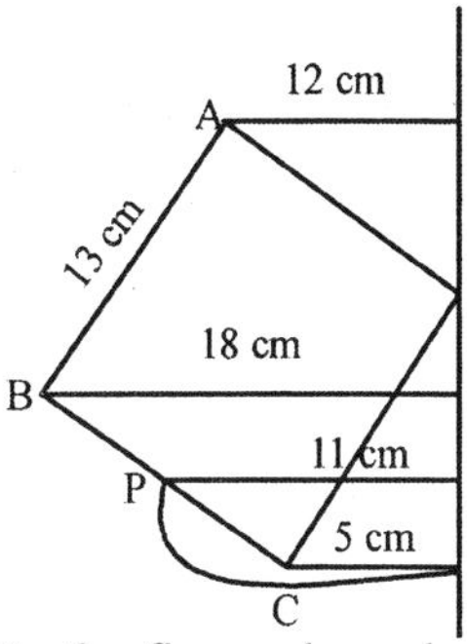

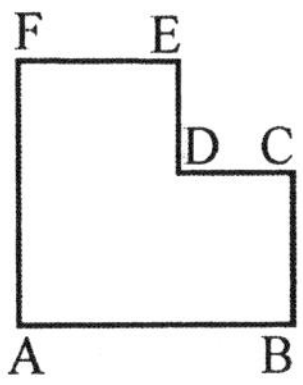

8. In the figure showed on the side. Draw A || line to edge AB so that it will divide the figure in 2 equal parts by area.

9. In a ΔABC a perpendicular AD is dropped on side BC from A, such that BD = x^2, DC = y^2 and AD = xy

   Prove that $\angle A = 90^\circ$

10. In the adjoining figure is that of crescent of moon. It is made of equal circles. CD = max width 3 cms. AD = max height 10 cms. Find the radius of the moon.

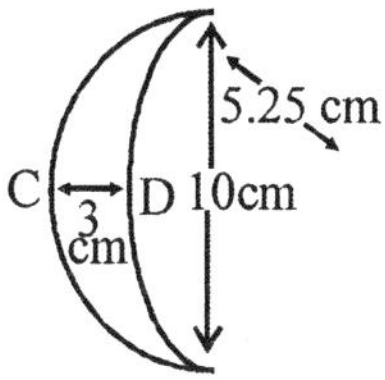

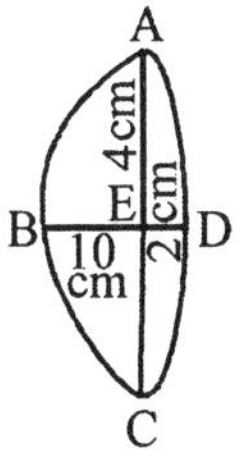

11. Here with is shown a lens. BD is the bisector of AC. ABC and ADC are unequal arcs of a circle. AE = 4 cms, BE = 10 cms, ED = 2 cms. Find the radii of the two circles.

12. In the figure given below BC is the circular arc of radius 8 cm. The centre of this circle is on AB produced. PAB is right angle. Find out the radius of the circle.

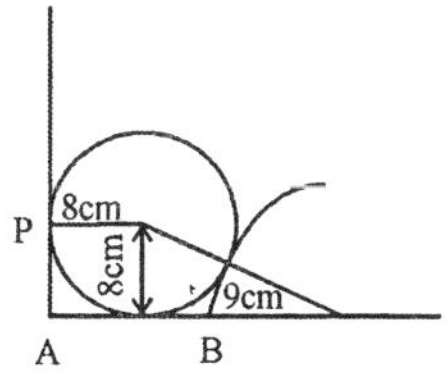

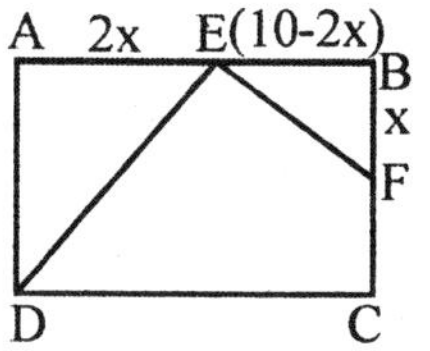

13. ABCD is a rectangle, AE = 2BF, 2BC = AB = 10cm. Area of rect ABCD = 29 cm^2. Find the length of AE.

14. A circle of 4 cms radius touches 2 perp. lines touching this circle and the two perp. lines two circles are drawn. Prove that their radii are $(12 + 8\sqrt{2}), (12 - 8\sqrt{2})$.

15. AB and CD are 2 chords of a circle at right angles to each other. Show that arc AC + arc BD = 90°, on the other hand if arc AC + arc BD = 90°. Then $AB \perp CD$.

***16. In a right $\angle d$ Δ the sum of radii of the in-circle and the circum-circle is equal to half the sum of sides containing the right angle. Prove this.**

***17. In a quadrilateral two adjacent sides are equal and the diagonal from the intersection of these sides bisects the angle opposite. Prove that either the quadrilateral in cyclic or else the remaining sides are equal.**

***18. Three equilateral Δs are drawn on the sides of a Δ opposite to the Δ itself then the three lines joining the vertex of the original Δ to the opposite vertex of the equilateral Δ are concurrent.**

***19. Four circular coins each of different radius are placed on the table in such a way that each touches the other two only. Prove that all the four pairs of contacts arc on the same circle.**

***20. In right angled triangles the difference of radius of the escribed circle with the diagonal and the radius of the incircle is equal to the diagonal.**

21. ABCD is a parallelogram. There is a point P inside so that $\angle CPD + \angle APB = 180^\circ$. Prove that $\angle PBC = \angle PDC$

22. ABCD is a parallelogram. There is a point P inside so that $\angle PBC = \angle PDC$ show that $\angle APB + \angle DPC = 180^\circ$.

23. C is any point on the arc of a circle less than a semi-circle with AB as the base. CA and CB are joined. AC is produced beyond C to D. CD = CB show that D is on a definite circle with AB as base.

24. ABCD is a quad. in a circle, if arc AB, arc BC, arc CD and arc DA are bisected at the points at P, Q, R, S then prove that $PR \perp QS$.

*25. Two circles are touching each other externally. Their radii are 'a' and 'b'. A direct common tangent is drawn to these circles. In between the direct common tangent and the two circles, a smaller circle touching the two circles and the direct common tangent is drawn. Its radius is 'c'. Prove that $\frac{1}{\sqrt{c}}=\frac{1}{\sqrt{a}}+\frac{1}{\sqrt{b}}$

26. A triangle which has the incircles and the circumcircle concentric is the equilateral Δ. Prove that the area between the 2 circles is $\frac{3\pi r^2}{4}$ if r is the radius of the circum-circle.

27. One circle touchecs another internally. From A the point of contact of these circles a line is drawn meeting the first one at P and the second one at F. From P a tangent is drawn to the first circle meeting the outer circle in C and D. Prove that CF = DF.

28. Two circles intersect at P and F. Through P a straight line is drawn meeting the circle in points A and B. Find the locus of incentre of AFB.

29. A rhombus is drawn round a rectangle. Prove that the four angular points are either in 2 perpendicular lines or otherwise on four definite arcs.

30. AB is a line segment M its middle-point. Circles are drawn with AB, MA and MB as diameter on the same side of AB. A circle is drawn so that it touches all the 3 circles internally, the circle as AB diameter and the other two externally.

Prove that its radius $= \frac{AB}{6}$.

31. PP_1 and QQ_1 are two direct common tangents drawn to two circles intersecting in points A and B. The common chord when produced intersects PP_1 in R and QQ_1 in S respectively.
Prove that $RS^2 = PP_1^2 + AB^2$

32. Out of Δs drawn on common side and the same vertical angle the perimeter of isosceles Δ is maximum.

33. Out of Δs drawn in a circle the perimeter of equilateral Δ is maximum.

34. Out of quadrilaterals drawn in a circles square is the largest possible.

35. The opposite sides of a cyclic quadrilateral are produced to meet in a point and the $\angle s$ so formed are bisected. Prove that they are at right angles.

36. On the circum-circle of the ΔABC, a point P is taken and the perpendiculars are drawn from it to the 3 sides of the ΔABC and the perpendiculars are allowed to meet the circle in D, E and F. Prove that ΔDEF is equiangular to ΔABC.

37. Two circles are at right angles. The square on each of the direct common tangent is equal to $\sqrt{2r_1r_2}$

38. Draw a Δ in a given circle such that 2 of its sides will pass two given points and the angle between these sides will be equal to a given angle.

39. From a fixed point A two straight lines APQ and AMN are drawn, intersecting two fixed lines in P and Q and M and N respectively. Show that the angle between the circle APM and the circle AQN is the same how so ever you draw the lines.

40. The incircle of ΔABC touches the sides BC, CA and AB at D, E and F respectively. Prove that the sides of the pedal Δ of DEF are parallel to the sides of the original Δ.

41. Two equal cirles are moving in two perpendicular lines in such a way that they touch each other and one of 2 perpendicular lines. What is the locus of the point of contact of these circles.

42. TAP is a fixed angle. X is a fixed point equidistant from TA and PA. Through A and X any circle is drawn. It cuts the chord TA and AP in points D and K. Prove that either the sum or difference of intercepts (AD + AK) made by them is fixed.

43. Two intersecting circles are such that one of them passes through the centre of another. Direct common tangents are drawn to these two circles. Prove that the line joining the points of contact of the first circle is a tangent to the 2nd circle.

44. The bisector of $\angle A$ of a ΔABC meets the base BC in D. AE is drawn so that $\angle ADC = \angle DAE$. Prove that AE is either tangential to the circle or is parallel to BC.

45. In ΔABC O is the orthocentre and r is the radius of the circum-circle then prove that

$OA^2 + BC^2 = OB^2 + CA^2 = OC^2 + BA^2 = 4r^2$

46. In the diagram on the side given below.

(a) XY is a st. line (b) $AC \perp XY$

(c) $PQ \perp XY$ (d) $\angle ABP = 90°$

(e) 2AB = BP (f) AC = 5 cms

(g) PQ = 24 cms

Prove that BC = 12 cms, AP = $13\sqrt{3}$ cms and QC = 22 cms.

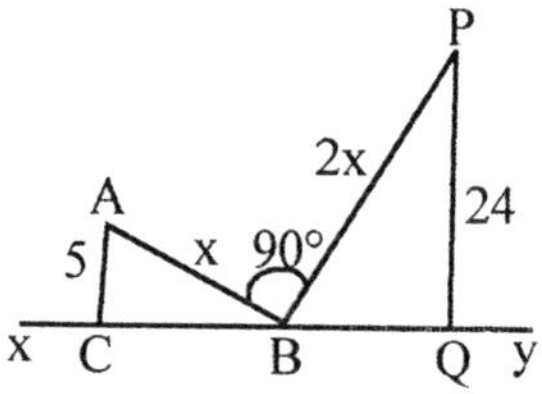

47. The base and the angle at the vertex is given, what is the locus of the (i) centroid (ii) Circum centre.

48. If in a quadrilateral the sum of one pair of opposite sides is equal to the other pair then we can intercept a circle in the quad. touching all the four sides.

49. Inside an angle PAF two externally touching circles are drawn. C is their point of contact. One of the 2 circles touches AP in P while the other touches AF in F.

Prove that $\angle PCF = 180 - \dfrac{\angle PAF}{2}$.

50. In a ΔABC AB > AC. From AB in measured AD= AC. AC is extended to AE making AE = AB. I and I^1 are the centres of the incircle and exterior circle's touching the base BC. Prove that the six points C, I, D, B, I^1, E are on the same circle.

51. Two circles intersect in A and B. Through A straight lines PAF and CAD are drawn meeting the two circles in P AND F and C and D respectively. CP and FD meet in O. Show that the angle between the circle in COF.

52. In a quad ABCD line BC = line AD. The midpts of the diagonals AC and BD are P and F. Show that PF makes equal $\angle s$ with AB and CD.

53. Two parallel lines and a point is given. From this point draw a line parallel line (use only a scale and pencil).

54. A side of a right angled Δ is double of the other side. With these sides as diameter semi circles are drawn. Prove that the common chord in $\frac{2}{5}\times$ diagonal.

55. P, Q, R, S is a quadrilateral drawn in a circle. At pts P, Q, R, S tangents are drawn. They meet in ABCD. If ABCD is cyclic then the chords PR and QS at rt. angles to each other.

56. XY is a chord of the circle at right angles to the diameter AB. YCD is another chord of the circle intersecting the diameter in C. Prove that XB and XA are the internal and external bisectors of ΔDXC.

57. Two circles of equal radii are drawn. They touch each other at A. A circle of double the radius is drawn touching the first Circle but cutting the second circle at P and F. Show that AB passes through P or F,

58. ABCD is a cyclic quad. O is the intersection of the diagonals. M is the centre of the circle about AOB. Prove that $MO \perp CD$.

59. ABCD is a square inscribed in a circle. PQRS is the largest square in the smaller segment bounded by AB. Prove that PQ = $\frac{1}{5}AB$.

60. Two circles intersect each other at right angles. Any diameter is taken in circle No. 1 and from the ends of the diameter two tangents are drawn to circle No. 2. Prove that sum of the square on the tangents is equal to square on the diameter of the circle No. 1.

4

Advanced Theorems

So far we have studied theorems that occur in the curriculum of Xth class of C.B.S.E. in Mathematics. But some more theorems have come up, as a result of further interest in the subject and have been put forward purely as an academic intersect. It is only for those who sustain interest in the subject and suggestion for improvement and criticism on any item is welcome.

Projection of one Line on Another Line.

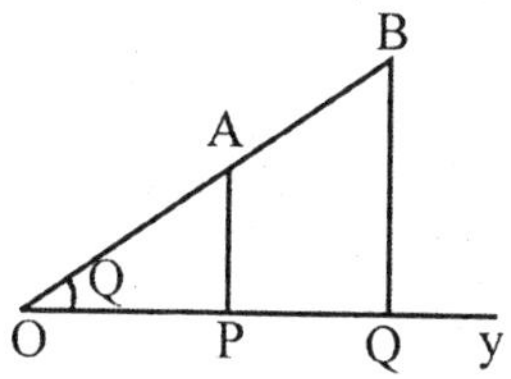

Given a line AB and a long line say XY inclined at an angle θ. From A draw $AP \perp XY$ and from B draw $BQ \perp XY$. PQ is the projection of AB on XY. According to Trignometry $PQ = AB \cos\theta$. The principle of projection is used in the next two theorems, which are known as extension of Pythagoras theorems.

Theorem No. 62

In an obtuse angled Δ the square on the side opposite the obtuse angle is equal to the sum of the squares on the other sides containing the obtuse angle together with twice the rectangle contained by one of these sides and the projection of the other side on it.

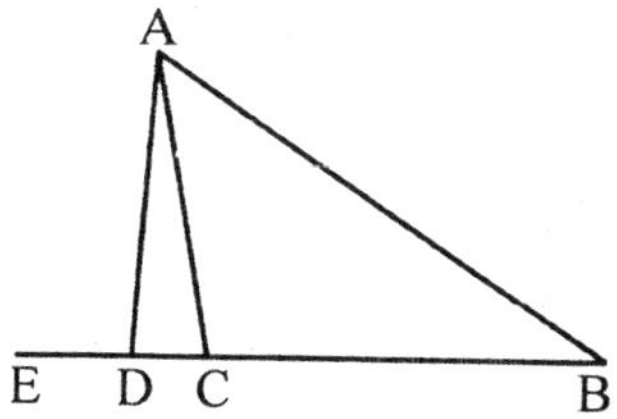

Given: ABC is an obtuse angled Δ . $\angle ACB$ is obtuse. Let $AD \perp BC$, so that BD is the projection of AB on BE and CD is the projection AC on CD.

To Prove: $AB^2 = BC^2 + CA^2 + 2BC.CD$

Proof: Because

$BD = BC + CD$

$BD^2 = BC^2 + CD^2 + 2BC.CD$

Adding AD^2 to both sides

$AD^2 + BD^2 = AD^2 + BC^2 + CD^2 + 2BC.CD$

But by Pythagoras theorem

$AB^2 = AD^2 + BD^2$

and $AC^2 = AD^2 + CD^2$

$\therefore$ $AB^2 = AC^2 + BC^2 + 2BC.CD$

The square on AB is equal to the sum of the squares on other sides (viz. BC^2 + AC^2) together with twice the rectangle contained by one of those sides viz. BC and the projection of AC on BD viz. CD. (H.P.)

Theorem No. 63

In every Δ the square on the side opposite the acute angle is equal to sum of the squares on the sides containing that angle diminished by twice the rectangle contained by one of those sides and the projection on it by the other.

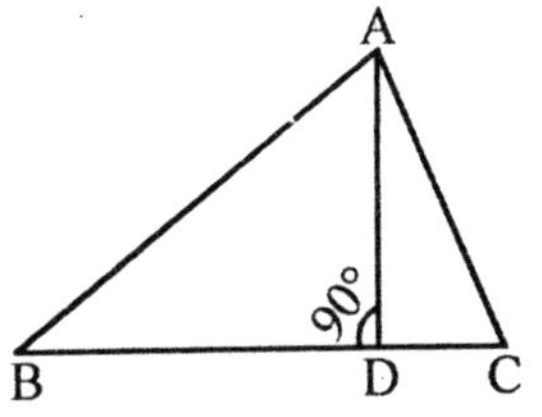

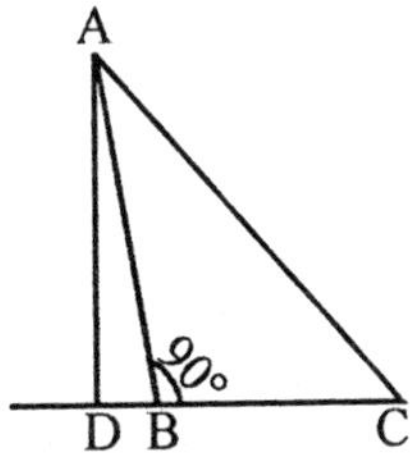

Given: ABC is a Δ in which $\angle C$ is acute. Let $AD \perp BC$ (Fig. 1) or BC produced (Fig. 2) so that CD is the projection of CA on BC.

To Prove: $AB^2 = AC^2 + BC^2 - 2BC.CD$

Proof: Since in both cases BD is the difference of BC and CD.

$BD = (BC \sim CD)$

$BD^2 = BC^2 + CD^2 - 2BC.CD$

Adding AD^2 to both sides

$$AD^2 + BD^2 = AD^2 + BC^2 + CD^2 - 2BC.CD$$

But by pythagoreas theorem

$$AD^2 + BD^2 = AB^2 \text{ and } AD^2 + CD^2 = AC^2$$
$$AB^2 = AC^2 + BC^2 - 2BC.CD$$
$$= AC^2 + BC^2 - 2BC.AC\cos\theta$$

The square on AB is equal to the sum of the squares on other sides (viz. $AC^2 + BC^2$) diminished by twice the rectangle contained by one of those sides viz. BC and the projection on it of the ofter viz. CD.

Theorem No. 64

In any Δ *the sum of the squares on two sides is equal to twice the square on half the third side together with twice the square on the median which bisect the third side.*

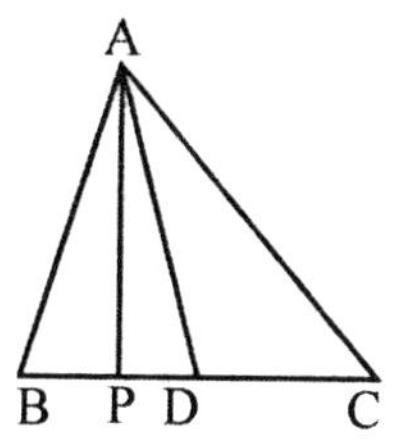

Given: A ΔABC. D is the middle point of the base. AD is the median which bisect the third side (BC) and $AP \perp BC$.

To Prove: $AB^2 + AC^2 = 2BD^2 + 2AD^2$

Proof: Draw $AP \perp BC$ out of $\angle s$ at D $\angle ADC$ is obtuse and $\angle ADB$ is acute.

$$AC^2 = AD^2 + DC^2 + 2DC.PD \qquad ...(1)$$

$$AB^2 = AD^2 + BD^2 - 2BD.PD \qquad ...(2)$$

But $\qquad DC = BD$

$\therefore \qquad AC^2 = AD^2 + DC^2 - 2DC.PD \qquad ...(3)$

Adding (1) and (3)

$$AB^2 + AC^2 = 2AD^2 + 2DC^2$$

$$= 2AD^2 + 2BD^2 \qquad \text{(H.P.)}$$

Corollary: In any Δ the difference of square on 2 sides is equal to twice the rectangle contained by the base and the projection of the media on the third side.

Proof: Subtract (3) from (1)

$$AC^2 - AB^2 = 4DC.DP = 2BC.DP \qquad \text{(H.P.)}$$

Exercise (I)

1. ABC is isosceles Δ in which AE=AC and BE $\perp$ AC. Show that $BC^2 = 2AC.CE$.

2. In a Δ ABC show that

(a) If C = 60°, then $c^2 = a^2 + b^2 - ab$

(b) If C = 120°, then $c^2 = a^2 + b^2 - ab$

3. If AB = 8 cm. O is the middle point A point P is taken on the circle with 5cm radius from O. Show that $AP^2 + BP^2 = 82$ sq cm

4. Base of Δ = 10 cm. Sum of the squares on the other 2 sides = 122 sq. cm. Find the locus of the vertex (circle of $\sqrt{72}$ cm radius).

5. In a Δ ABC, the base BC is bisected at X. a = 17 cms and b = 15 cms and c = 8 cms. Calculate the median of BC and deduce $\angle A$.
(8.5 cms, 90°)

6. The medians of a Δ are as follows

(a) 5 cms, 6 cms, 7 cms

(b) 13 cms, 14 cms, 15 cms. find out the sides.

$$\left(\frac{2}{3}\sqrt{145}, \frac{2}{3}\sqrt{73}, \frac{2}{3}\sqrt{112}\right)$$

$$\left(\frac{1}{3}\sqrt{505}, \frac{1}{3}\sqrt{592}, \frac{1}{3}\sqrt{\frac{1545}{2}}\right)$$

7. Base = 14 cm, Sum of the squares on the other side = 148 sq cm. What is the locus of the vertex. (Circle of 5 cm. radius)

8. A and B are 2 points, PF is any line. Find out a point O on PF so that $OA^2 + OB^2$ will be minimum possible?

9. Prove that the sum of squares on the sides of a parallelogram is equal to the square on the diagonals.

10. The sides of a rhombus and the shorter of the two diagonals

measure 3 cms. Find the second diagonal to the nearest hundredth of a centimeter. (**Ans.** 5.30 cm)

11. In any quadrilateral the squares on the four sides are together equal to the sum of the squares on its diagonals together with four times the square on the straight line which joins the middle point of the diagonals.

12. The sum of the squares on the sides of a quadrilateral is greater than the sum of the squares on its diagonals by four times the square on the straight line which joins the middle points of the diagonals.

***13. In any Δ ABC a point X is taken on BC so that m. BX = n. CX. Show that $mAB^2 + nAC^2 = mBX^2 + nCX^2 + (m+n)AX^2$.**

***14. In any Δ a point X is taken on BC produced so that mBX = nCX. Prove that $mAB^2 - nAC^2 = mBX^2 - nCX^2 + (m-n)AX^2$.**

***15. In any Δ the square on the three sides is equal to 4 times the square on the medians.**

16. In a trapezium ABCD, AB $\parallel$ CD and AD = BC. Prove that AB.CD = $AC^2 - AD^2$.

17. P is the middle point of AB. There is a point F in AB. Prove that $AF^2 + FB^2 = 2FA.FB + 4PF^2$.

18. The three sides of a Δ are 3 cms 5 cms and 7 cms, what is the angle opposite 7 cms side? (**Ans.** 120°)

19. In a Δ ABC, $\angle C$ is obtuse. AD $\perp$ BC and is further extended to a points E and F so that FD = AB and DE = AC.

    Show that BE = CF.

20. The sum of the squares on the diagonal of a trapezium = square on the inclined sides + twice the rectangle on parallel sides.

21. In a quadrilateral the sum of the squarcs on the diagonals is twice the squares on the lines joining the middle points of opposite sides.

22. In a Δ ABC, C is the centroid and D is the middle point of AG. Prove that $DB^2 + DC^2 = AB^2 - 10AD^2$

23. AB is a diameter of the circle and CD $\parallel$ AB. A point X is taken on AB. Prove that $AX^2 + BX^2 = CX^2 + DX^2$

24. ABCD is a quadrilateral in which

    AB = 13 cms, BC = 13 cms, CD = 14 cms, DA = 4 cms, BD = 15cms.

    Prove that ABCD is cyclic.

25. ABC is the Δ and G is the centroid. Show that $AB^2 + BC^2 + CA^2 = 3(GA^2 + GB^2 + GC^2)$

26. If a st. line is bisected at X and is also divided

(i) internally

(ii) externally into unequal segments at Y then in either case show that $AY^2 + YB^2 = 2(AX^2 + XY^2)$.

27. ABC is a isosceles Δ in which AB = AC. AY is drawn dividing BC internally or externally at Y.

    Then show that $AY^2 = AC^2 - BY.YC$ (for internal division)

    $AY^2 = AC^2 - BY.YC$ (for external division)

28. In any quadrilateral the sum of the squares on the sides is equal to the sum of the squares on the diagonals together with four times the square on the line joining the middle point of the diagonals.

Theorem No. 65

If two chords of a circle intersect each other within the circle, the rectangle contained by their segments are equal.

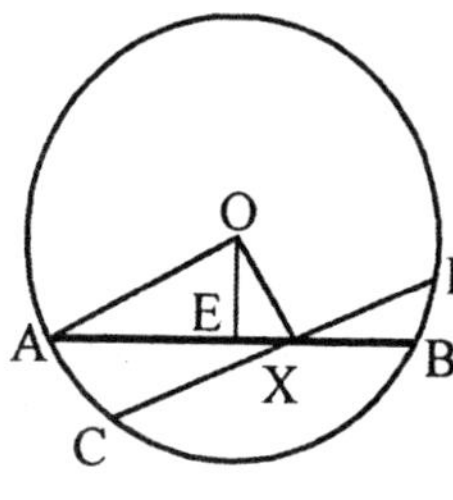

Given:- AB and CD are two chords of a circle of center o and radius r. They intersect each other at X within the circle.

To Prove:- AX . XB = CX . XD

Construction:- Let E be the centre of AB.

Join OE and OA and OX.

Proof:- Rectangle AX . XB

$= (AE + EX).(EB - EX)$

But $= AE = EB$ $\quad = (AE + EX)(AE - EX)$

$= AE^2 - EX^2 = AE^2 + OE^2 - OE^2 - EX^2$

$= (AE^2 + OE^2) - (OE^2 + EX^2)$

$= AE^2 - OX^2 = r^2 - OX^2$

Similarly $\quad CX.XD = r^2 - OX^2$

$AX.XB = CX.XD = r^2 - OX^2$ (H.P.)

Alternative Proof

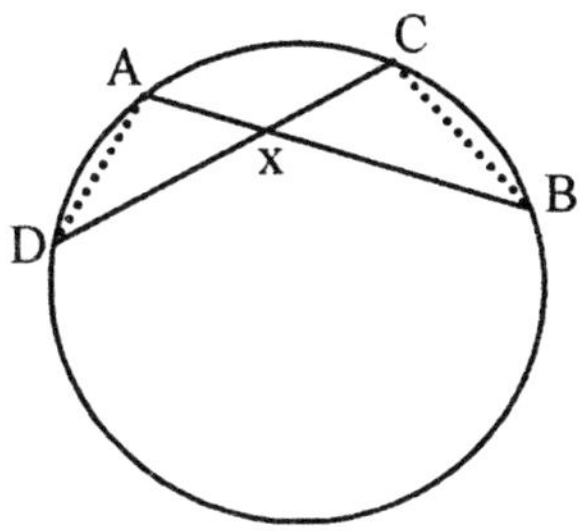

Given:- AB and CD are 2 chords of a circle. They intersect at X within the circle.

To Prove:- $AX.XB = CX.XD$

Construction:- Join AD and BC.

Proof:- In Δs AXD and CXB

$\angle DAX = \angle BCX$ and $\angle ADX = \angle XCB$ ($\angle s$ in the same segment)

$\therefore$ There Δs are equiangular

$$\frac{AX}{CX} = \frac{DX}{BX}$$

$\therefore \quad AX.BX = CX.DX$ (H.P.)

Theorem No. 66

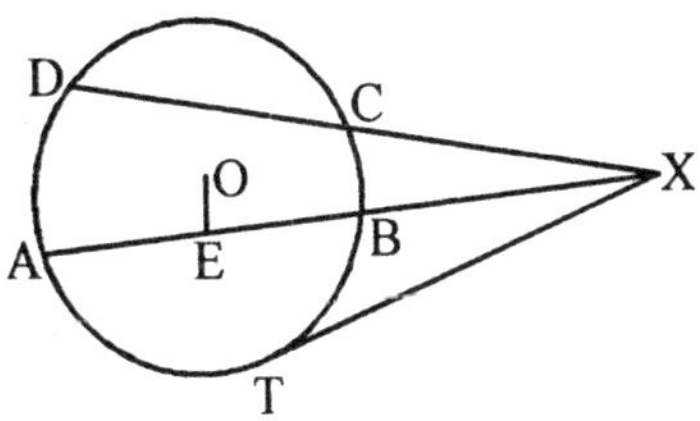

If 2 chords of a circle, when produced intersect each other outside the circle, then the rectangle contained by the segments of the one are equal to the rectangle by the segments of the other and each rectangle is equal to the square of the tangent from the point of intersection.

Given:- ABCD is a circle with centre O and radius r. AB and CD are chords intersecting each other when produced outside the circle in point X. Let XT be the tangent to the circle from X the point of intersection.

To Prove:- Rectangle AX . XB = Rectangle CX . XD = XT^2

Construction:-Draw OE ⊥ AB.

Join OE, OB, OC and OT.

Proof:- Rectangle AX.XB

$= (XE + EA)(XE - EB)$

$= (XE - EA)(XE - EA)$ $\because$ EA = EB

$= XE^2 - EA^2$

$= (XE^2 + OE^2) - (OE^2 + EA^2)$

$= OX^2 - r^2$ ...(1)

Similarly CX . XD

$= OX^2 - r^2$ ...(2)

Also X T $= OX^2 - OT^2$

$= OX^2 - r^2$ ...(3)

$\therefore$ AX.XB = CX.XD = XT^2 (H.P.)

Alternative Proof

Given:- ABCD is a circle. Let AB and CD be the 2 chords meeting each other in X outside the circle. XT is the tangent.

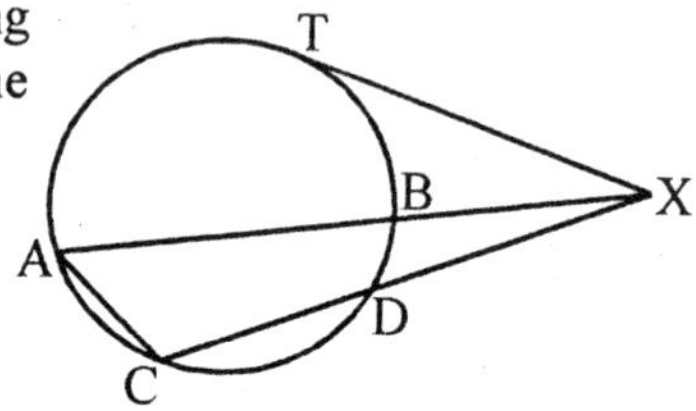

To Prove:- AX . XB = CX . XD

Proof:- $\angle BAC + \angle BDC = 180°$ (Th 53)

$\angle BDX + \angle BDC = 180°$ (L.P.A.)

$\therefore$ $\angle BAC = \angle BDX$

Similarly $\angle DCA = \angle DBX$

$\therefore$ Δs ACX and Δ BDX are equiangular.

$$\therefore \qquad \frac{AX}{DX} = \frac{CX}{XB}$$

$$\therefore \qquad AX \, . \, XB = CX \, . \, XD$$

Theorem No. 67

(Converse No. 66)

If from a point two line regments are drawn and two points are taken on each, such that the rectangle contained by the whole line and the line upto which the point is taken on that line in both cases, then the four points i.e. two points on each line are concyclic.

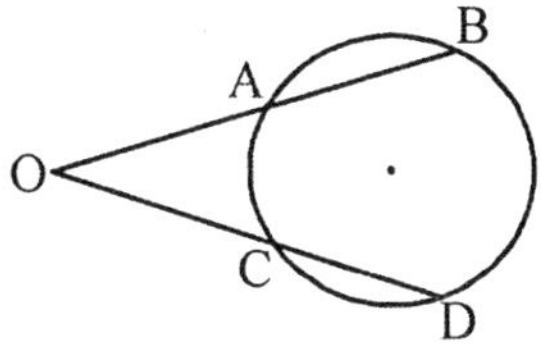

Given:- Two line segments OAB and OCD are drawn and the four points taken viz. A, B on one line and C, D on the other so that OA.OB = OC.OD.

To Prove:- That A, B, C, D are concyclic.

Proof:- Suppose the circle drawn through A, B and C passes through D^1 instead of D.

$$\text{Then} \qquad OA \times OB = OC \times OD^1$$

$$\therefore \qquad OD^1 = \frac{OA \times OB}{OC}$$

but by hypothesis

$$OD = \frac{OA \times OB}{OC}$$

This means D and D^1 meet coincide

$\therefore$ A, B, C, D are concyclic.

Exercise (J)

1. ABC is a Δ in which $\angle A = 90°$ and AD $\perp$ BC then

(a) $BD.CD = AD^2$

(b) $AD.BC = BA^2$ and

(c) $CB.CD = AC^2$

2. Prove that the tangents from any point in the extended common chord are equal.

3. A line segment XY is drawn of unlimited length and a point P is taken outside it. A line PK is drawn to the line and a point D is taken on it such that PK.KD is fixed value. What is the locus of D as K moves on the whole line.

4. AB is a diameter of the circle On AB or on extended AB a perpendicular is drawn. A line is drawn from A, meets this perpendicular in P and E. Prove that AP.AF is always fixed.

5. What is the locus of points from which tangent to intersecting circles are equal.

6. From a fixed point in a circle perpendicular chords are drawn. The sum of the squares on these chords is always the same in whatever direction these chords are drawn.

7. M is the centre of the circle and MK is the radius. From a point A outside the circle a perpendicular AF is drawn to the radius. AK is joined and if necessary produced to meet the circle in point B. Prove that 2 MK.KF = AK.KB.

8. In a Δ ABC, $\angle B$ and $\angle C$ are acute. BD and CF are drawn perps from B and C to the opposite sides.

   Prove that $BC^2 = AB.BF + AC.CD$.

   If $\angle C$ is obtuse what will be the change in the final results.

9. ABC is a Δ in which $\angle A$ is acute. BC^2 is less than $AB^2 + BC^2$ by twice the square of the tangent that would be drawn from A to the circle with BC as diameter.

10. Taking advantage of the above theorem draw a secant to the circle from a point outside the circle so that it will be (a) bisected or (b) trisected by the near point of the circle.

11. ABCD and EFBA are two circles intersecting in A and B. DE and DF are two direct common tangents. By producing AB it meets ED and CF in G and H respectively. Prove that $GH^2 = ED^2 + AB^2$.

12. A man is standing on a 2 km high mountain. The diameter of the earth is 12800 km. Show that he can see up to a distance of 160.012 kms approximately around uninteruptedly.

13. A circular arc over a gate 16 meters wide has a radius of 8 meters. What is the chord of half the arc. (9.8 meter)

14. The Empress mill at Nagpur has a chimney 40m high. Assuming

the earth's diameter is 12800 km. Find the distance at which top of chimney is visible. (22.56 km appor.)

15. The minimum distance of a point from the circle is 1.2 m. The tangent to the circle is 2.4 m from the point. What is the radius of the circle. (1.8 m)

16. If the horizon is visible to an observer 20.00 meters about sea-level is 16.0 kms distant, find roughly the diameter of the earth. **(Ans.** 12800 km)

17. Hence find the approximate distance at which a bright light raised 60 m. above seaevel is visible. (Appox. 27.72 km)

18. PAB is a secant line drawn from outside the circle. PM is perp. from P on to a diameter CD of the circle. If M falls within the circle then $PM^2 = PA.PB + CM.MP$.

    If M falls outside the circle then $PM^2 = PA.PB - CM.MD$.

19. If the height of Qutub Minar is 750 meters then from what distance its top can be seen.

20. From a fixed in a circle two perp. chords are drawn then the sum of the squares on these chords is the same wherever you draw them.

Theorem No. 68

If one angle of one Δ is equal to one angle of another Δ, then the areas of each of the 2 Δs are proportional to the respective product of the two sides.

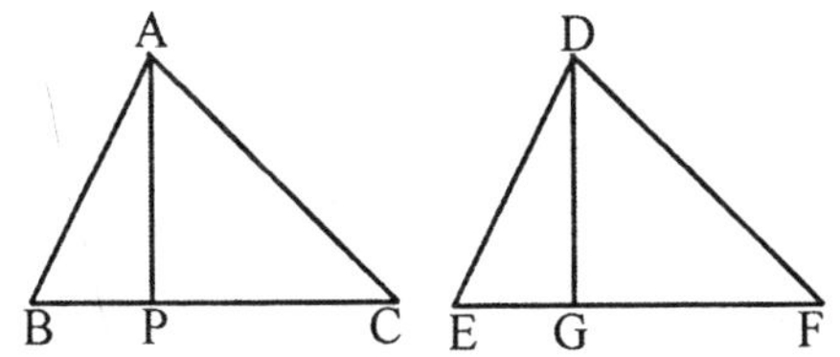

Given:- ABC and DEF are 2 Δs in which $\angle B = \angle E$

To Prove:- $\dfrac{Area\ of\ \Delta ABC}{Area\ of\ \Delta DEF} = \dfrac{AB.BC}{DE.EF}$

Construction:- Draw $AP \perp BC$ and $DG \perp EF$

Let $AP = m$ and $DG = n$

Proof:- $$\frac{Area\ of\ \Delta ABC}{Area\ of\ \Delta DEF} = \frac{\frac{1}{2}BC.AP}{\frac{1}{2}EF.DG} = \frac{BC.AP}{EF.DG}$$

But Δ ABP is equiangular to Δ DEG

$$\therefore \qquad \frac{AP}{DG} = \frac{AB}{DE} = \frac{m}{n}$$

$$\therefore \qquad \frac{Area\ of\ \Delta ABC}{Area\ of\ \Delta DEF} = \frac{AB.BC}{DE.EF} \qquad \text{(H.P.)}$$

Theorem No. 69

If a line is bisected at a point and is further divided at another point in unequal division, then the rectangle contained by unequal divisions is equal (square on half the line)-(square on the line between the point of bisection and the point of division into dissimilar divisions.

The is also true if the point of division is external.

Given:- A line is bisected at a point M and is also divided into 2 parts (AX and XB) for internal Division and (AY and YB) for external Division.

To Prove:- $AM^2 - MX^2 = AX.XB$

$MY^2 - AM^2 = AY.YB$

Proof:- Let $AM = a$

$MS = b$

$MY = c$

A M X B

Fig. I

A M B Y

Fig. II

Then Fig. I (i) Since M is the mid pt of AB

$AM = MB = a$

$AX = (a + b)$

$XB = (a - b)$

$AX.XB = (a + b)(a - b)$

$= a^2 - b^2$

$= AM^2 - MX^2$

and Fig. II $AY = AM + MY = a + c$

$BY = MY - MB = c - a$

$$AY.BY = (c + a)(c - a)$$
$$= c^2 - a^2$$
$$= MY^2 - AM^2 \qquad \text{(H.P.)}$$

Exercise (K)

1. If M is the mid point of AB and X is a point on MB. Prove that

   $MX = \frac{1}{2}(AX - BX)$

2. If M is the mid point of AB and the line is extended to a point X then prove that

   $MX = \frac{1}{2}(AX + BX)$ and $AX^2 + BX^2 = 2MX^2 + 2AM^2$

3. ABCD is a st. line and X and Y are the centre of AB and CD. Prove that

   $AD + BC = 2XY$.

4. M is the mid pt. of AB. AB is extended to a point X. Prove that

   (a) $AM.AX = MB.\ BX + 2AM^2$

   (b) $AX^2 = 4\,AM.\ MX + BX^2$

5. ABCD is a straight line. AB = CD.

   $AD^2 + BC^2 = 2AB^2 + 2BD^2$.

6. On a line AB, X is a point. Prove that

   $AB^2 + AX^2 = 2AB.AX + XB^2$.

7. On a line AB, X and Y are the pt. of trisection. Prove that

   $AB^2 = AX^2 + 2XB^2$

8. M is the mid-pt of a line AB and is further extended to a point X so that $MB.MX = BX^2$

   then prove that $XA^2 = 5BX^2$.

9. AB is extended to a point X, so that $XA^2 = 4XB^2 + AB^2$

   Prove that $XA = \frac{5}{2}XB$

10. By extending a given line AB, find a fixed point D in the extended part, so that

    AD.DB = a given square in area.

11. M is the mid point of AB. Find a point C by extending it, so that $AC^2 - BC^2 = CM^2$

12. On a given side draw an isosceles obtuse angled Δ so that the square on its maximum side will be 3 times the square on the equal side.

13. By extending AB find a point P in the extended portion, so that $AB^2 + AP^2 = 2AP.BP$.

14. Line AB has its middle point at C. Line AB is extended and a point D is taken on the extended portion. Prove that $AD^2 - DB^2 = 4AC.CD$.

15. Divide the diagonal of a st. angled Δ in 2 parts so that the difference on the 2 parts will be = square on either of 2 sides.

Practical Geometry - 5

Construction No. 30

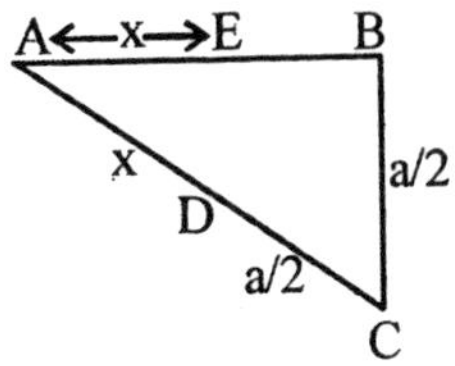

To divide a line of given length so that the rectangle contained by the whole line and its part from the point of division upto its end is equal the square on the remaining part.

Construction:

Draw AB of given length a. Draw a right angle to AB at B. Make $BC=\frac{a}{2}$. Join AC. Make $CD=\frac{a}{2}$. From A cut off CD = AE. E is the required point.

Proof: Let AD = x

ABC is a rt. angled Δ

$\therefore \quad AC^2 = AB^2 + BC^2$

$$\left(x+\frac{a}{2}\right)^2 = a^2 + \frac{a^2}{4}$$

$$x^2 + ax + \frac{a^2}{4} = a^2 + \frac{a^2}{4}$$

$\therefore \quad x^2 = a^2 - ax = a(a-x)$

or $\quad AE^2 = AB.EB$

This is known as medial section of AB at E.

Construction No. 31

To draw an issosceles Δ having each of the base angles = twice the angle at the vertex or (To draw an isosceles Δ with vertical $\angle = 36^\circ$ and base $\angle s = 72^\circ$ each)

Construction:

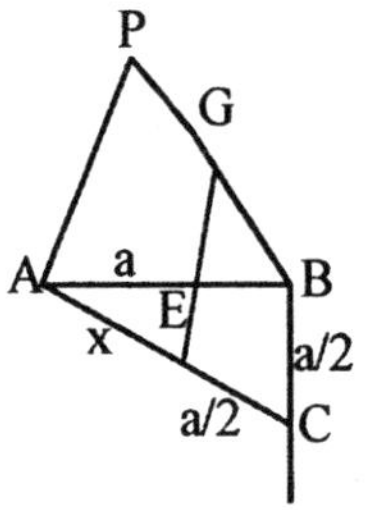

Let AB be the given side. Divide AB in medial section i.e. $AB^2 = AB.BX$. With XB as base draw an isosceles Δ so that XG = BG = AX.

Δ XGB is the required Δ. From A draw AP $\parallel$ XG and produce BG to meet AP in P. PAB is another Δ with base angles = 72° and $\angle$ at the vertex = 36°.

Proof: Because XG = XA

$\angle XAG = \angle XGA$

$\angle GXB = \angle XGA + \angle XAG = 2\angle XAG$

Because XG = GB

$\angle GXB = \angle GBX + 2\angle XAG$

$\angle GBX = 2\angle XAG$

The $\angle s$ of the Δ GXB

$= A + 2A + 2A = 180^\circ$

$\therefore$ $5A = 180^\circ$

$A = 36^\circ$

$\therefore$ $\angle XGB = 36^\circ$

and $\angle GXB$ and $\angle GBX$ are each = 72°

As GX is $\parallel$ AP.

APB is equiangular to GXB. (H.P.)

Exercise (L)

1. In a ΔABC, $\angle B = \angle C = 2\angle A$ prove that $\frac{BC}{AB} = \frac{\sqrt{5}-1}{2}$

Proof: Let $AD = BD = BC = x$

Let $DC = y$

$AB = AC = x + y$

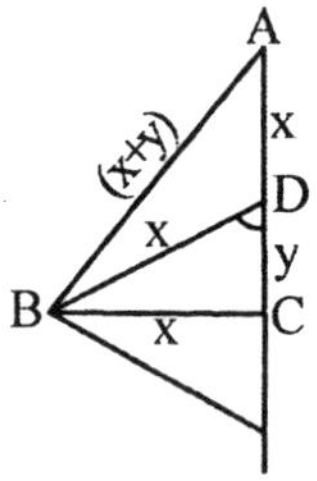

ΔBDC is similar to ΔABC

$$\therefore \quad \frac{x+y}{x} = \frac{x}{y}$$

$$xy + y^2 = x^2$$

or $$y^2 + xy - x^2 = 0$$

$$y = \frac{-x \pm \sqrt{x^2 + 4x^2}}{2} = \frac{x(\sqrt{5}-1)}{2}$$

$$\therefore \quad \frac{y}{x} = \frac{\sqrt{5}-1}{2} = \frac{x}{x+y} = \frac{BC}{AB}$$

2. If a st. line is divided in medial section at point X, then show that

$$AB^2 + BX^2 = 3AX^2$$

$$AX^2 = AB.BX = AB(AB - AX)$$

$$\therefore \quad 2AX^2 = 2AB^2 - 2AB.AX$$

$$3AX^2 = 2AB^2 - 2AB.AX + AX^2$$

$$= AB^2 + AB^2 - 2AB.AX + AX^2 = AB^2 + (AB - AX)^2$$

$$= AB^2 + BX^2$$

3. In a ΔABC

$$\angle B = \angle C = 2\angle A$$

$$AC = 2R, \sin ABC$$

$$AB = 2R, \sin ADC$$

But $AC = AB$ and $\angle ABC = 72°$ and $\angle ADC = 108°$

$\therefore$ $\sin \angle ABC = \sin \angle ADC$

$\therefore$ $R = R_1$

$\therefore$ Circle about ABC is same as circle about ABD.

4. BC is a side of a regular decagon.

5. If a st. line is divided into medial section. Then difference of squares of the segment AX and BX is equal to rectangle contained by the segments

$$AX^2 - BX^2 = AB.BX - BX^2$$
$$= BX(AB - BX) = AX.BX$$

6. If a $\Delta ABC \angle A = 36°$, $\angle B = 72°$

BP $\perp$ AC and AE $\perp$ BC prove that

(1) $5AC^3 = RBP^2 + 4AE^2$

(2) $AC^2 + BC^2 = 4BD^2$

(3) $PC = \sqrt{5} - 1$

(4) $\frac{AC}{CD} + \frac{CD}{AC} = 3$

Construction PD = CD. Join BP $\perp$ AC

(i) $4AC^2 = 4AE^2 + 4EC^2 = 4AE^2 + BC^2$ ...(1)

Now $AC = 2CP + AD$

$AC - 2CP = AD$

$AC - 2CP = BC$

$AC^2 - 4AC.CP + 4CP^2 = BC^2$

$AC^2 - 2AC.CD + 4CP^2 = BC^2$

But $AC.CD = AD^2 = BC^2$

$AC^2 - 3BC^2 + 4CP^2 = 0$

$AC^2 = 3BC^2 - 4CP^2$ ...(2)

Adding (1) and (2)

$5AC^2 = 4BC^2 + 4AE^2 - 4CP^2$

$= 4BP^2 + 4AE^2$

(ii) $AC^2 = 3BC^2 - 4CP^2$

$AC^2 + BC^2 = 4BC^2 - 4CP^2 = 4BP^2$

$$AC^2 = 3BC^2 - 4CP^2 \qquad ...(1)$$

(iii) $AC^2 = 3BC^2 - 4CP^2$

Suppose BE = 1 and CD = x then BC = 2

$$AC = (2+x) \text{ and } CP = \frac{x}{2} m$$

then (1) gives $(2+x)^2 = 3.2^2 - 4\frac{x^2}{4}$

$$4 + 4x + x^2 = 12 - x^2$$

$$2x^2 + 4x - 8 = 0$$

$$x^2 + 2x - 4 = 0$$

$$x = \frac{-2 \pm \sqrt{4+16}}{2} = \sqrt{5} - 1$$

$$CD = \sqrt{5} - 1 \,.\; CD = \sqrt{5} - 1$$

(iv) $\frac{AC}{CD} + \frac{CD}{AC} = \frac{\sqrt{5}+1}{\sqrt{5}-1} + \frac{\sqrt{5}-1}{\sqrt{5}+1}$

$$= \frac{\left(5+1+2\sqrt{5}\right) + \left(5+1-2\sqrt{5}\right)}{5-1} = \frac{12}{4} = 3$$

Exercise (M)

1. ABC and DBC are $2\Delta s$ standing on BC and on the same side. E is a point on AB. Parallels are drawn to AC and DA. They meet CB and DB at F and G. Prove that FG $\parallel$ CD.

2. ABC is a Δ. DEF is a line meeting BC, AC and AB in D, E, F there by making $\angle AFD = \angle AEF$. Prove that $\frac{BD}{CD} = \frac{BF}{CE}$.

3. ABC is a Δ in which the incircle touches the base in D and the other sides in E and F. EF is joined and it meets BC in P. Prove that PD divides BC harmonically.

4. In a ΔABC a point X is taken on the base BC. AX is joined and a point O is taken on AX. Through O parallel is drawn through the

base OC. It cuts the line AB, AX and AC in P, O and Y. Show that PO : OY = BX : CX.

5. AB, AC and AD are 3 lines emanating from a common point A. P is a point in AD. perpendiculars are drawn to AB and AC respectively. Show that PM : PN and AM : AN are always fixed.

6. In an isosceles Δ about which a circle is drawn. Prove that

$$\frac{\textit{diameter of the circle}}{\textit{equal side}} = \frac{\textit{equal side}}{\textit{perpendicular to the vertex from the base}}.$$

7. In a ΔABC, AP is the bisector of $\angle A$ at the vertex meeting the base in P. If M is the centre of the incircle then show that

$$\text{AM/MB} = \frac{AB+AC}{BC}.$$

8. ABCD is a parallelogram. From a line AXY is drawn cutting BC and AD in X and Y. Prove that the area of BX. BY is always fixed.

9. The rectangle contained by 2 perpendiculars drawn from any point on the circle to any of two tangents is equal to the square of perpendicular drawn from the same point to the chord of contact.

10. The distance of a point from the base of an isosceles Δ is in middle proportion to the distance from the sides. What is its locus.

11. A line AB is divided internally and externally in the same ratio of X:Y, the prove that

$$\frac{2}{AB} = \frac{1}{AX} + \frac{1}{AY}$$

12. In a ΔABC, points P and F are taken on AB and AC such that AP $= \frac{1}{5}$AB and AF $= \frac{1}{5}$AC. BF and CP intersect in M.

Prove that

(a) M is on the median of the base.

(b) The median is divided in the ratio of 1:2 and smaller part is on the side of A.

(c) BM $= \frac{5}{6}$ BF and CM $= \frac{5}{6}$ CD.

13. Line AB is harmonically divided in X and Y. Then AB divides XY harmonically.

14. In a Δ inset a square. How many squares are possible.

15. One brick is placed over another. The bricks are of identical size. Its dimensions 8 cm x 5 cms x 4 cms. How high are the corners of the brick II.

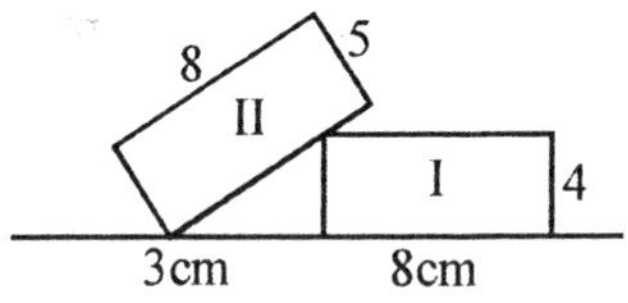

16. An electric bulb is 12cms above the centre of table 9cms x 5cms x 4cms. What would be the shadow of the table.

17. ABC and PQR are 2 similar Δs. AD and PS are the two heights. AB = 15 cms, BC = 14 cms, AD = 12 cms, PS = 5 cms. Prove that the areas are in the ratio of 9:1.

18. Two circles intersect in A and B. Through A straight line CAD is drawn meeting the two circles in C and D.

Prove that BC : BD = $\dfrac{\textit{diameter of circle CAB}}{\textit{diameter of circle DAB}}$

19. In a ΔABC, a point is taken on BC so that BD = $\frac{1}{4}BC$ and on AC a point E is taken so that AE = $\frac{1}{3}$AC. Prove that AD passes through the mid point of BE.

20. PF is a line cutting equal sides AB and AC of an isosceles ΔABC in P and F. If 4BP.CF = BC^2 then prove that AB, AC and PF touch a fixed circle of ABC.

21. If ABCD is a quadrilateral and if the bisectors of $\angle s$ of A and B meet the corners of C and D then prove that $\angle C = \angle D$.

22. In a ΔABC the bisector of the vertical $\angle A$ meets the base in D and the circumcircle in E. Prove that AD.AE = AB.AC.

23. There is only 1 point which when joined to all the three vertices and produced to meet the opposite sides then it divides each line in the ratio of 2:1.

24. In a ΔABC AD is the median of the base BC. On AD is any point

taken. BP and CP are joined and produced to meet the opposite sides in P and E then prove that BP : PF = CP : PE.

25. In a ΔABC a line DE is drawn $\|$ to BC and intersecting in AB and AC in D and E. The distance between DE and BC is 5 cms. The area of ΔADE= 9 sq cms and that of quad. DECB = 40 sq. cms. Find the length of DE.

26. Inocribe a square in a semi-circle with AB as diameter. Prove that the area of the square = $\frac{1}{5}$ of the area of square on AB.

27. Draw a parallel to the base and bisect it in 2 parts.

28. Draw 2 parallels to the base of a Δ so that it is divided into 3 equal parts.

29. Draw a square in the sector of a circle. What is the ratio of the radius to the side of the square if central angle is (a) 60° and (b) 45°.

30. In a ΔABC AB = AC = 2BC. D is a point on AC such that $\angle DBC = \angle BAC$. From D a parallel is drawn to BC. Find the ratio of $\Delta ABC : \Delta BCD$ and $\Delta BED : \Delta EDA$.

31. M is the centre of the circle. MG is the radius A and B are 2 points on MG so that MA. MB = MG^2. Prove that the ratio of the distances of any point on the circles. from A and B is always the same.

32. DEF is the pedal Δ of the ΔABC. D is an BC and D is on AC.

    Show that $\Delta ABC : \Delta DEF = \dfrac{BE^2}{AE^2}$

33. Find out a point inside a Δ so that its distances from each of the vertex are in the ratio of opposite sides.

34. A, B, C are points in one straight line. Find the locus of points which make the same $\angle$ with AB and AC.

35. A, B, C, D are 4 points in one straight line. Find out a point P so that AB, BC,CD make the same angle with P.

36. The common tangent of 2 externally touching circles is middle proportion of two diameters.

Theorem No. 70

(Simpson's line)

The feet of perpendiculars drawn from any point on the circmference of a Δ on the three sides are collinear.

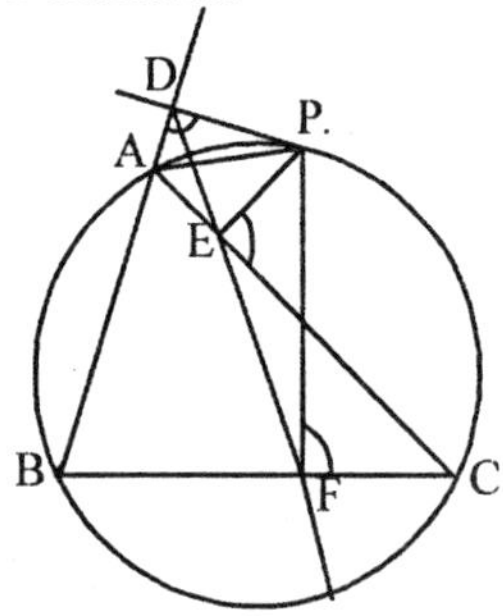

Given:- A Δ ABC about which a circle is drawn. P is any point on the circumference of the circle. PD, PE and PF are the three perpendiculars drawn from P to AB, AC and BC respectively.

To Prove :- That D, E, F are in one straight line.

Proof:- PD $\perp$ AB and PE $\perp$ AC

$\therefore$ $\angle PDA + \angle PEA = 180°$

$\therefore$ P, D, A, E are cyclic

$\therefore$ $\angle PAD = \angle PED$ ($\angle s$ in the same segment)

Similarly PE $\perp$ AC and PF $\perp$ AC

$\therefore$ $\angle PEC = \angle PFC = 90°$

$\therefore$ P,E, F, C, are cyclic

$\therefore$ $\angle PCF + \angle PEF = 180°$ ...(1)

But $\angle PCF = \angle PAD$

($\angle$ of cyclic quad. = opposite $\angle$ obtained by producing the side)

and $\angle PAD = \angle PED$

and from (1) we have

$\therefore$ $\angle PED + \angle PEF = 180°$

$\therefore$ D, E, F are in one st. line (L.P.A.)

This St. line is usually known as Simpon's line. (H.P.)

Exercise (N)

1. In the figure of theorem No. 70, PF produced meet the circle in G. Prove that AG || DF.
2. Find the points on the circle so that the line coincides with a side of the Δ. How many such points are there.
3. ABCD is a quadrilateral. DA and CB meet in F. AB and DC meet in P. The circumcircles FAB and PCB meet in L. Prove that quadrilaterals FDCL and ADPL are cyclic.
4. Find a point on the sides of quad. such that the feet of perpendiculars on the sides of a quad. are in one st. line. How many such points are possible.

***5. The centre of the inscribed circle is the orthocentre of the 3 centres of escribed circle.**

6. Base, vertex angle, and the point where the incircle touches the base are given. Draw the triangle.
7. The three points in which the incircle of Δ touches the 3 sides are given. Draw the Δ.
8. The three centres of three escribed circles are given. Draw the triangle.
9. The angle at the vertex, the perimeter and the radius of the incircle are given. Draw the Δ.
10. The angle at the vertex, the perpendicular from the vertex to the base and the radius of the inscribed circles. Draw the Δ.

*11. If a point P on the circumference of a Δ is joined to the orthocentre O, prove that the centre of OP is on the Simpson line to circle from O.

12. The base, $\angle$ at the vertex and the position of Simpson's line are given. Draw the Δ.
13. The vertex, the circumradius and the position of Simpson line and the point from which the Simpson line is drawn, are given. Draw the Δ.

***14. A point M is taken on the circle and 3 chords MA, MB and**

MC are drawn. 3 circles are drawn with the chords as diameter then the other 3 points of intersection with circles are in one st. line. (Cambridge Mathematical Vol I Page 169)

Theorem No. 71

(The Nine points circle)

In any Δ, the middle points of the three sides, the feet of perpendicular from the 3 vertices to the opposite sides, the middle points of the three lines joining the vertices to the orthocentre, are concyclic.

These nine points through the circle passes is known as the *NINE POINTS CIRCLE.*

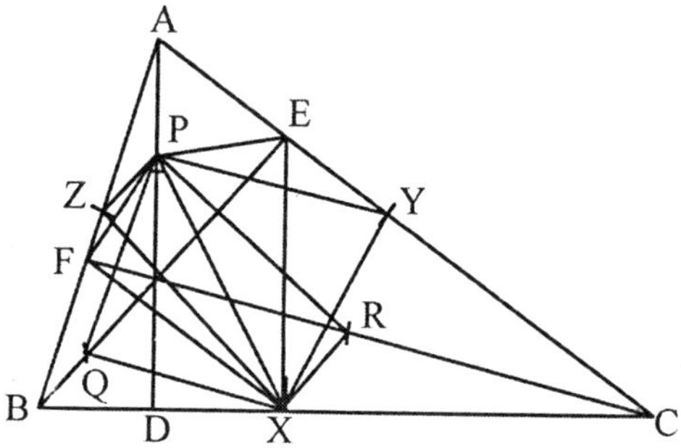

Given:- In a ΔABC, X, Y, Z are the midpts of the sides, D, E, F are the feet of perpendiculars from the 3 vertices, O is the orthocentre P, Q, R are centres of OA, OB, OC respectively.

To Prove:- That X, Y, Z, D, E, F and P, Q, R are concyclic.

Construction:- Join XY, XZ, XP, YP, and ZP.

Proof:- 1. PDX is a rt. angled Δ

PX is the hypotenuse. With PX as diameter a circle passes through the 3 points P, X and D. This is the nine points circle.

2. AP = PO and AY = YC.

$\therefore$ PY $\parallel$ OC

AY = YC and CX = XB

$\therefore$ YX $\parallel$ AB

$\angle$ between OC and AB = 90°

$\therefore$ Y is on the circle with PX as diameter.

3. AP = PO and AZ = ZB

$\therefore$ PZ $\parallel$ BE

AZ = ZB and BX = XC

$\therefore$ ZX $\parallel$ AC

$\angle$ between BE and AC = 90°

$\therefore$ Z is on the circle with PX as diameter.

4. $\angle PEX = \angle PEO + \angle OEX$

As P is the mid point of AO and AEO is rt angled Δ.

$\angle PEO = \angle POE$ and $\angle POE = \angle BOD$

As X is the mid Point of BC and BEX is rt. angled Δ.

$\angle BEX = \angle XBO$

$= \angle POE + \angle BOX = \angle BOD + \angle OBX = 90°$

$\therefore$ E is on the circle with PX as diameter

5. Now consider

$\angle PFX = \angle PFO + \angle OFX$

$\angle PFO = \angle POF$ (as P is the mid pt. of AO and AFO is rt angled Δ)

$\angle POF = \angle DOC$ (Verically opposite)

$= \angle OFX = \angle OCD$

(as X is the mid pt. of BC and BFC is rt angled Δ)

$\therefore \angle PFX = \angle DOC + \angle OCD = 90°$

$\therefore$ F is on the circle with PX as diameter.

6. Now PQ $\parallel$ AB and QX $\parallel$ OC

as P and Q are centres of OA and OB respectively and

Q and X are centres of OB and BC rcspcctively

$\therefore \angle$ Between PQ and QC = $\angle$ Between AB and FC = 90°

Hence Q is on the circle with PX as diameter

7. Further PR $\parallel$ AC as P is mid pt. of AO

and R is the mid pt. of CO

and XR || BE as X is the mid pt of BC

and R is the mid pt. of CO.

∴ ∠ Between PR and XR = ∠ Between AC and BE = 90°

∴ R is on the circle with PX as diameter

We have proved that Y, Z,; E, F, Q and R are on the circle with PXD with diameter as PX and D as rt angle.

∴ X, Y, Z, P, Q, R, and D, E, F are on one circle.

Corollary 1

The centre of the nine pts circle is the center of the line joining the circum centre to the ortho-centre.

Let S be the circumcentre and OS be the line joining the circum centre to the orthocentre. DX and EY are the chords of nine pts. circle. They meet at the centre of SO. Which is the centre of nine pts circle.

Corollary 2

The radius of the nine pts circle is half the radius of circum-circle. PX is the diameter of nine pts circle. The middle pt of SO is also the centre of the nine pt. circle.

∴ PX and So intersect in N.

AP X S is a parallelogram

One side PX is the diameter of the nine pt. circle

= As radius of the circmcirc.

∴ Radius of nine pt circle = $\frac{1}{2}$ radius of the circumcircle

Corollary 3

The centre of gravity is on the line joining the orthocentre to the circumcenter.

Exercise (O)

1. Given the base and the vertical angle A. Find the locus of the nine pts circle.

2. The nine pts circle of any Δ ABC whose orthocentre is O is also the nine pts circle of the Δ s AOB, BOC and COA.

3. If I, I_1, I_2, I_3 have the usual meaning then the circle about the main Δ ABC is nine pts circle of the Δ s formed by 3 out of the 4 pts formed by I, I_1, I_2, I_3.

4. Given the base and the vertical angle of the Δ the nine pt circle always touches a fixed circle.

Theorem No. 72

The rectangle contained by 2 sides of a Δ is equal to the square of the bisector of the angle contained by the 2 sides together with the rectangle contained by the segments of the third side with which the bisector divides it.

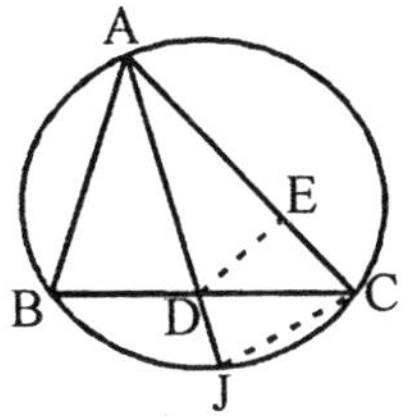

Given:-ABC is a Δ. AD is the bisector of the $\angle A$. AD dividers the base in segment BD and CD.

To Prove:- AB.AC = AD^2 + BD.DC

Construction:- Draw the circum-circle of the Δ ABC. Produce AD to meet the circumferenec in J. Join JC. From AC cut off AE = AB assuming AC > AB.

Proof:- In Δ s ABD and AED

because $\begin{cases} AB = AE & (Const) \\ AD \text{ is common} \\ \angle BAD = \angle CAD \end{cases}$

These Δ s ABD and AED are congruent.

$\therefore$ $\angle ABD = \angle AED$

But $\angle ABD = \angle AJC$ (Th. 52)

$\therefore$ $\angle AED = \angle AJC$

$\therefore$ DECJ is a cyclic quad.

$\therefore$ AJ.AD = AC.AE (Th. 66)

$\therefore$ (AD + JD). AD = AC.AE

$\therefore$ AC.AE = AD^2 + AD.JD

$\therefore$ AC.AE = AD^2 + BD.CD (Th 65) (H.P.)

Corollary :- In any Δ the difference between the rectangle contained by 2 sides and the rectangle contatined by the 2 segments into which the third side is divided by the bisector of the exterior angle.

Example:- Find out the internal bisector and the external bisector opposite the side 7 cm of the sides of the Δ are 5 cms, 6 cms and 7 cms.

Theorem No. 73

In any Δ the rectangle contained by 2 sides is equal to the rectangle contained by the diameter of the circumcircle of the Δ and the perpendicular from vertex to the opposite side.

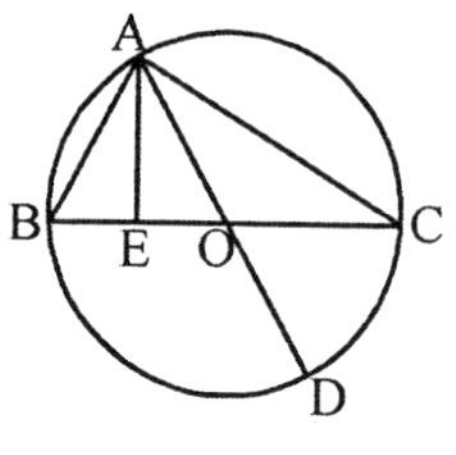

Given:- ABC is Δ. AD is the diameter of the circum circle. AE the perp. from the vertex.

To Prove:- AB $\times$ AC = AE $\times$ AD

Proof:- Δ ABE and Δ ACD are equiangular as $\angle AEB = \angle ACD = 90°$,

and $\angle ABE = \angle ADC$ as these are $\angle s$ in the same segment

$\therefore$ $\frac{AB}{AE} = \frac{AD}{AC}$

$\therefore$ AB $\times$ AC = AD $\times$ AE (H.P.)

Theorem No. 74

(Ptolemy's Theorem)

In a cyclic quadrilateral, the rectangle contatined by the two diagonals is equal to the sum of the rectangles contained by the opposite sides.

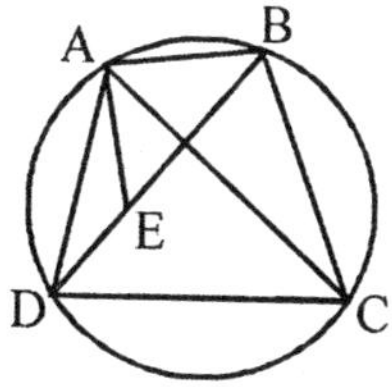

Given:- Quadrilateral ABCD

To Prove:- AC $\times$ BD

$=(AB \times CD)+(AD \times BC)$

Construction:- Draw a line AE so that $\angle BAE=\angle CAD$

To Prove:- In Δs ABE and CAD

because $\angle ABE=\angle ACD$ [$\angle s$ in the same segment]

$\angle BAE=\angle CAD$ (construction)

These Δs ABE and CAD are similar

$\therefore \quad \frac{AB}{BE}=\frac{AC}{DC}$

$\therefore \quad$ AB.DC = AC. BE ...(1)

Then $\angle BAE=\angle CAD$ (construction)

Subtracting $\angle EAC$ from both sides

$\angle BAE-\angle EAC=\angle CAD-\angle EAC$

$\angle BAC=\angle DAE$

and $\angle BCA=\angle ADE$ ($\angle s$ in the same segment)

The Δs BCA is equiangular to ΔADE

$\therefore \qquad \frac{AD}{DE}=\frac{AC}{BC}$

$\therefore \qquad AD.BC=AC.DE$ ...(2)

Adding (1) and (2)

$$AB.DC+AD.BC=AC.DE+AC.BE$$

$$=AC(DE+BE)=AC \times BD \qquad \text{(H.P.)}$$

(Ptolcmy AD 90 to AD 168 Born in Egypt. He was a Mathematician, geographer, Astronomer and Astroger. He lived in Egypt under Roman Empire The above theorem goes by his name.

Theorem No. 75

(Ceva's Theorem)

If the bisctors of the $3\angle s$ *of a* Δ *viz. A, B, C meet the opposite sides in D, E, F respectively, then* $\frac{BD}{DC}\times\frac{CE}{EA}\times\frac{AF}{FB}=1$

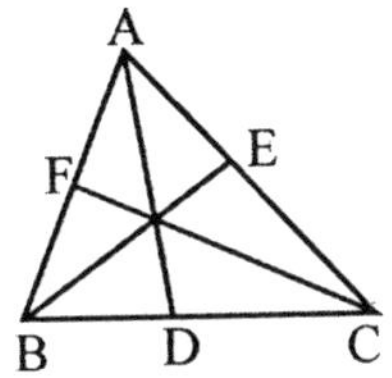

Given: ABC is Δ. AD, BE, and CF are respectively the bisectors of $\angle A, \angle B, \angle C$.

To Prove: $\frac{BD}{DC}\times\frac{CE}{EA}\times\frac{AF}{FB}=1$

Proof: $\frac{BD}{DC}=\frac{AB}{AC}$ [Th. 38]

$$\frac{CE}{EA}=\frac{BC}{AB}$$

$$\frac{AF}{FB}=\frac{AC}{BC}$$

$$\therefore \quad \frac{BD}{DC}\times\frac{CE}{EA}\times\frac{AF}{FB}=\frac{AB}{AC}\times\frac{BC}{AB}\times\frac{AC}{BC}=1 \qquad \text{(H.P.)}$$

Theorem No. 76

The three points of intersection of the adjacent trisectors of the angles of any triangle form an equilateral triangle.

(Enunciated by Frank Morley (1860 - 1937) chairman of the Mathematics Dept. of John Hopkins University. He stated this theorem in 1899)

The following proof is given by Shri V.G. Tohekar.

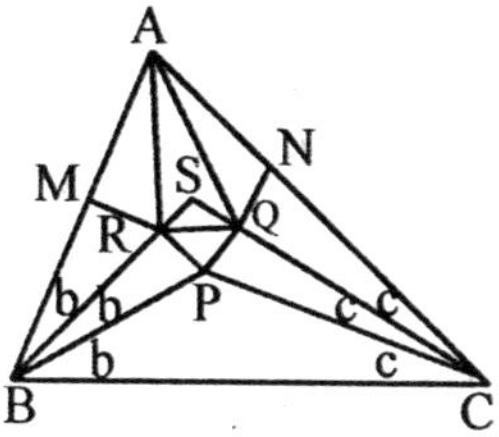

Given:. Arbitrary ΔABC.

To Prove: The three points of intersection of the adjacent trisectors of $\angle A$, $\angle B$, $\angle C$ from an equilateral Δ. i.e. PQR is equilateral.

Proof: Let

(i) $\angle A = 3a$, $\angle B = 3b$, $\angle C = 3c$

(ii) BP and CP, trisectors adjacent to BC of $\angle B$ and $\angle C$ respectively meet in P as shown in Fig. I.

(iii) The remaining trisectors of $\angle B$ and $\angle C$ meet in S.

$\therefore$ In ΔSBC, BP and CP will be bisectors of $\angle SBC$ and $\angle SCB$ respectively.

$\therefore$ P is the incentre of ΔSBC

$\therefore$ SP is the bisector of $\angle BSC$ ...(2)

At P draw on both sides of PS, an angle each of 30°

i.e. $\angle RPS = \angle OPS = 30^\circ$

(arms of the $\angle s$ meeting BS and CS at R and Q respectively) ...(3)

2. Now $\Delta PRS \equiv \Delta PQS$ $\left[\begin{array}{l} \therefore \angle RPS = \angle QPS = 30^\circ \\ \text{PS is common side} \\ \angle RSP = \angle QPS \text{ in view of (2)} \end{array}\right]$

$PR = PQ$

We have $\angle RPQ = \angle RPS + \angle QPS$ ...(4)

$= 30^\circ + 30^\circ = 60^\circ$

$\therefore$ ΔPQR is equilateral [as PR = PQ giving $\angle PRQ = 60^0$]

$\therefore$ We will be done if we show that AR and AQ are trisectors of $\angle A$ ...(5)

For this purpose take points M and N on BA and CA respectively such that BM = BP and CN = CP

$\therefore$ $\Delta MBR \equiv \Delta PBR$ ...(6)

($\because$ BM = BP by construction, BR is common $\angle MBR = \angle PBR = b$)

$\therefore$ $MR = RP$

Similarly $PQ = QN$

$\therefore$ $MR = RQ = QN$

$\because$ ΔPQR is equilateral $RP = PQ = RQ$

3. Now in ΔSBC, $\angle B = \angle C = 2b + 2c$

$\therefore$ $\angle BSC = 180^\circ - 2b - 2c = 180^\circ - 2(a + b + c) + 2a$

$$= 180^\circ - \frac{2}{3}(3a + 3b + 3c) + 2a = 60^\circ + 2a$$

$\therefore$ $\angle PSR = 30^\circ + a$ (in view of 4) ...(7)

$\therefore$ $\angle BRP$ = exterior angle of PSR $= \angle PSR + \angle SPR$

$= (30^\circ + a) + 30^\circ$ using (3) and (7) $= 60^\circ + a$

$\therefore$ $\angle MRB = 60^\circ + a$ [in view of 6]

$\therefore$ $\angle MRQ = 360^\circ - \angle MRB - \angle BRP - \angle PRQ$

$= 360^\circ - (60^\circ + a) - (60^\circ + a) - 60^\circ = 180^\circ - 2a$

Similarly $\angle RQN = 180^\circ - 2a$

4. Thus equal line segments MR, RQ, QN are such that

$\angle MRQ = \angle RQN$

∴ Their end points are expected to lie on a circle.

We establish this fact as follows.

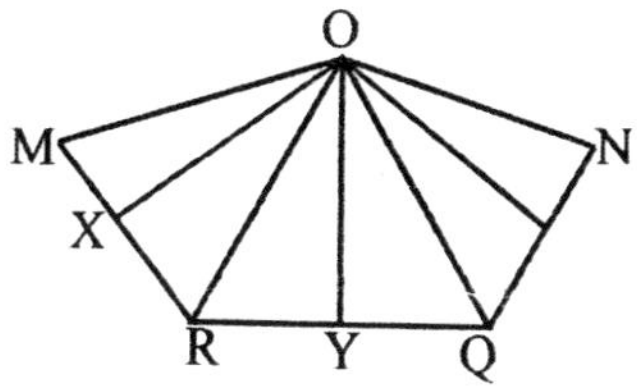

Let XO be ⊥ bisector of MR and YO be ⊥ bisector of RQ.

It can be readily seen that

$$\Delta OMR \equiv \Delta ORQ \equiv \Delta OQN$$

∴ $$ON = OQ$$

∴ Circle with centre O and radius MO passes through the four points M, R, Q, N (Fig. 3)

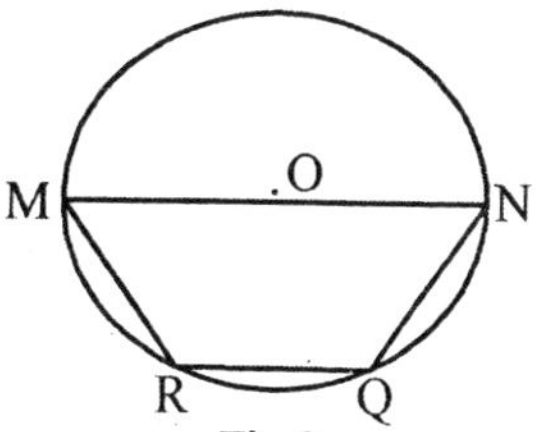

Fig. 3

Further radii OR and OQ bisect $\angle MRQ$ and $\angle RQN$ respects'vely and line segment XO and YO bisect $\angle MOR$ and $\angle ROQ$ respectively.

∴ We know that $\angle MRO = \frac{1}{2}\angle MRQ$

$$= \frac{1}{2}(180 - 2Q) = (90° - a)$$

∴ $$\angle XOR = a,\ \angle MOR = 2a$$

This implies that $\angle MON = 6a$

In the circle with centre O and radius MO and shown in fig. 3 chord MN subtends $\angle 6a$ at centre O.

∴ at the circumference MN subtends $\angle 3$ a. But MN subtends $\angle 3$ a at the vertex. A giving the ΔABC

∴ The circle must through A Now ∵ MR, RQ, QN subtend $\angle 2a$ at O each of these subtends $\angle$ a at A on the circumference.

∴ AR and AG trisect $\angle$ A and this is what was to be established as stated in (5). Hence the theorem. (H.P.)

Exercise (P)

1. A, B, C are 3 points in this order. Find a point P in this line so that PB is in the middle proportion to PA and PC i.e.

   PA : PB = PB : PC

2. P is a point in the base of a ΔABC. From this point parallels are drawn to AB and AC meeting AC and AB in D and F respectively. Prove that the area of ΔADF is in mid proportion to the areas of ΔBFP and ΔPDC.

3. ABC is a Δ. AD is bisector of a $\angle A$ meeting the base in D meeting the base in D. Prove that the circum radius of ABD circumradius of $\Delta BCD = BD : CD$

4. PA and PB are two chords of a circle from a point P. A tangent is drawn to the circle at the point P. A line is drawn parallel to the tangent meeting PA and PB in C and D respectively. Prove that PA. PC = PB.PD.

5. (a) Two intersecting circles have got their line of centres A, B, C, D, meeting them A and C and B and D respectively. P is the centre of the common chord. Prove that $\frac{PA}{PB} = \frac{PB}{PC}$.

   (b) On a line AB two points P, Q are taken so that the sequence is A, P, Q, B. Find a point between P and Q so that MP : MQ = MA: MB.

6. By extending a given line find a point in its extended portion so that the sum or difference of the squares of its disfferences from the ends of the line is equal to a given square.

7. Three points A, P, C are in a straight line. Find the locus of a point which makes equal angles with AP and PC.

8. What is locus of middle points of the chords which meet at right angles at a given point in a circle?

9. A point in taken inside or outside a circle and from it two lines are drawn to the circumference of the circle inside or outside and the lines forming a diameter so obtained are divided in a particular ratio. What will be the locus of the points so obtained?

*10. Two pairs of straight lines meet each other in 2 points. Prove that the four triangles so formed have their circum-circles all passing through one point.

*11. Two intersecting pairs of straight lines meet each other in 2 points. Prove that four triangles so formed have their middle points of the line joining the common point of the 4 circum circles to each of the othocentre are in one line.

12. The ratio of 2 perpendiculars from two given points on to a moving line is the same. Prove that the lines all pass through some fixed point. What is the point?

13. In any Δ the circum radius is greater than diameter of the in-circle. In what Δ these are the same?

*14. ABC is an isosceles Δ in which the angle at the vertex is half each of the base angles. A point X is taken on AB so that AX = BC.

Prove that the projection of BX on AC is $\frac{BC}{2}$.

15. AB is a chord of the circle. C is a point on the circumference. AC and BC are joined. They meet on a diameter perpendicular to AB in points D and E. If M is the centre of the circle prove that

MD.ME = $(\text{Radius})^2$

16. From the intersection of 2 circles draw a line terminated by the 2 circumferences so that intercept made by the 2 circles is in a given ratio.

17. The area of a quadrilateral, one side and the middle point of the opposite side are given. Prove that locus of the angular points are only on parallel lines.

18. In a parallelogram P is such a point $\angle APB = \angle CPD$. Prove that the circles of Δs APB, BPC, CPD, DPA are all equal.

19. In a Δ I is the incentre and I_1 is the centre of the escribed circle corresponding to the side BC.

Prove that $AI.AI_1 = AB.AC$.

20. Two points are given on a circle. Find a point on the circle between A and B so that the projection of AP and BP on the tangent at P will be in a given ratio.

*21. From the ends of two mutually perp. fixed radii are drawn two chords parallel to each other and the centres of these chords are joined and circle are drawn over them. Prove that the circles pass through one special point.

22. In a trapezium a parallel is drawn to the parallel sides through the inter-section of the diagonals. Prove that it divides the inclined sides in the ratio of parallel sides and is bisected at the inter-section of the diagonals.

23. On AB is drawn a semi-circle and a tangent is drawn to it at point C. It meets the tangent at A and B in D and E. If CF is drawn perpendicular to AB, prove that it meets AE and BD in point M such that MC = MF.

24. A, B, C, D are 4 points in one line. P is a point above A, B, C, D such that the circles ABP and BDP touch each other at P. Prove that $\angle BPD$ is supplementary to $\angle APC$.

25. ABC is a right angled Δ, right angle at C. CD is bisector of the angle C and meets the diagonal in D. From D is drawn a parallel to BC. Prove that

    $DE(AC+BC) = AC.BC.$

26. In Δ ABC, BE $\perp$ AC and CF $\perp$ AB. BE and FC meet in O. Prove that circles BEFC and AFOE meet at right angles.

27. AB is a diameter of the circle. AF and BE are two chords of the circle meeting in R. Prove that the circle ERF intersect the main circle at 90°.

28. AB and CD are 2 chords of a circle meeting at right angles in point P. O is the centre of the circle. Prove that

    $PA^2 + PB^2 + PC^2 + PD^2 = (\text{diameter})^2$

    and $AB^2 + CD^2 + 4\,PO^2 = 2(\text{diameter})^2$

29. Two parallel tangents are drawn to a circle. They make an intercept on a third tangent which is divided at its point of contact with the circle such that the rectangle contained by the 2 parts is equal to $(\text{radius})^2$.

*30. Two circles touch each other at A. From A, two straight lines are drawn at right angles to each other meeting circle in P and F and B and D respectinvely. The line of centres meet the circumferences in C and E. Prove that

    $CE^2 = PF^2 + BD^2.$

31. In a Δ ABC $\angle A = 90°$. The side BA and CA are either produced or not produced and a line PF is drawn parallel to BC meeting BA in P and CA in F. Prove that

    PF.BC = AB. AP + AC. AF.

*32. A flagstaff is erected on a 30 M high mountain. It is found that the flagstaff makes equal angles at points 35M and 54M distant in a straight line. Find the height of the flagstaff.

33. On the high seas a man is standing on the deck of a ship x meters above sea-water. Assuming the earth's diameter to be 12800 km, show that he can see a distance y kms from his eye surrounding him upto the horizon so that $y = \sqrt{.03(12800 + .03)} = 19.596$ kms

    where x = 30 mts.

*34. ABCD is a quadrilateral drawn in a circle. DA and CB meet in P. AB and DC meet in F. Prove that square on PF is equal to the sum of the squares from P and F.

*35. ABC is an equilateral Δ with A as centre and AC as radius, a circle is drawn. Further an AB as diameter another circle is drawn on AC a point P is taken and a circle is drawn with P as centre touching the first circle internally and the second circle externally. Prove that $AP = \frac{4}{5}AC$.

36. ABCD is a cyclic quadrilateral in which BC = AD. Prove that

    $AC^2 = AB.AD + BC^2$.

*37. ABCD is a cyclic quadrilateral drawn in a circle. Circle A X B and circle C X D touch each other at X. Show that locus of x is a circle whose centre is the meeting point of AB and DC.

38. CD is a diameter of the circle and a point X is taken on it. AB is parallel to CD. On the other side of AB away from CD, M is the highest point. Prove that

    $XA^2 + XB^2 = XC^2 + XD^2 = 2MX^2$.

*39. AB and CD are 2 chords of a circle meeting in point P at right angles to each other. O is the centre of thc circle. OP and AC have their centres at M and N respectively. Prove that

    $2(MN^2 + PN^2) = (\text{Radius})^2$

40. AB is a diameter of the circle. PTB is a chord at right angles to the diameter meeting AB in T. A circle is enclosed within the space AT, PT and the arc AP. It touches the diameter in point C. Prove that

(a) BF = BC and that PT bisects $\angle APT$ and

(b) The rectangle contained by the radii of the two circles $= \frac{1}{2}$ AC.CB.

41. In a Δ ABC, AB = AC. The perp. bisector of AB and BC meet in point D. Prove that

(a) $AC^2 = BD.DC$ and

(b) $BD^2 \sim AC^2 = BD.ED$.

42. ABCD is a square circumscribed by a cricle. P is a point on the circumference. Prove that

$2\,PA.\,PC = PB^2 \sim PD^2$.

43. In the segment of a line AB a point P is taken such that the square on AP i.e. AM×NP is equal to the rectangle contained by AB and BP. If C is the midpoint of AP. Show that

CM = CN = CB.

*44. Squares are drawn on the 3 sides of a Δ and the three adjacent ends of the squares are joined forming a hexagon. Show that the sum of the squares on the 3 lines joined is equal to three times the squares on the sides of the Δ.

45. In a Δ ABC, $\angle B = 90°$ the perpendicular bisector of AC meets BC in E and AB in F. Prove that

$AC^2 = 2BC.CE = 2AB.AF$.

46. AB and AC are 2 radii of a circle. A tangent is to be drawn to the arc between AC and AB produced, so that it will be of a given length.

*47. PA and PC are two tangents drawn to a circle from P. The middle point of the chord of contact is G. From P is drawn a chord meeting the circle in X and Y. Prove that

(a) AC bisects the angle XGY,

(b) If M is the certer of the circle then XYMG is cyclic.

(c) If MG meets the circle in B and D. Then CX bisects the $\angle PXG$ and DY bisects the $\angle PYG$.

48. ACB is the segment of a circle, AB being the chord and C is the middle point. Draw a line from C intersecting the chord AB in E and the remaining part of the circle in so that ED is of a given length.

49. In a quadrilateral ABCD, the length AD = BC and AC = CD and $\angle ACB + \angle ADC = 180°$. Prove that

    $AB^2 = BC^2 + CD^2 + DA^2$.

50. PA and PB are 2 tangents to a circle. A and B are the 2 points of contact. In $\overrightarrow{PA}$ a point D is taken and in $\overrightarrow{PB}$ is extended and a point is taken on it so that BE = AD. If ED meets the circle in F and G show that the centre of FG and BE is on AB.

51. AB is a diameter of a circle. CD is a chord making an angle of 45° with AB. It meets AB in P. Prove that

    $2CP^2 + 2DP^2 = AB^2$.

*52. ABC is a Δ and G is the centroid, X is any other point. Prove that

$$AX^2 + BX^2 + CX^2 = \frac{AB^2 + BC^2 + CA^2}{3} + 3CX^2.$$

53. Three circles are given to you. Find out the locus of a point from which the sum of the squares of the tangents to these circles is equal to a given square.

    In case the no. of circles is four or more, what will be the locus of the point in the above case.

54. Three points A, B, C. are in a straight line. Different circles are drawn through B and C and tangents are drawn to these circles from A, what is the locus of the points of contact to these circles.

55. Inside a Δ ABC, a point M is taken and $\angle AMB$ and $\angle AMC$ are bisected. They meet AB and AC in points P and F. If PF is parallel to BC. What is the locus of M?

*56. ABC is an equilateral Δ inside a circle. AB meets the circles in PQ, BC meets the circle in GF and CA meets the circle in NM

    AM + BQ + CF = AP + BG + CN.

*57. ABC is a rt. angled Δ, rt angled at B. If AK is the bisector of $\angle A$ and I is the In-centre of the Δ.

If $\frac{AI}{IK} = \frac{\sqrt{3}+1}{\sqrt{3}-1}$

Find $\angle A$ and $\angle C$.

*58. O is the circum-centre of the Δ ABC, If AO, BO, CO meets the BC, AC, AB in P, Q, R, and $\angle BAC = \angle AQR = \angle RQB$

Prove that $\angle BAC = 50°$.

59. Two circles are intersecting in A and B. P is moving on one circle. PA and PB are joined. They meet the second circle in C and D. Prove that the Cirum centre of Δ PCD moves on a circle that is concentric with the second circle and has a fixed radius.

*60. Two equal circles are intersecting at right angles to each other. The area which is common to both the circles, + area on their radius is equal to half the area of these circles.

61. Two intersecting circles are drawn in such a way that one passes through the mid point of another. Both the common tangents are drawn. Then the line which joins the points of contact of the first circles touches the second circle.

Exercise (Q)

1. Draw a Δ inscribed in a given circle through 2 given points inside the circle and whose angle at the vertex is equal to a given angle. How many Δ s are possible.

*2. Sum of the base of an isosceles Δ and the perpendicular from the vertex is given and the equal side is also given. Draw the isosceles Δ.

*3. Difference of the base of the isosceles Δ and the perpendicular from the vertex and the equal side is given. Draw the isosceles Δ.

4. Base, ratio of the remaining two sides is given and the third parameter is one of the following is given, draw the Δ.

(a) Difference of the base angles

*(b) Median bisecting one of the sides other than the base

(c) Difference of the squares on the other sides

(d) The Radius of the circum-circle

(e) Points where the in-circle touches the base.

5. Base, sum of the squares on the other sides and the difference of squares on the other sides is given. Draw the Δ.

6. If the diagonal of the square is 2 cm. larger than each of the side, how will you draw the square.

*7. All the four sides of a cyclic quadrilateral are gvien, how will you draw the same.

*8. The distance of 3 angular points of a point inside a square are given. How will you draw the square.

9. Draw an equilateral Δ whose e.g. is at a given point and whose two sides pass through 2 points.

6

Hints to Difficult Problems

Exercise (g)

Ex. 11

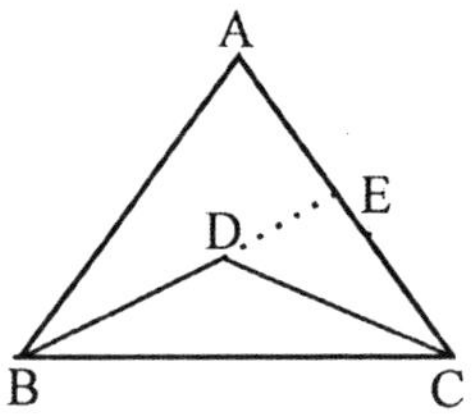

$AB + AE > BE$

$AB + AE > BD + DE \quad ...(1)$

$DE + EC > DC \quad ...(2)$

Adding (1) and (2)

$AB + AE + DE + EC > BD + DE + DC$

Cancelling DE which is common

$AB + AE + EC > BD + DC$

$\therefore \quad AB + AC > BD + DC$

Ex. 12

Apply Ex. 11 thrice round the Δ. First A as vertex then B and C as vertices and add the 3 equations.

Ex. 17

Produce AD to E so that AD = DE

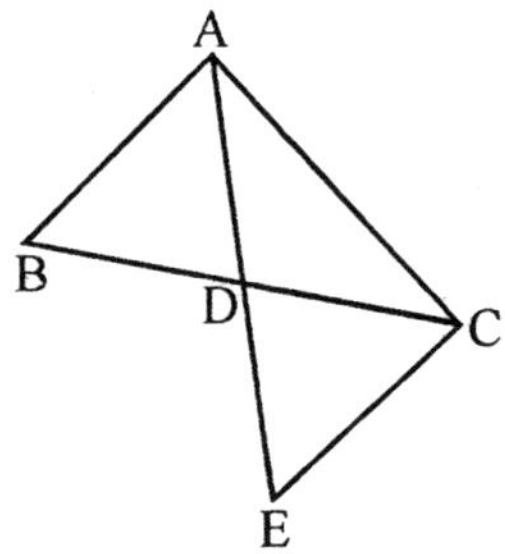

Join CE, $\quad \Delta ABD \equiv \Delta CDE$

$\therefore \quad AB = CE$

$AC + CE > AE = 2AD$

$AC + AB > 2AD \quad ...(4)$

Similarly $\quad AB + BC > 2BE \quad ...(5)$

$AC + BC > 2CE \quad ...(6)$

On adding equations (4), (5) and (6)

We get AB + BC + AC > AD + BE + CF

Ex. 21

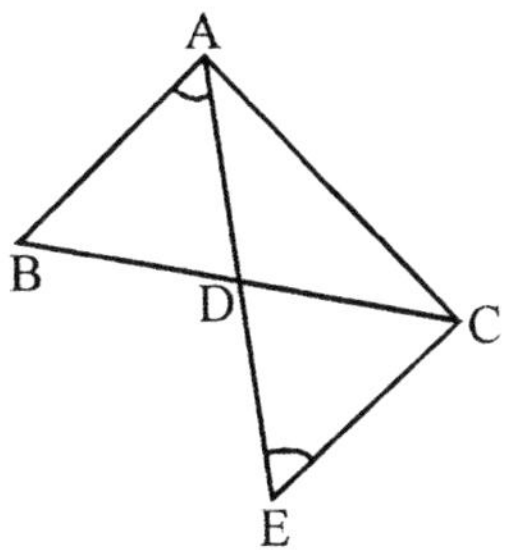

Produce AD the median to E,

Prove Δs ABD ≡ Δ DCE

$\angle BAD = \angle CED$

AC > AB > CE

$\therefore \quad \angle CEA > \angle DAC$

$\therefore \quad \angle BAD > \angle DAC$

$\therefore$ AD is between the bisector of $\angle A$ and AC.

Exercise (l)

Ex. 26

First draw the ΔABD with AB = 3 cms, AD = 2 cms and $\angle A = 120^\circ$.

Then separately draw ΔBDE. BE = 7 cms = BC + CE. $\angle E = 20^\circ$ and BD as drawn from the Δ drawn earlier.

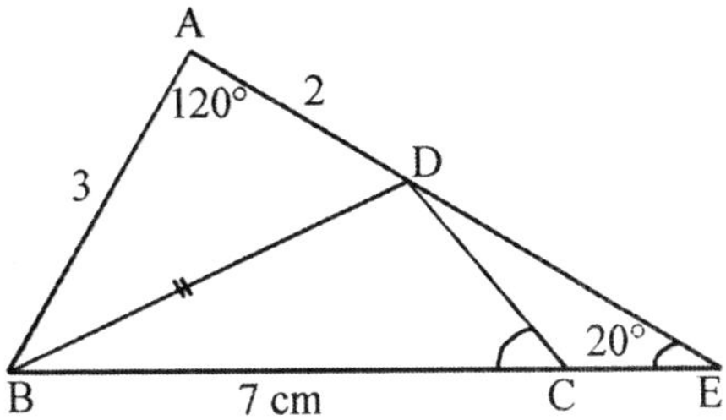

We get the ΔBDE. Then on this construct the ΔABD again. From D make $\angle CDE = \angle CED$. We get the quadrilateral ABCD.

Ex. 27

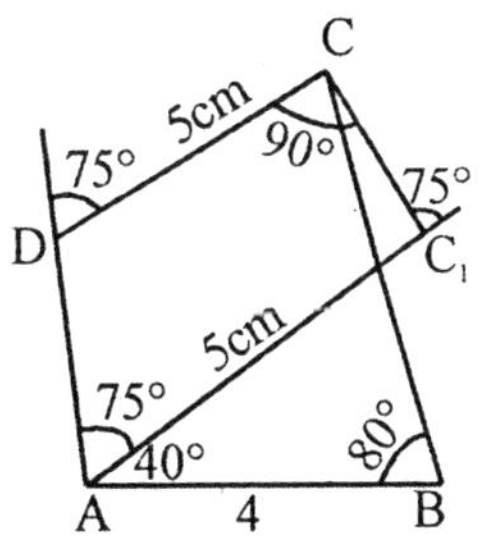

Construction: Make AB = 4.00 cms. At A make an angle of 40° and take $AC_1 = 5.0$ cms. At B make an angle of 80°, At C_1 make an angle of 75° CC_1 and CB meet at C. From C draw CD at right angles to CC_1 Make CD = 5 cm. Join AD. Quadrilateral ABCD.

Ex. 28

Construction: The DC = 5.5 cms and $\angle PDC = 50°$, Make PD = 4 cms and make with DP an angle = 75°, From C cut off a distance of 6 cms to B From B draw a parallel to DP = 4 cm. Join AD

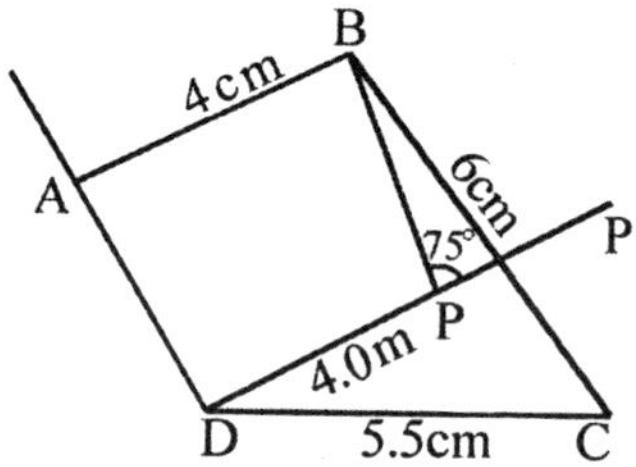

At P draw PB || AD

Exercise (m)

Ex. 3

Use Theorem 26.

Ex. 13

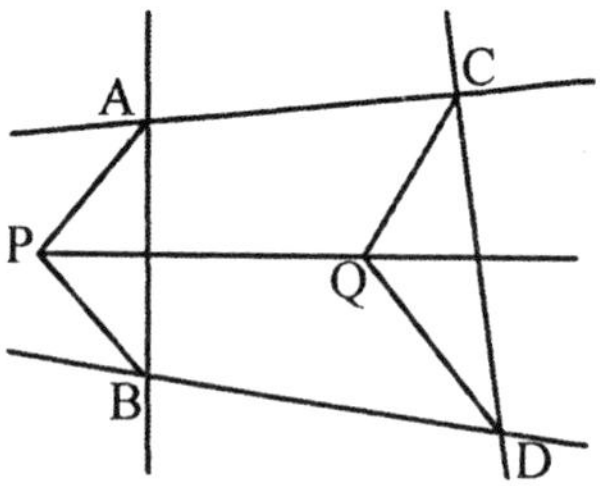

AC and BD are the two lines that meet outside the paper. Draw any two lines AB and CD. Bisect the $\angle S$ at A and B and let them meet in P. Bisect $\angle S$ at C and D. Let them meet in Q PQ is the line.

Exercise (n)

Ex. 9

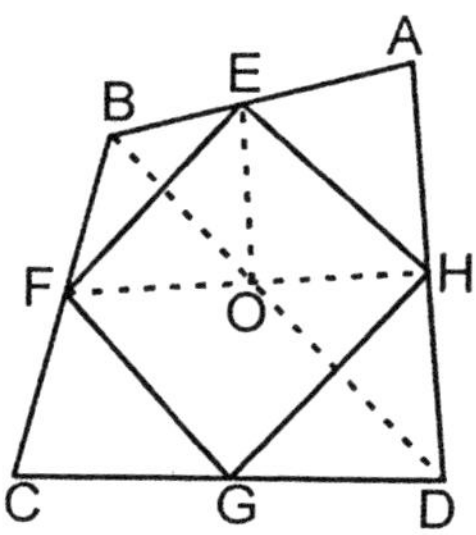

E, F, G, H are the midpts of the four sides

(1) ΔAEH is folded about EH

AE falls along EO

AH falls along HO

Then ΔAEH occupies the position ΔOEH

(2) Similarly ΔDGH if folded about HG

(3) ΔCFG is folded about FG.

(4) BFE is folded about EF

$\therefore$ AH = OH = HD

$\therefore$ AOD = 90°

Ex. 22

ABCD is a quad. Denote the sides by a, b, c, d and the diagonals by e, f. We are given that

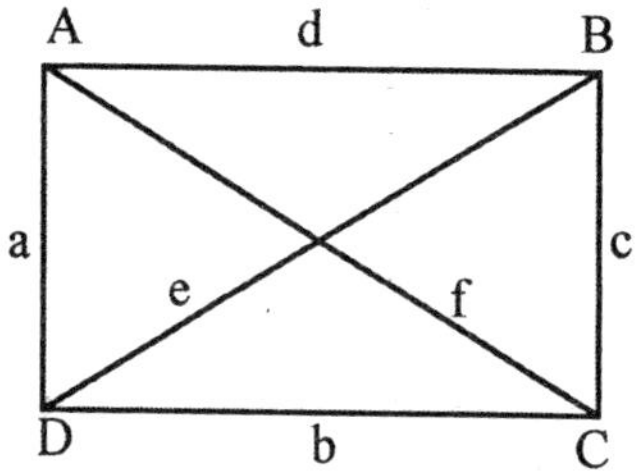

$a + b + c = b + c + f = c + d + e = a + d + f$

(1) + (3) = (2) + (4)

$a + b + e + c + d + e = b + c + f + a + d + f$

$\therefore$ $2e = 2f$

$\therefore$ $e = f$

substituting in (1) and (2)

$a + b + f = b + c + f$

That means $a = c$

and $b = d$

The diagonals are equal and the opposite sides are equal. That means it is a rectangle.

Ex. 23

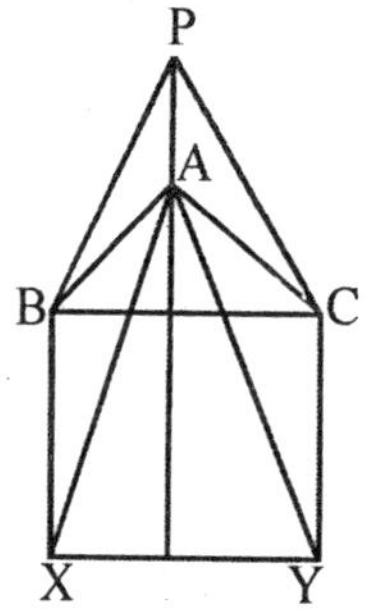

Prove: $\Delta PBC \equiv \Delta AXY$

$\therefore$ PB = AX and PC = AY

PB || AX and PC || AY

BX || PA

$\therefore$ PA $\perp$ BC.

Ex. 24

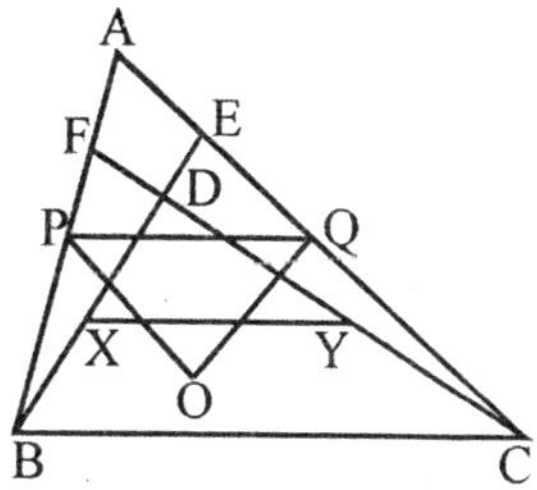

DX = BX; DY = YC

$XY = \frac{1}{2} BC$

$PQ = \frac{1}{2} BC$

$\therefore$ $\Delta DXY \equiv \Delta OPQ$

$$OP = \frac{1}{2} DC$$

$$2OP = OC$$

Similarly $\quad 2OQ = BD$

Ex. 29

1. BP and CF are both perpendiculars to AF the bisector of $\angle A$ of a ΔABC
2. We have to prove DP = DF
3. $PD = \frac{1}{2} OC = \frac{1}{2}(b - c)$
4. $DF = \frac{1}{2} BT = \frac{1}{2}(b - c)$
5. DP = DF.

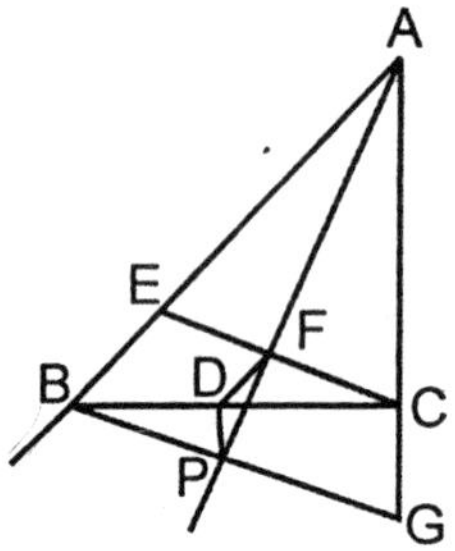

Ex. 35

Quadrilateral: AB = CD

Let X and Y be the middle pts of AD and BC and S and T of BD and AC respectively

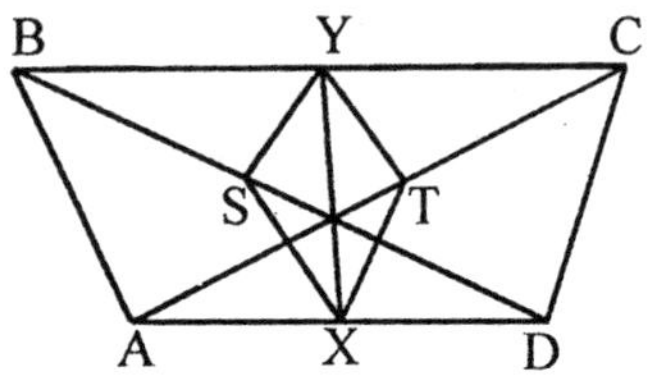

$SY = \frac{1}{2} CD \qquad TX = \frac{1}{2} CD$

$SX = \frac{1}{2} AB \qquad YT = \frac{1}{2} AB$

But AB = CD

SXTY is a rhombus

$\therefore$ XY bisects $\angle SXT$

Ex. 36

Given: AB = BC = CD

BPFC is a parallelogram

BP = 2BC

To Prove: PD $\perp$ AF

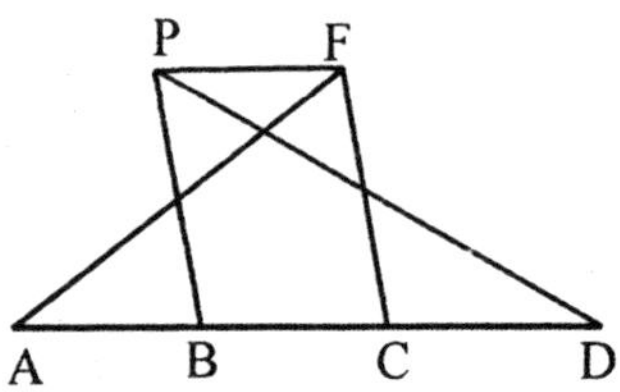

Proof: $\angle CAF = \angle AFC = \dfrac{180 - \angle FCA}{2}$

$$\angle BDP = \angle BPD = \frac{180 - \angle PBD}{2}$$

$$\therefore \quad \angle CAF + \angle BDP = \frac{180 - \angle FCA + 180 - \angle PBD}{2}$$

$$= \frac{360 - \angle FCA - PBD}{2} = \frac{360 - 180}{2} = 90^\circ$$

Ex. 39

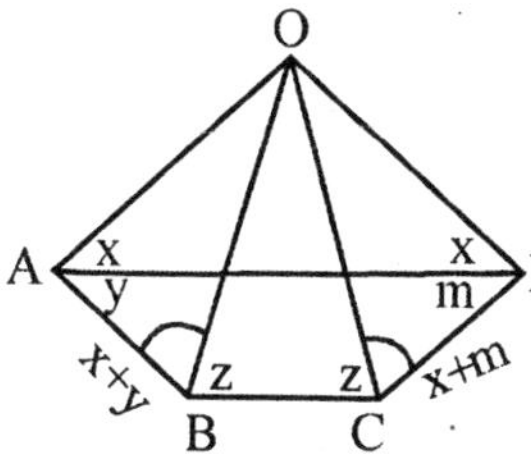

Given: OA = OB = OC = OD

$\angle OAB = \angle OBA = (x+y)$

$\angle OBC = \angle OCB = z$

$\angle ODC = \angle OCD = x + m$

$\angle OAD = \angle ODA = x$

$\angle DAB = y$

$\angle DCB = z + x + m$

$$\therefore \quad \angle DCB + \angle DAB = x + m + z + y$$

$$= \angle ABC + \angle ADC = 180^\circ$$

Ex. 40

Sum of interior $\angle s$ of a polygon = (2S – 4) rt angles of S sides.

Sum of interior $\angle s$ of a polygon with (3S – 4) sides = [2 (3S – 4) – 4] rt $\angle s$

= (65 – 12) rt angles

= 3 (25 – 4) rt angles

Ex. 43

(1) $g = x + i$

$= x + y + d$

$\therefore \; g - d = x + y$

(2) $b = h + y$

$= g + x + y$

$b - g = x + y$

(3) $b - g = g - d$

$2g = b + d$

$$g = \frac{b+d}{2}$$

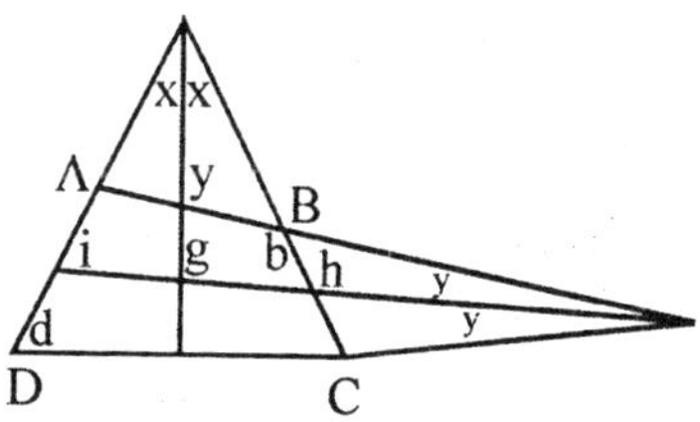

Ex. 45

Given : ABCD is a square HG and LM are inclined at right angles to each other.

To Prove: HG = LM

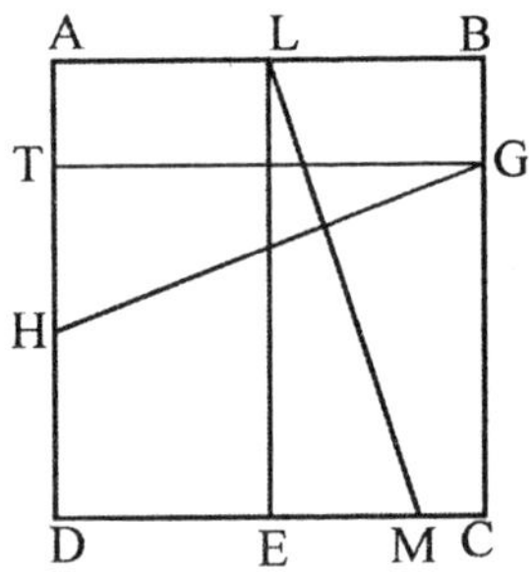

Proof:

$\angle AHG + \angle ALM = 180^\circ$

$\angle AHG = \angle BLM = \angle LMD$

$\angle HGT = \angle ELM$

because they are complement of equal $\angle s$

LE = TG

are sides of the same square.

$\therefore$ HG = LM

Ex. 46

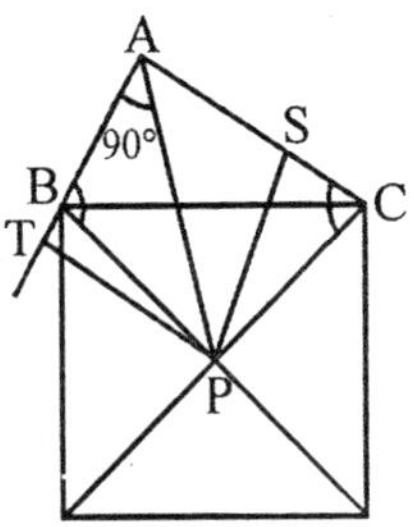

$\angle BPC = 90^\circ$,

$\angle A = 90^\circ$

$\angle ABP + \angle ACP = 180^\circ$

$\angle APB + \angle PBT = 180^\circ$

$\therefore \quad \angle ACP = \angle PBT$

In Δs PTB and PSC

because $\begin{cases} PB = PC \\ \angle PBT = \angle ACP \\ \angle PTB = \angle PSC = 90^\circ \end{cases}$

These Δs are congruent

$\therefore$ PT = PS

$\therefore$ AP bisects $\angle A$

Ex. 47

$\angle AFE + \angle AEF = 120^\circ$

$\angle AFE + \angle BFD = 120^\circ$

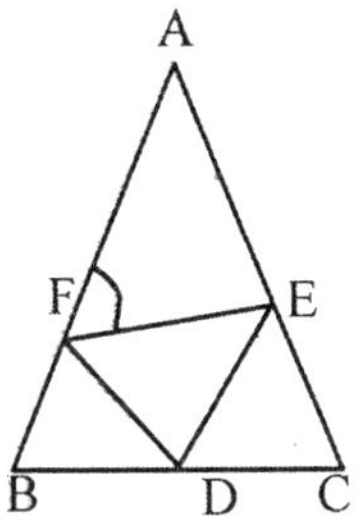

$\therefore$ $\angle AEF = \angle BFD$

and $\angle BFE = \angle AED$

$\therefore$ The outside Δs are congreunt

$\therefore$ AF = BD

AE = DC

$\therefore$ AF + AE = BC

Ex. 50

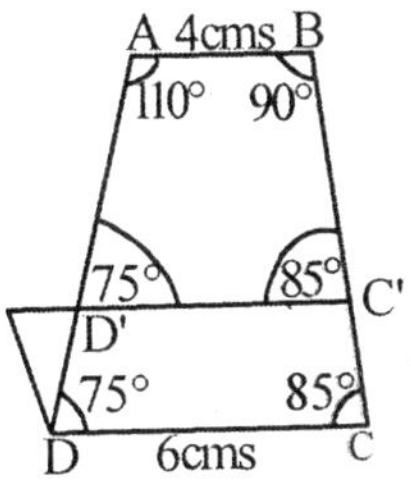

(1) Draw AB = 4 cms

(2) Make at A 110°

(3) Make at B 90°

(4) Draw any line C' D' making angle of 85° □ with BC

(5) Draw a || line with CD= 6 cms at a point below.

(6) ABCD is the quad.

Ex. 51

First draw a rough figures AB = 3.8 cms, BC = 4.5 cms, CD = 3 cms. $\angle AEB$ = 28° and CFB = 35°

Draw AG || CDF and CG || EDA at 35° and 28° respectively.

We can draw ΔBAG as BA, AG and $\angle BAG$ are known. Then we have to draw BCG

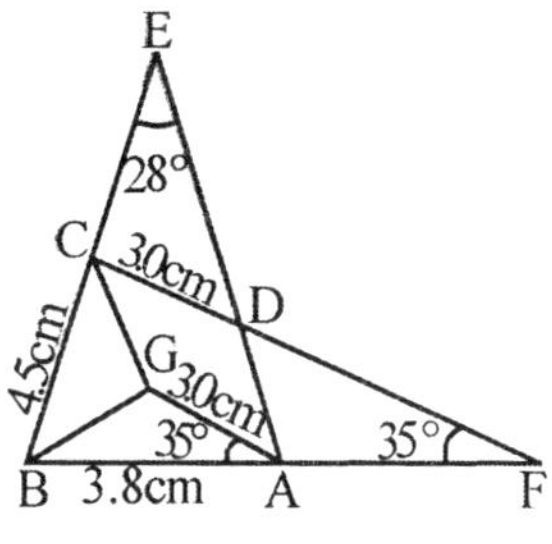

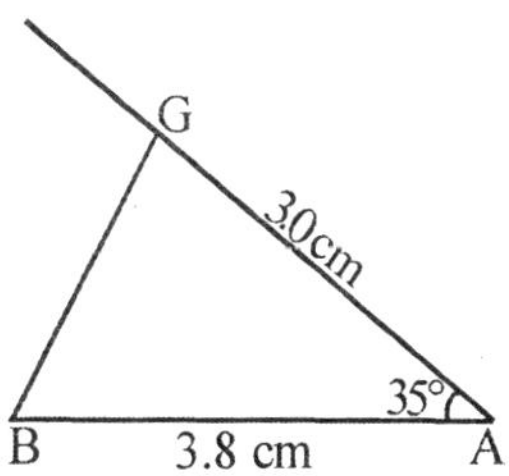

There are 2 methods

(1) Draw separately ΔCGB in which case CB is known, $\angle GCB$ = 28° and BG as found out from ΔBAG drawn already. Then complete on that ΔBAG draw CD || GA and take CD = 3 cms. Thus A, B, C, D are known, or

(2) On BG draw a segment of circle of 28° and cut off from the circle BC = 4.5 cms and then draw CD || to GA we get the quadrilateral ABCD. (For this refer construction No. 26)

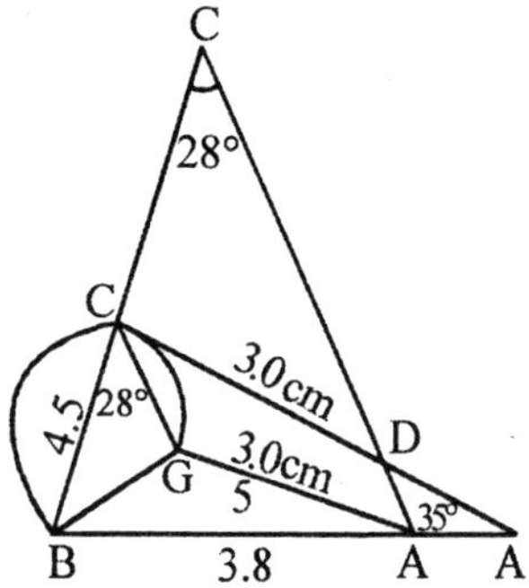

Exercise (U)

Ex. 8

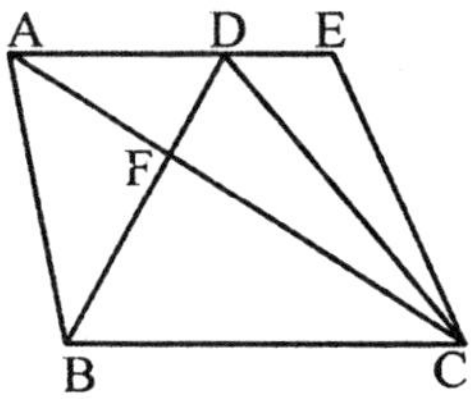

Area of ΔDCE

$= \text{Ar } \Delta AEC - \text{Ar } \Delta ADC$

$= \text{Ar } \Delta ABC - \text{Ar } \Delta ADC$

$= \text{Ar } \Delta DBC - \text{Ar } \Delta ADC$

$= \text{Ar } \Delta BFC - \text{Ar } \Delta FAD$

Ex. 9

$AD || CF || BE$

$\Delta AFC = \Delta CFD$ in area

$\Delta BFC = \Delta FCE$ in area

$\Delta ABC = \Delta CDE$ in area

$2\ \Delta ABC = \Delta FDE$ in area

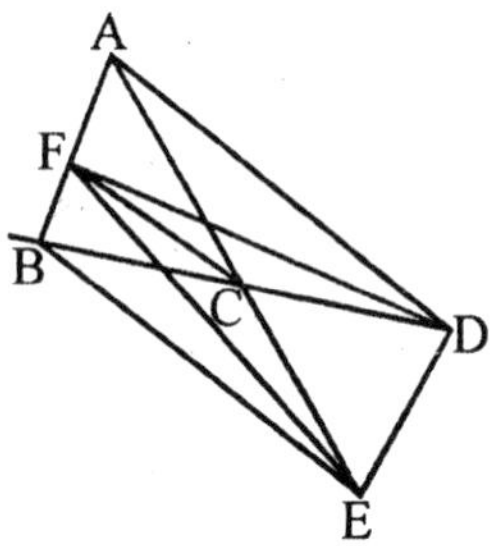

Ex. 11

$\Delta BED = 2\text{AB} \times 3\text{CN} = 6\ \Delta ABC$

$\Delta CEF = 6\ \Delta ABC$

$\Delta AFD = 6\ \Delta ABC$

$\Delta DEF = 18\ \Delta ABC + \Delta ABC$

$= 19\ \Delta ABC$

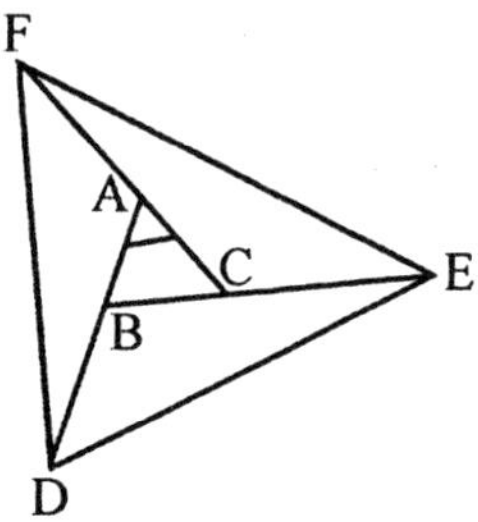

Ex. 12

Regular octahedron consis of 8 Δs

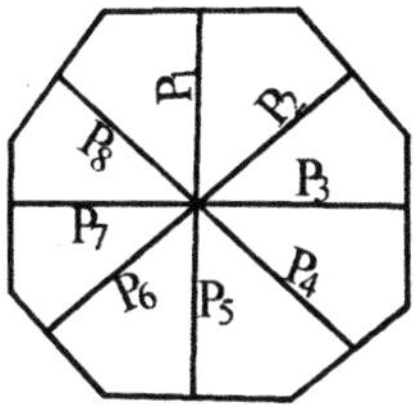

$\frac{1}{2}$ side $\times$ (8 perpendiculars $p_1 + p_2 + p_3 + p_4 + p_5 + p_6 + p_7 + p_8$)= Area of octahedron

$(p_1 + p_2 + p_3 + p_4 + p_5 + p_6 + p_7 + p_8) = \frac{2\,Area\ of\ octahedron}{side}$ is always the same where you take the point.

Ex. 13

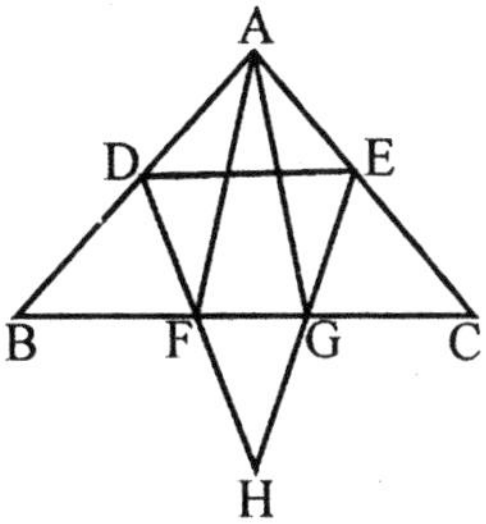

Join AF and AG

AF ‖ GH

AG ‖ FH

AFGH is a parallelgram

FG bisects the parallelogram

$\therefore$ ΔFGH equal in area to

$\Delta AFG = \frac{1}{3}\Delta ABC.$

Ex. 14

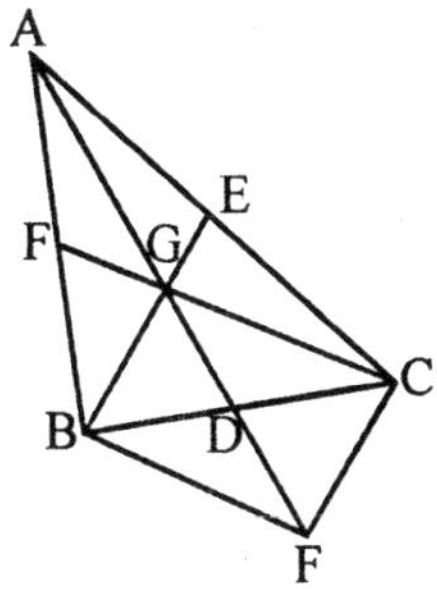

Triangle formed with $\frac{2}{3}$ AG, $\frac{2}{3}$ BG, $\frac{2}{3}$ CF is GFC.

$\therefore$ Its area $= \frac{1}{2}$ paralellogram BGCF

$= \frac{1}{2} \times \frac{2}{3}\ \Delta ABC = \frac{1}{3}$

ΔABC

Ex. 15

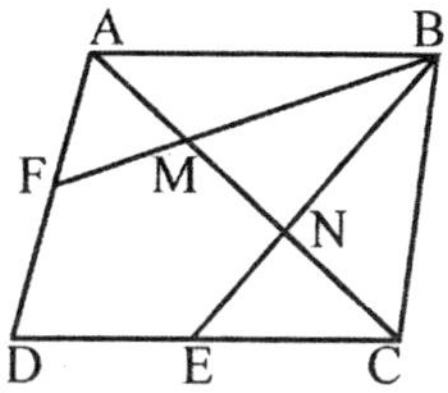

Pentagon = FM NED

$$= \square \text{gm ABCD} - \left[\frac{1}{4}+\frac{1}{4}+\frac{1}{6}\right] \square$$

gm ABCD

$$= \frac{1}{3} \square \text{gm ABCD}$$

$$\because \Delta AFB = \Delta BFC = \frac{1}{3} \square \text{ gm ABCD}$$

$$\Delta BMN = \frac{1}{6} \square \ gm\,ABCD$$

Ex. 18

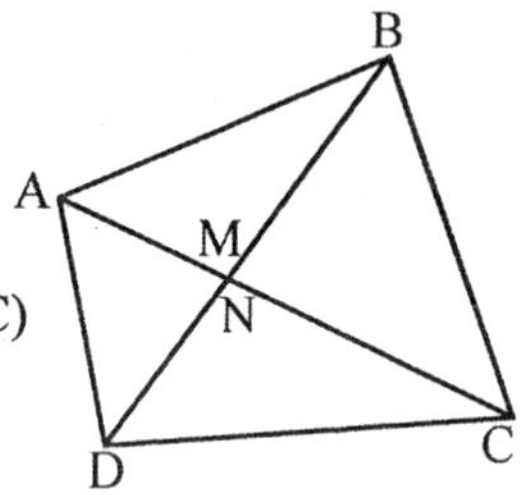

$$AB^2 + DC^2 = AD^2 + BC^2$$

$$AB^2 - BC^2 = AD^2 - DC^2$$

$$AM^2 - MC^2 = AN^2 - NC^2$$

$$(AM + MC)(AM - MC) = (AN + NC)(AN - NC)$$

$$(AM - MC) = (AN - NC)$$

$$(AM + MC - 2MC) = (AN + NC - 2NC)$$

$$MC = NC$$

M and N must coincide

$\therefore$ AB and CD intersect at rt. angles.

Ex. 19

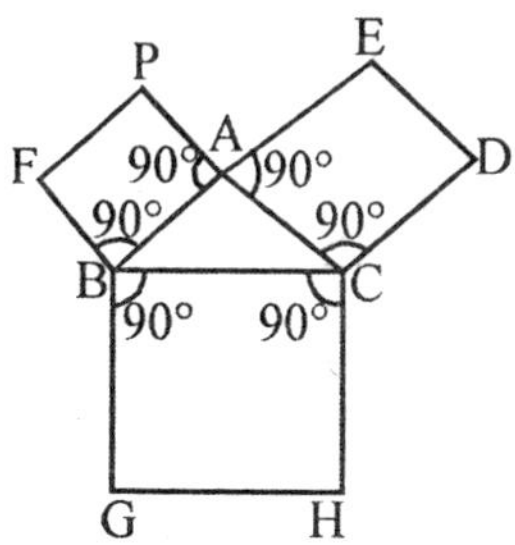

$\Delta ABC = \Delta APE$ in area

because AB = AP

AC = AE

and $\angle PAE + \angle BAC = 180^\circ$

Like this $\Delta FBC = \Delta ABC$ in area

$\Delta DCH = \Delta ABC$ in area.

Ex. 20

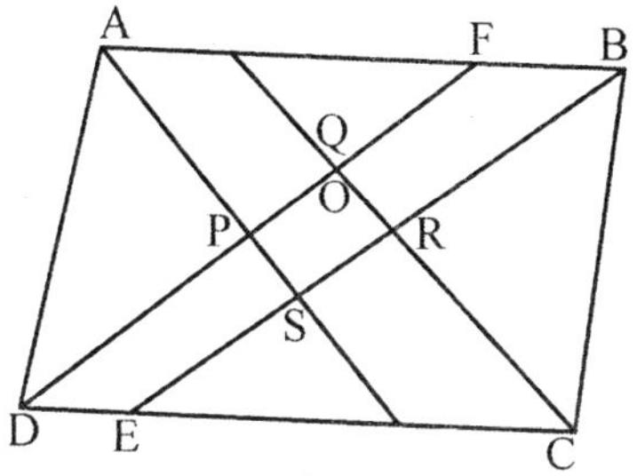

(1) $\angle A + \angle D = 180^\circ$

Angle at P = 90°

The same hold good for the angles at Q, R and S

$\therefore$ PQRS is a rectangle

(2) Diagonal PR and QS of the rectangle are equal to ED and FB = difference in the sides of the main parallelogram.

(3) Centre of the main parallelogram O is also the centre of the parallelogram PQRS.

Ex. 21

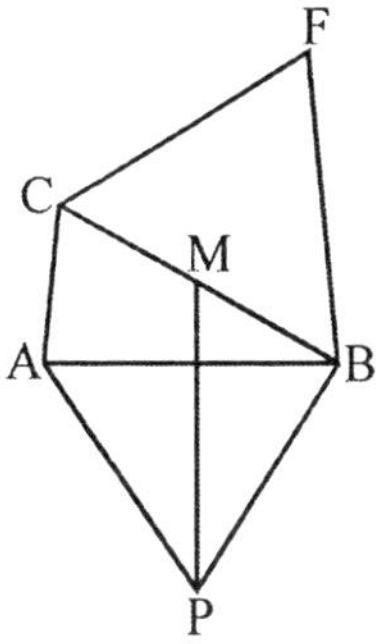

(1) APB and CFB are equilateral ΔS .

(2) $\Delta AFB \equiv \Delta CBP$

because BP = AB

CB = BF

$\angle CBP = \angle ABF = 60^\circ + \angle ABC$

(3) Let M be the midpoint of BC

(4) $\Delta CBP = 2\Delta MBP$

= quad AMBP = $\Delta ABP + \Delta AMB$

$= \Delta ABP + \frac{1}{2}\Delta ABC$

Ex. 22

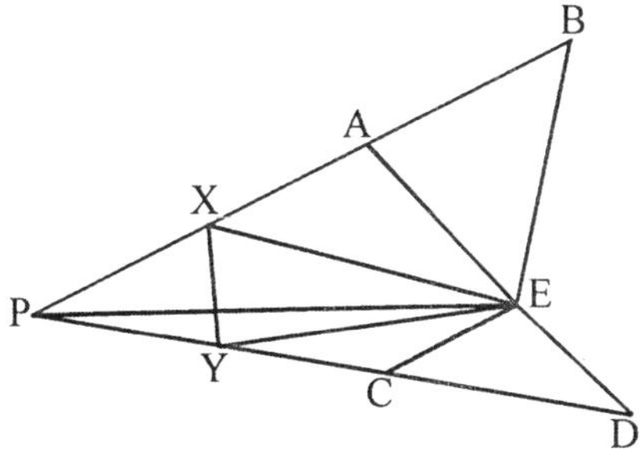

Let AB and CD meet in P.

Let PX = AB and PY = CD

$\Delta PXE = \Delta ABE$ in area.

also $\Delta PYE = \Delta CED$ in area.

but $\Delta ABE + \Delta CED$ is fixed

$\therefore$ $\Delta PXE + \Delta PYE$ is fixed.

$\therefore$ Area of quad. PXEY is fixed.

out of which area of PXY is fixed already.

$\therefore$ Area of XYE is fixed.

E is parallel to XY.

Ex. 23

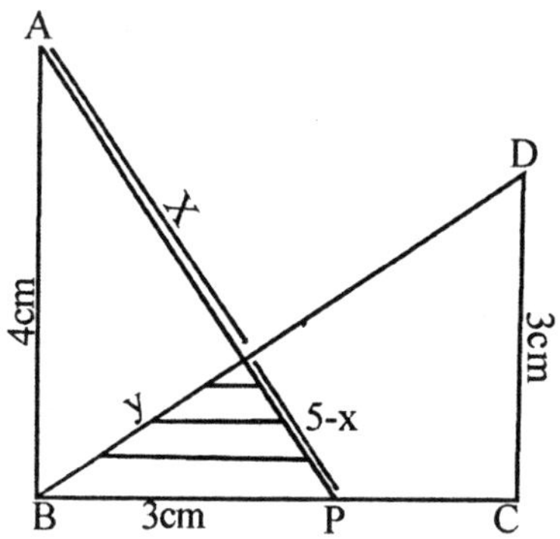

$y^2 = 4^2 - x^2 = 3^2 - (5 - x)^2$

$16 - x^2 = 9 - (25 - 10x + x^2)$

$16 - 9 + 25 = 10x$

$\therefore$ $x = 3.2$

$\therefore$ $y^2 = 16^2 - 3.2^2$

$y = \sqrt{5.76}$

BD and AC meet at right angles

Area required $= \frac{1}{2} y (5 - x)$

$= \frac{1}{2} \sqrt{4.76} \times 1.8 = 9 \times \sqrt{4.76}$

Ex. 24

$s(s - b)$

$= \frac{(a+b+c)}{2} \frac{(a+c-b)}{2}$

$= \frac{a^2 + c^2 + 2ac - b^2}{4}$

$= \frac{1}{2} ac = \text{Area of } \Delta .$

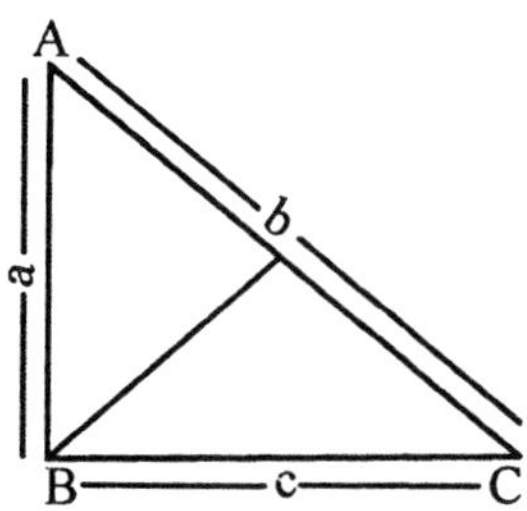

Exercise (B)

Ex. 8

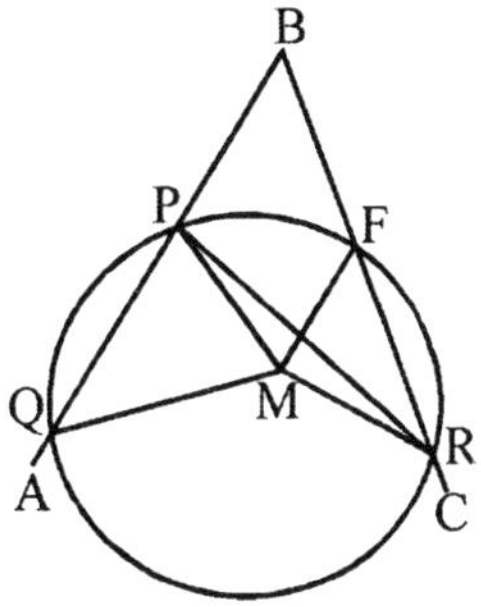

(1) $\angle ABC = \angle QPR - \angle PRF$

$$= \frac{1}{2}\angle QMR - \frac{1}{2}\angle PMF$$

$$= \frac{1}{2}(\angle QMR - \angle PMF)$$

$\therefore$ $2\angle ABC = \angle QMR - \angle PMF$

(2) $\angle PSF = \angle SPR + \angle SRP$

$$= \frac{1}{2}\angle QMR + \frac{1}{2}\angle PMF$$

$$= \frac{1}{2}(\angle QMR + \angle PMF)$$

$2\angle PSF = \angle QMR + \angle PMF$

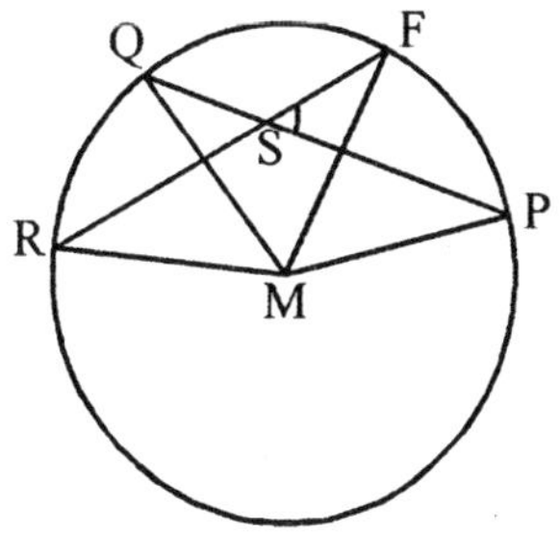

Ex. 15

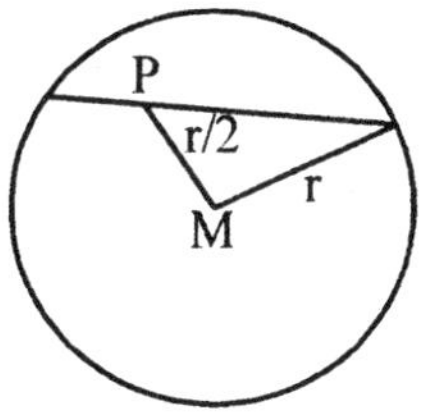

Join PM, Let M be the Centre of the circle Bisect it

From the point of bisection a with radius $\frac{r}{2}$ draw a circle.

Ex. 16

Join PM. On PM as diameter draw a circle.

That will be locus of mid points of chords.

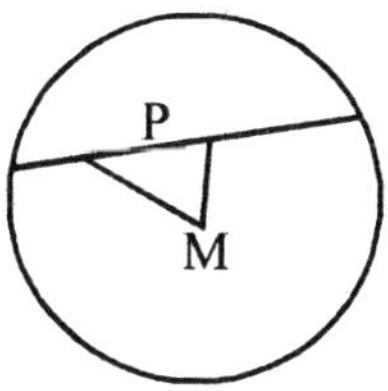

Ex. 17

Join PM

Find the point of trisection say 'C' with C as centre draw a ⊙ with $\frac{r}{3}$ radius

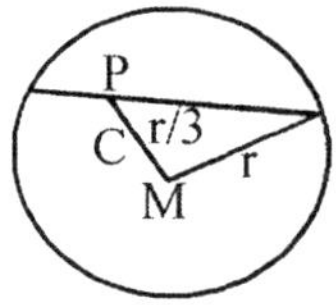

That will be locus of lines which will be trisected.

Ex. 24

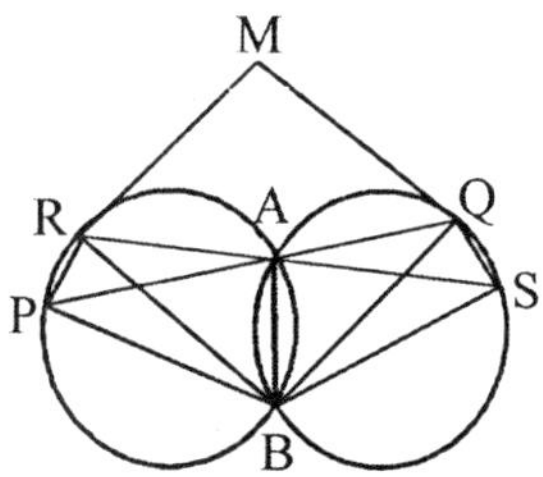

$\angle APB + \angle AQB$ is a fixed angle.

$\therefore$ $\angle PBQ$ is also fixed.

$\angle PBA + \angle ABQ$ is fixed.

$\angle PBA = \angle ARM$

$\angle ABQ = \angle ASQ$

$\therefore$ $\angle ARM + \angle ASQ$ is fixed.

$\therefore$ $\angle RMQ$ is fixed.

Ex. 26

P and B are the intersections of 2 circles. Through P lines AB and CD are drawn making $\angle APQ = \angle DPQ$.

We have to prove that AB = CD.

Draw ⊥s QM, QN on AB and CD. They are equal as $\angle QPM = \angle QPN$

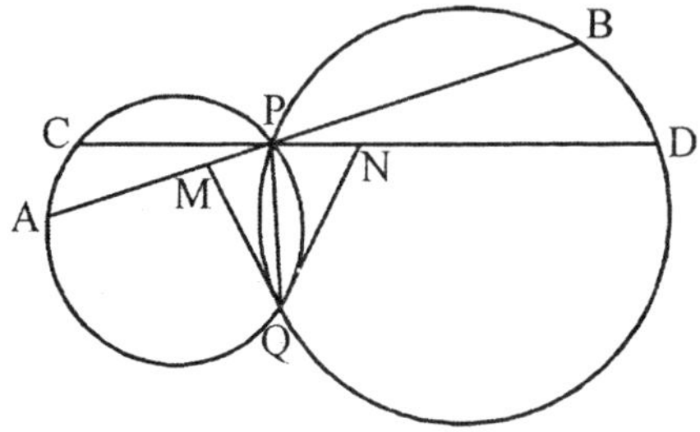

$\therefore$ $\Delta QMB \equiv QND$ and

MB = ND ...(1)

$\Delta QCN \equiv \Delta QAM$

$\therefore$ MA = CN ...(2)

Adding (1) + (2) CD = AB

Ex. 30

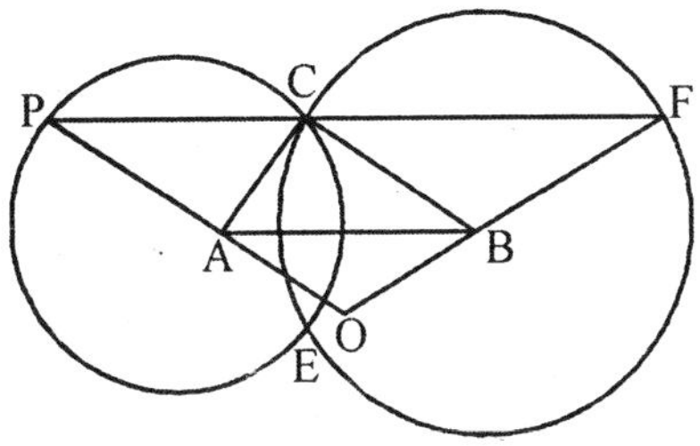

(A) POF is a Δ

$\therefore \angle OPF + \angle OFP + \angle POF = 180°$

3 $\angle s$ of a Δ = 180°

$\angle ACP + \angle BCF + \angle ACB = 180°$

(Because 3 $\angle s$ at the st. line PCF = 180°)

But $\angle OPF = \angle ACP$

and $\angle OFP = \angle BCF$

$\therefore \quad \angle POF = \angle ACF$

(B) Comparing $\angle s$ of the ΔPEF and the $\angle s$ of ΔABC

$$\angle EFC = \frac{1}{2}\angle EBC = \angle ABC$$

and $\angle EPC = \frac{1}{2}\angle EAC = \angle CAB$

$\therefore$ remaining $\angle PEF = \angle ACB$

$$\therefore \quad \angle PEF = \angle ACB = \angle POF$$

Ex. 32

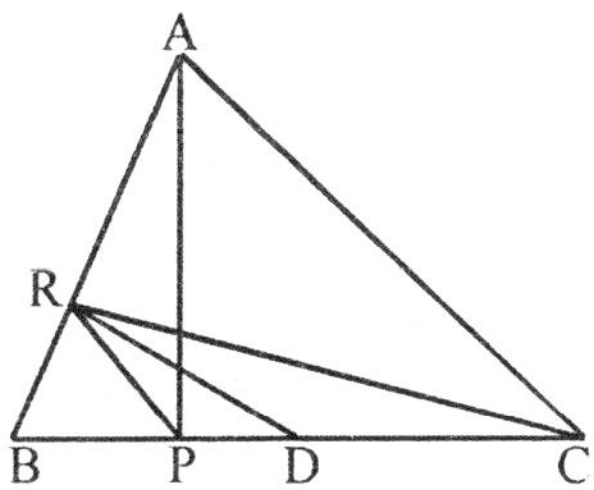

D is the mid point of BC

CR $\perp$ AB and AP $\perp$ BC

BRC is a rt. triangle

$\therefore$ DR = DB = DC

$\angle DRB = \angle RBD = \angle B$

RP joins two feet of perps

$\therefore$ $\angle PRB = \angle C$

$\therefore$ $\angle PRD = \angle B - \angle C$

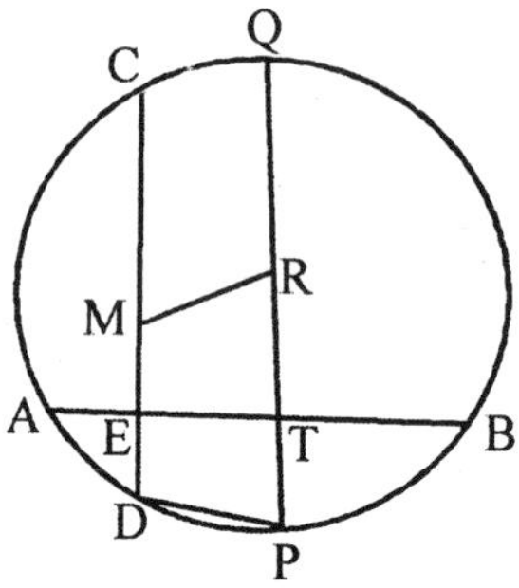

Ex. 33

1. AB and CD are 2 perp. chords meeting in E.
2. Let QP be the diameter || CD. Let it meet the chord AB in T.
3. Let TP = TR and Let ED = EM and join MR
4. MRPD is a trapezium
5. CE – ED = CE - ME = CM
6. QT – TP = QT – TR = QR
7. $\therefore$ CM = MR
8. Difference (CE - ED) = MR

   Wherever C is located.

Ex. 40

1. X is a pt. given outside a circle whose centre is M.
2. Join MX and divide it into 2 equal parts at P.
3. With P as centre and half the radius of the given the main circle.
4. Circle cutting the main circle with centre M in P and Q. Produce XP, XS to meet the main circle in S and T.

   XS and XT are the lines bisected at P and Q respectively.

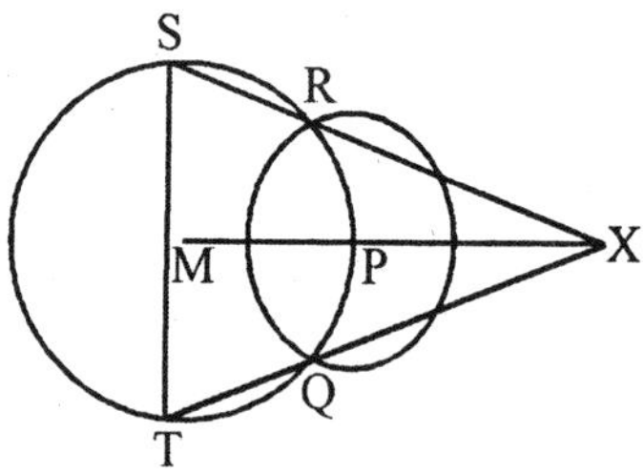

Exercise (C_1)

Ex. 10

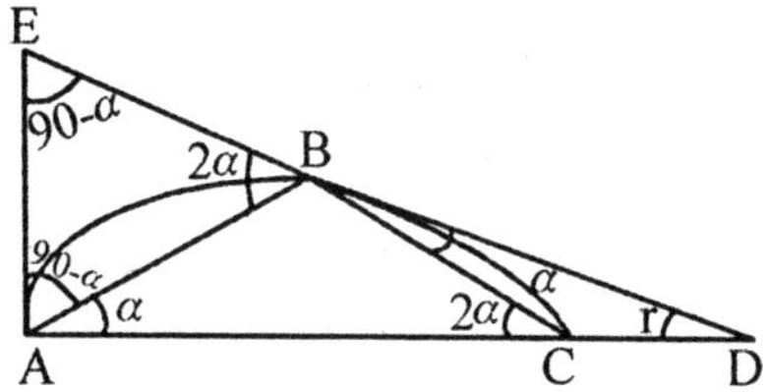

ABC is an arc of a circle

Join AB and BC

Arc AB = 2Arc BC

It means

$\angle ACB = 2\alpha$ and $\angle BAC = \alpha$

AE $\perp$ AC

$\angle EAB = (90-\alpha)$

$\angle AEB = (90-\alpha)$

AB = EB

Also $\angle CDB = \angle CBD = \alpha$

AB = EB

EB = BD

Ex. 20

(1) AB and CD are two pazalld chords of the ⊙

(2) Tangent at D meets AB in T

(3) TP is 2nd Tangent from T.

(4) PC meets AB in X

(5) M is the Centre of the circle

(6) MPTD is cycle quad of which MT is dic.

(7) $\angle PDT = \angle PCT$

(8) $\angle PXT = \angle PCT$

(9) ∴ MXPTD is only the same circle.

10) MX $\perp$ TD X is only the ⊙

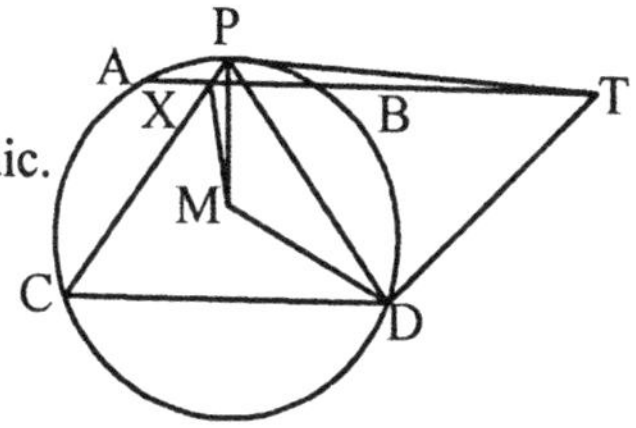

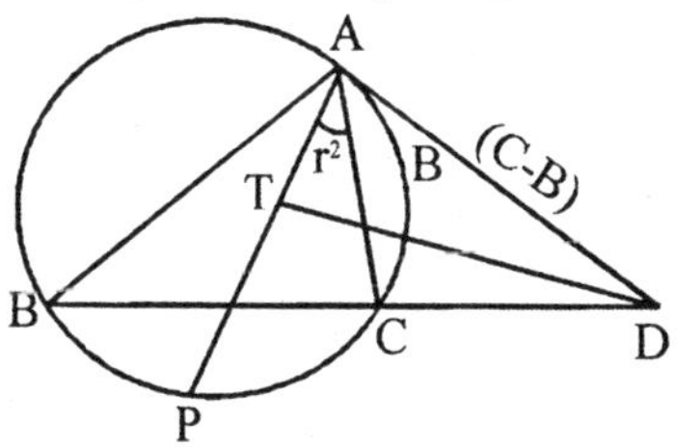

Ex. 21

1. AD is a tangent at A

∴ $\angle CAD = \angle B$

AP is a bisector of $\angle A$

∴ $CAP = \frac{\angle A}{2}$

$$\therefore \quad \angle TAD = \frac{\angle A}{2} + \angle B$$

2. $\angle ADC = \angle ACB - \angle CAB = (C - B)$

$$\therefore \quad \angle ADT = \frac{C-B}{2}$$

3. $\angle ATB = \left(\frac{A}{2} + B\right) + \left(\frac{C-B}{2}\right)$

$$= \frac{A+B+C}{2} = 90^\circ.$$

Ex. 24

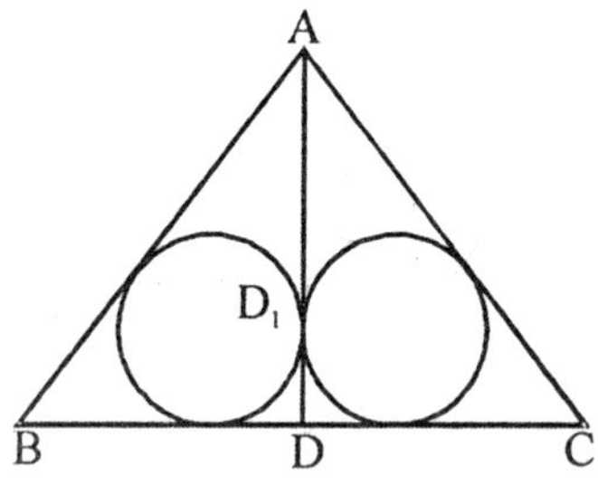

The incircle of ADB will touch AD at point D_1 if $\frac{AD+BD-AB}{2} = DD_1$

The incircle of ACD will touch AD at point D_2 if $\frac{AD+CD-AC}{2} = DD_2$

In order that $DD_1 = DD_2$

$$BD - AB = CD - AC$$
$$AB - BD = AC - CD$$
$$AF = AE.$$

Ex. 26

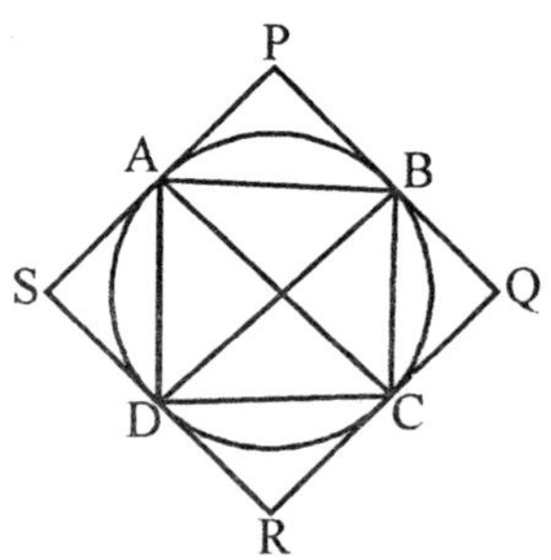

PQRS is a cyclic quad.

$$\therefore \quad \angle P + \angle R = 180^\circ$$

$$\angle Q + \angle S = 180^\circ$$

$$\angle PAB = \angle PBA = \frac{180 - \angle P}{2} = \angle ACB$$

$$\angle RCD = \angle RDC = \frac{180 - \angle R}{2} = \angle BDC$$

$$\angle ACB + \angle BDC = \frac{360-(P+R)}{2} = 90°$$

Ex. 29

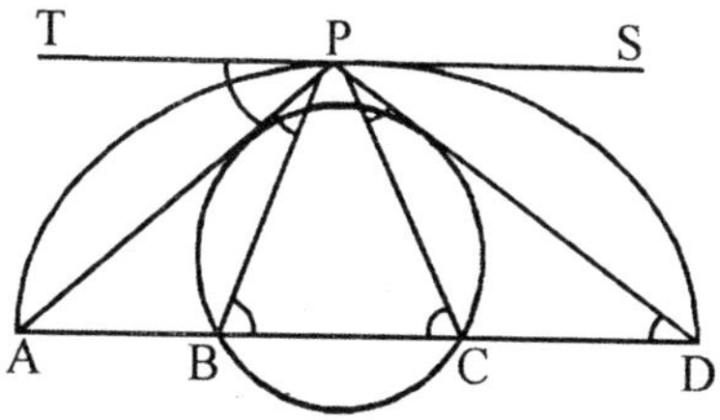

Draw a tangent TS to the circle APD

$$\angle TPA = \angle PDA$$

But we are given that

$$\angle APB = \angle CPD$$

$\therefore\ \angle TPA + \angle APB = \angle PDA + \angle CPD$

$\therefore\qquad \angle TPB = \angle PCB$

$\therefore$ TS touches circle of ΔABC also.

Ex. 30

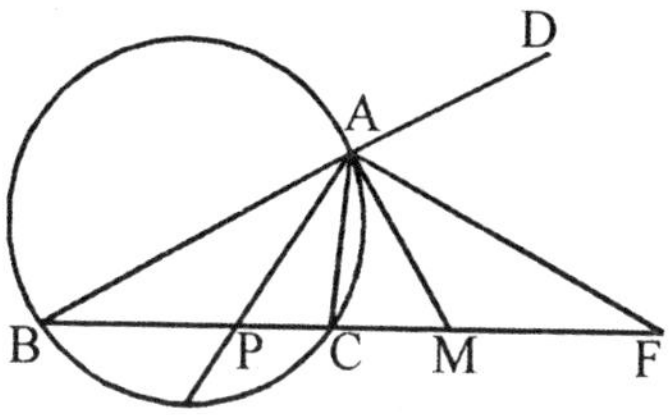

AP is the internal bisector of $\angle A$

AF is the external bisector of $\angle A$

AF makes an angle $= \left(90 - \frac{A}{2}\right)$ with AC

AM is tangent to the circle about ABC.

$\therefore\qquad \angle PAM = \frac{A}{2} + B$

and $\qquad \angle APM = \frac{A}{2} + B$

$\therefore\qquad$ AM = PM $\qquad$ ...(1)

Consider the ΔMAF

$$\angle MAF = \angle CAF - \angle CAM$$

$$= 90 - \frac{A}{2} - B$$

$$\angle MFA = 180 - \angle AFC$$

$$= 180 - (\angle ABC + \angle BAC + \angle CAF)$$

$$= 180 - \left(B + A + 90 - \frac{A}{2}\right)$$

$$= 90 - B - \frac{A}{2} = \angle MAF$$

$\therefore$ AM = FM ...(2)

From (1) and (2) we have PM = FM

Ex. 34

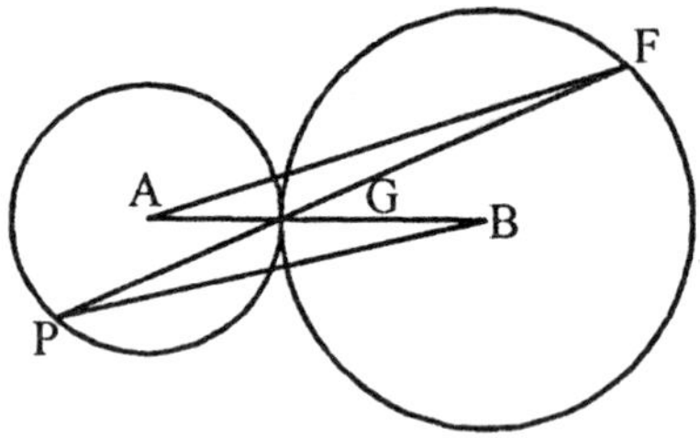

A, G, P are in one st. line.

Join BP and AF

$\Delta ABP = \Delta AFP$ in area

$\because BF \parallel AP$

$\Delta BGP = \Delta AGF$ in area

Ex. 35

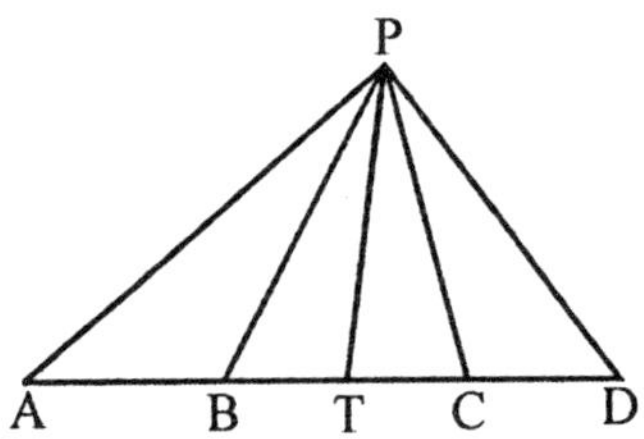

$\angle BPT = \angle PAB$

$\angle CPT = \angle PDT$

$\angle BPT + \angle CPT = \angle PAB + \angle PDT$

$\angle BPT + \angle CPT = 180^\circ - \angle APD$

$\angle BPT + \angle CPT + \angle APD = 180^\circ$

$\angle BPT + \angle CPT + \angle APT + \angle DPT = 180^\circ$

$\angle APC + \angle BPD = 180^\circ$

Ex. 39

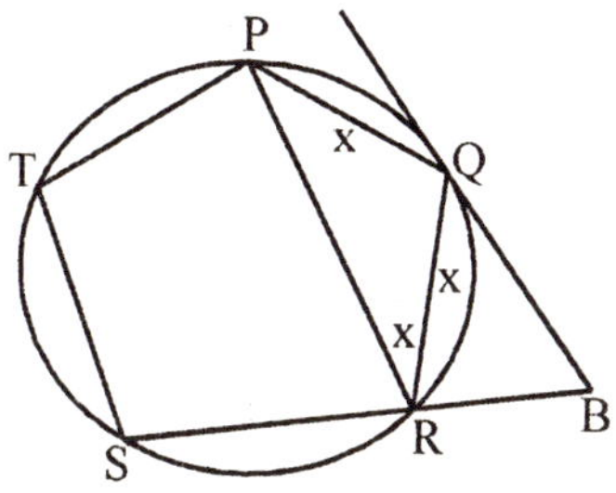

Arc PQ = Arc QR

$\therefore$ chord PQ = chord QR

$\angle QPR = \angle QRP$

$\angle RQB = \angle RPQ$ (Th. 59)

$\therefore$ PR || OB

$\angle PRS = \angle QBS$

But $\angle PRS + \angle PTS = 180^\circ$

$\therefore$ $\angle QBS + \angle PTS = 180^\circ$

Ex. 40

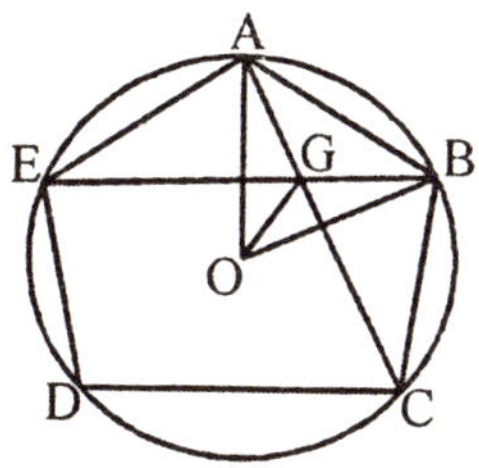

$\angle ABC = 108^\circ$

$\angle BAC = 36^\circ$

$\angle AOB = 72^\circ$

$\angle GOB = 36^\circ$

$\therefore$ AB meets the circle AGO as tangent

Exercise (D_1)

Ex. 10

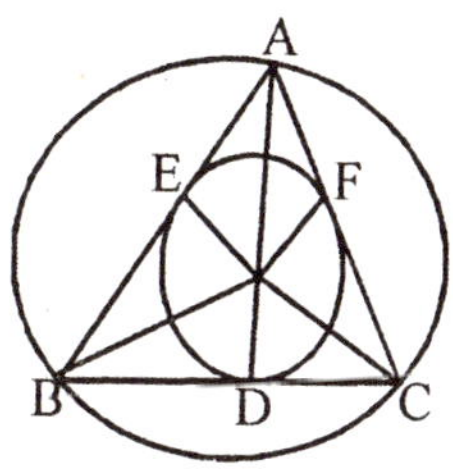

1. Draw an equilateral Δ inside the circle. If the radius of the circle is r the side of Δ will be $r\sqrt{3}$
2. Draw its in circle DEF $\left(radius \frac{r}{2}\right)$
3. From any point inside the circle we cannot draw an equilateral Δ.
4. From any given point we can draw 2 tangents to circle DEF
5. There will be 2 Δs for each point.

Ex. 11

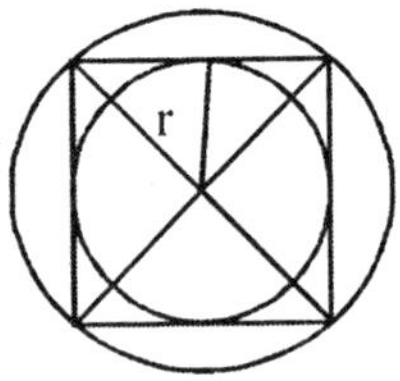

1. A square is to be drawn inside a circle radius r.
2. Draw a circle with radius $\frac{r}{\sqrt{2}}$
3. No square can be drawn from any point inside the circle.
4. Squares from each point can be drawn from any pts. outside the circle.

Exercise (D_2)

Ex. 2

One median = 3 cms second median = 6 cms. Angle at the vertex = 45°. Draw the Δ.

First draw the median BE = 3 cms. Find out the centroid on this median which is 1cm from E. On BE draw a segment of 45° which will also contain the vertex A of the Δ. Join M the centre of 45° segment to B. On BM draw a semi circle.

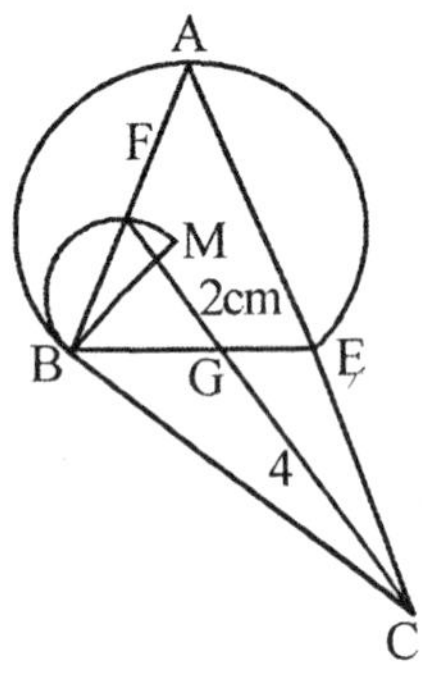

On this is the centre of AB which is located at $\frac{1}{3}$ of the second median from G. 2cm from G Find out the centre of AB. Join it to B to get A. Find out C which is at a distance of 4 cm from G. Join AC. This should contain E. ABC is the Δ.

Exercise (F)

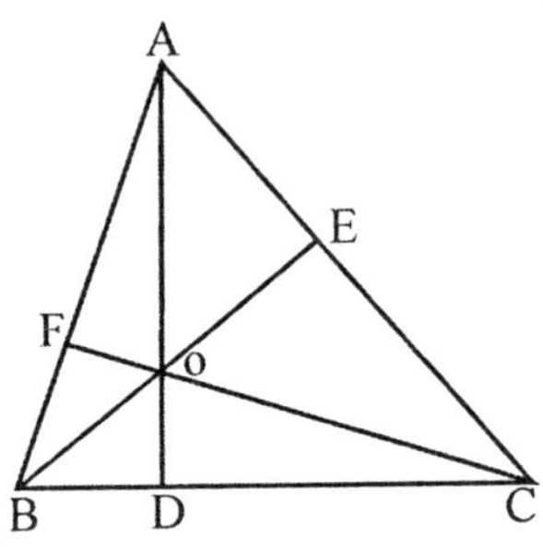

1. If OBC is the main Δ, A is the orthocentre

   [BA ⊥ OC, CE ⊥ OB, OD ⊥ BC]
2. $\angle EDC = \angle FDB = \angle A$
3. OD, OE, OF are the bisectors of the pedal Δ.

4. A, B, and C are centres of the bisectors of D, E, F. etc. respectively.

5. $\angle OBC = (90^{\circ} - C)$ and $\angle OCB = (90^{\circ} - B)$

$\therefore$ BOC $= \angle B + \angle C$

6. The midpt. of AB, OC and DE are in one line [AB is diameter ABDE, OC is diameter of OECD]

7. $\angle D = (180 - 2A)$

8. $\angle BOC = (180 - A)$

9. For a ΔBOC, the orthocentre is A

10. $\angle BOD = C$ and $\angle BPD = C$

$\therefore$ BO = OP

11. $\angle EAO = \angle EBC = \angle MEB$

$\therefore$ ME is a tangent to the circle AFOE

12 Let L be the circum centre

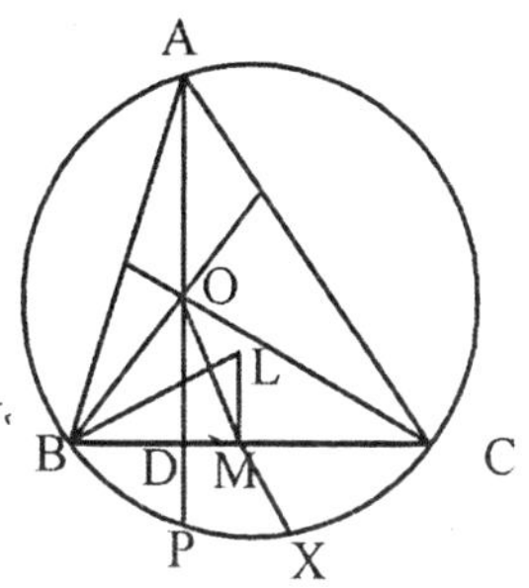

M be the centre of BC

$\therefore$ $\angle LMB = \angle LMC = 90^{\circ}$

Now $\angle PBD = \angle PAC = \angle OBC = (90° - c)$

$\therefore$ $\Delta BDP \equiv \Delta BDO$ (ASA)

$\angle OMD = \angle PMD$

and $\angle OMD = \angle XMC$

$\therefore$ $\angle PMD = \angle XMC$

$\therefore$ $\angle LMP$ = obtuse $\angle LMX$

Now in Δs LMP and LMX

LP = LX (both radii)

LM is common

and $\angle LMP = \angle LMC$ first pt.

$\therefore$ There ΔS are congruenat

$\therefore$ MP = MX

but MP = OM

$\therefore$ OM = MX

and BM = MC

$\therefore$ BOCX is a parallelogram

OM = MX

and AL = LX

$LM = \frac{1}{2} AO$

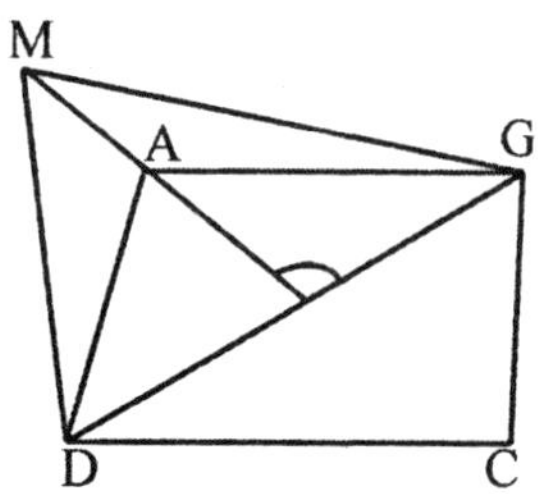

13. $BM \perp BC$

$MD \perp DC$

$\therefore$ $MA \perp BD$

14. Proved in example 12.

15. (a) Vertex (A), ortho centre (O) and the circum centre (L) is given.

Draw the circum circle. Draw LM || AO and make $LM = \frac{1}{2}$ AO and draw a perp. at M.

Exercise (G)

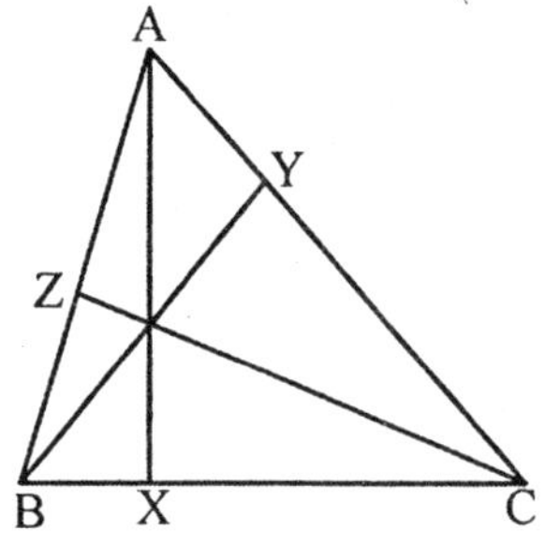

Ex. 4 1. AX + BX > AB

2. AX + XC > AC

3. BY + CY > BC

4. BY + AY > AB

5. CZ + AZ > AC

6. CZ + BZ > BC

Adding the 6 equations

$$2(AX + BY + CZ) + (AB + BC + CA) > 2(AB + BC + CA)$$

$$\therefore \quad AX + BY + CZ > \frac{1}{2}(AB + BC + CA)$$

Ex. 19 Given:- XE and XD are common tangents between circle 1 and circle 2, and between the circles 1 and 3. They meet in X join XF.

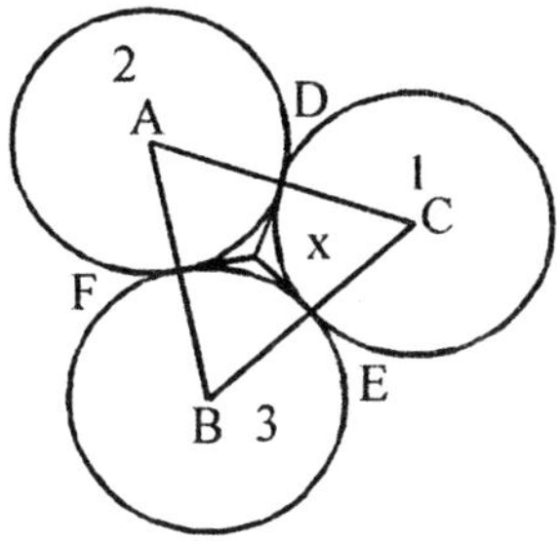

To Prove:- XF is a tangent between circle 3 and circle 2 that means we have to prove $XF = XD = XE$ and $\angle AFX = 90°$

Proof:- $\angle DXE = 180° - C$

$$\angle AFE = 90° - \frac{A}{2}$$

$$\angle DFB = 90° - \frac{B}{2}$$

$$\therefore \angle DFE = 180° - \left(90 - \frac{A}{2}\right) - \left(90° - \frac{B}{2}\right) = \frac{A}{2} + \frac{B}{2}$$

That means $\angle DXE = 2\angle DFE$

That means with X as centre and XE as radius, the circle will pass through F and D. That means XE = XF = XD

$\therefore \qquad \Delta AFX \equiv \Delta ADX$

and $\qquad \angle AFX = \angle ADX = 90°$

Ex. 20 If the side of the square is x, and radius of the circle is r then

$$2x^2 = 4r^2, \text{ or } r^2 = \frac{x^2}{2}.$$

If the radius of the circle is r its diameter is 2r then the side of the equrilateral Δ is $\sqrt{3}r = y$

$$r^2 = \frac{y^2}{3} = \frac{x^2}{2}$$

or $\qquad 3x^2 = 2y^2$

Ex. 26 AFPC is cyclic

$\therefore \qquad \angle PFB = \angle ACP$

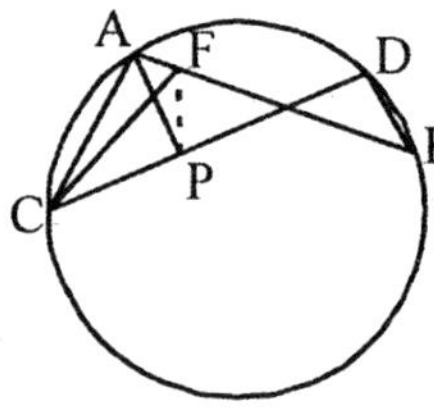

$$\angle ACP = \angle ABD$$

as they are $\angle s$ in the same segment

$\therefore \quad \angle PFB = \angle ABD$

$\therefore \quad PF \parallel BD$

Ex. 27 $\angle AOP = \theta$

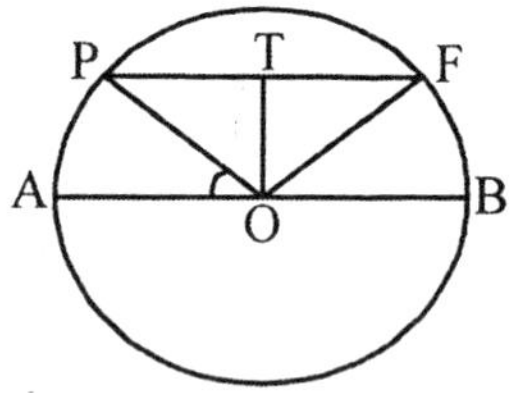

$\angle TOP = (90° - \theta)$

$\angle OTP = (90° - \theta)$

$\angle OPT = 2\theta$

$\angle OFP = 2\theta$

$\angle POF = 180 - 4\theta$

$\angle AOP + \angle POF = 180 - 3\theta$

$\therefore \quad \angle FOB = 3\theta$

$\therefore$ and $\angle AOP = \theta$

Ex. 28 1. ABC is an equilateral Δ drawn in a circle

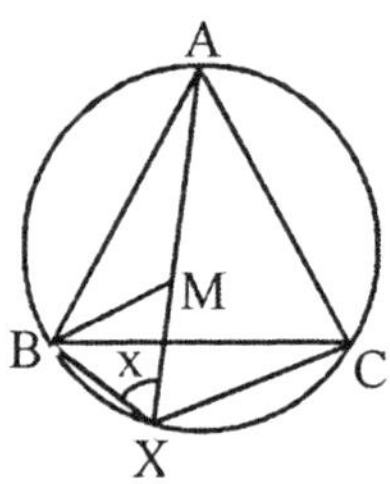

2. AX is a chord which cuts BC in D
3. To prove that AX = BX + CX
4. $\angle AXB = \angle ACB = 60°$
5. From AX cut off XM = BX
6. $\angle BXM = \angle ACB = 60°$
7. $\therefore$ XBM is an equilateral Δ.

8. $\angle ABM = \angle ABC - \angle MBD$

$= 60° - \angle MBD$

$\angle CBX = \angle MBX - \angle MBD$

$= 60° - \angle MBD$

9. $\Delta CBX \equiv \Delta ABM$

as $CB = AB$

$BX = \angle BM$

$\angle ABM = \angle CBX$

10. $\therefore$ $AM = XC;$

$XM = BX$

11. $AM + XM = BX + XC$

$\therefore$ $AX = BX + XC$

Ex. 29 (a) $\text{Arc AP} = \text{Arc AQ} = \dfrac{\text{Circumference}}{3}$

$\text{Arc AB} = \text{Area AD} = \dfrac{\text{Circumference}}{4}$

$\text{Arc DQ} = \text{Arc BP} = \left(\dfrac{1}{3} - \dfrac{1}{4}\right) \text{circumference}$

$= \dfrac{\text{Circumference}}{12}$

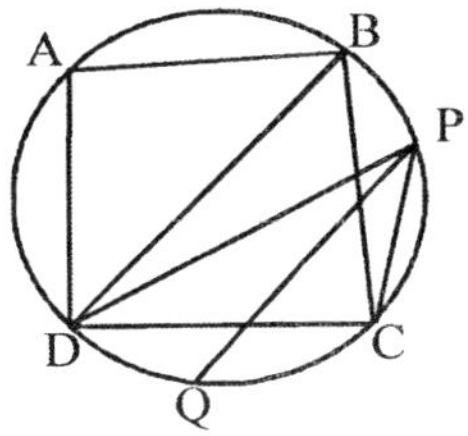

$\therefore$ $\angle BDP = \angle DPQ$

$\therefore$ $PQ \parallel BC$

(b) $\angle s$ of Δ PCD

CD subtends = $\angle 45°$; PC subtends = $\angle 22\frac{1}{2}^{\circ}$

Remaining $\angle F \parallel 112\frac{1}{2}^{\circ}$

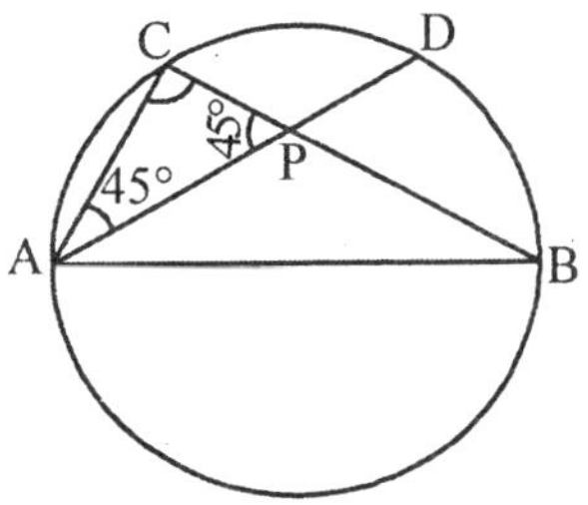

Ex. 30 AB diameter subtends $\angle 90°$

$CD = \dfrac{\text{diameter}}{2}$ subtends $\angle 45°$

$\therefore$ CD makes 45° with AD

$\therefore$ $AC = CP$

Ex. 31 Because equal $\angle s$ QBO and RBO stand on common baswe BO.

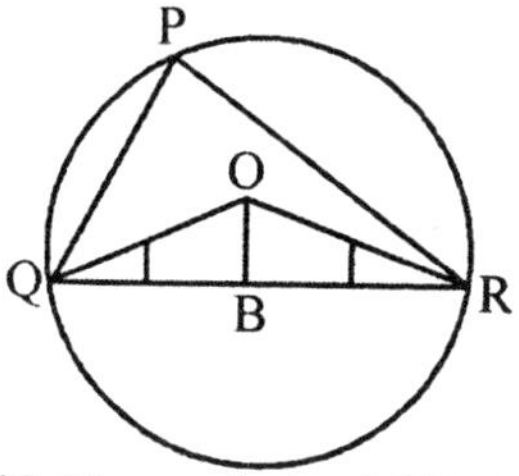

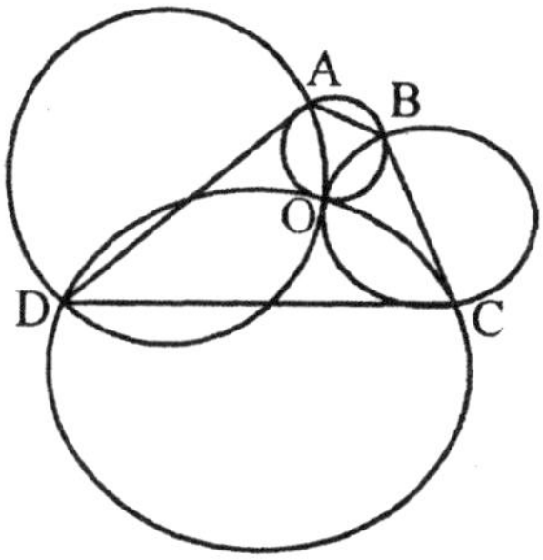

Ex. 32 Circumcircle AOB, BOC, COD, DOA will have AO, BO, CO and DO as common chords.

Therefore the line of centres will have the angles between their centres the same as between the diagonals.

Hence a parallelogram.

Ex. 33 AB is at rt. angle to the chord PQ. In this position Δ PQB will have maximum area as PQ and AB are maximum.

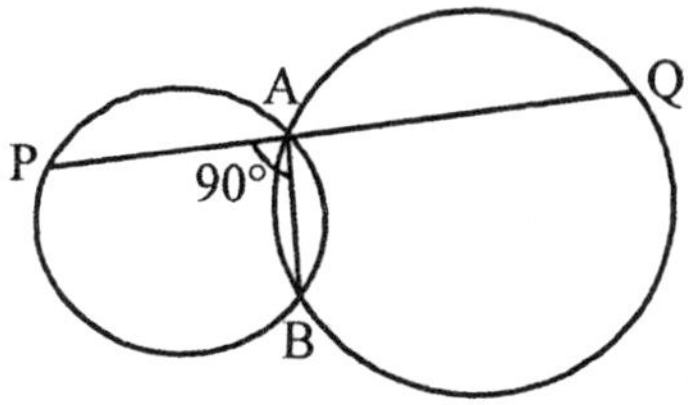

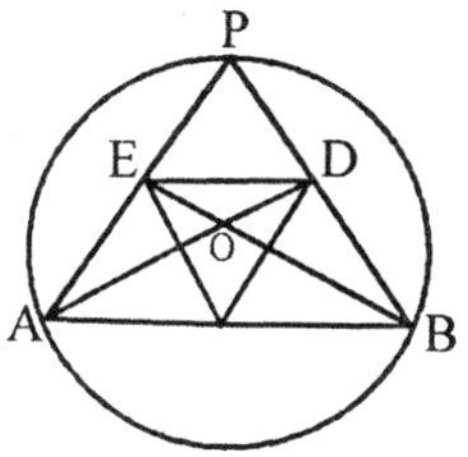

Ex. 34 AB is of fixed length P is a pt. moving on the circle Perps. drawn from A and B on to both sides of the circle. Circle AEDB will always be fixed. The locus of centre of ED will always be at a fixed dist. from the centre of AB.

Hence will be a circle.

Ex. 35 OP will be equal to OQ. The circle with O as centre will be the locus.

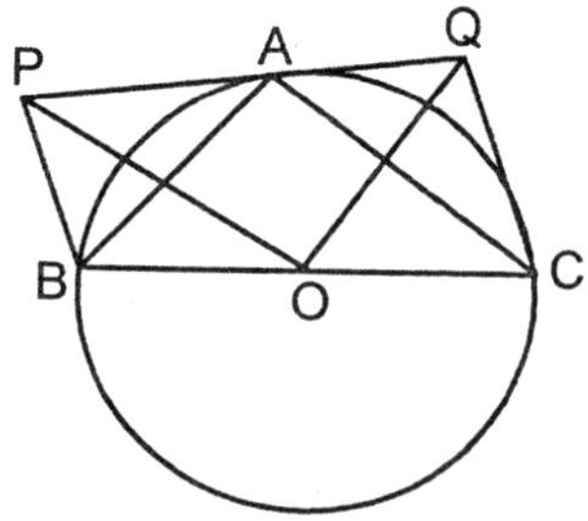

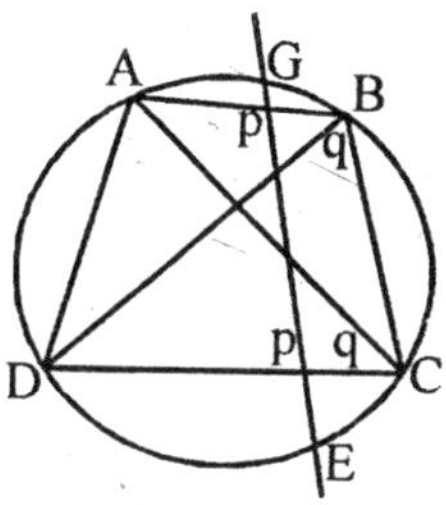

Ex. 36 ABCD is a cyclic quad. Line GE makes with AB and DC angle = p. Then it will make with diagonals (p - q)

Ex. 37 Join P to the centre of the square M and produce PM to meet the opposite side in R. Draw a perp. to PR at M. Let it meet the sides in Q and R. PQRS is the required square.

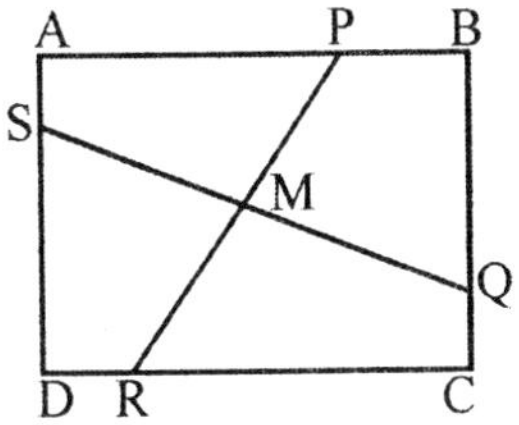

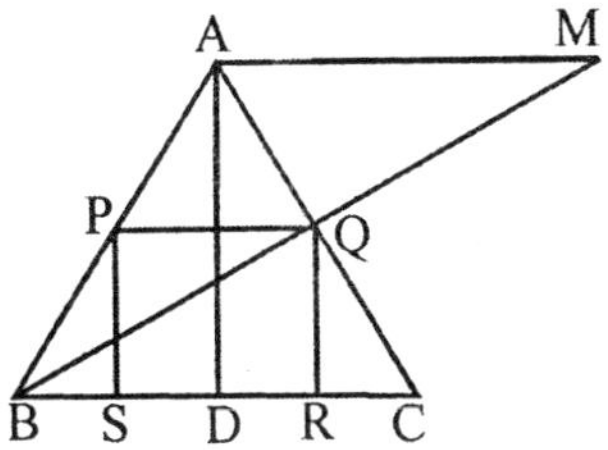

Ex. 38 1. Draw AD $\perp$ BC.

2. Draw AM || AD and = AD
3. Join M to B
4. It meets AC in Q
5. Q is one of the corners of the square
6. Prove $PQ = \frac{30}{11}$
7. 3 squares can be drawn corresponding to 3 sides of the Δ.
8. One square PQRS is drawn.

Ex. 39 1. Given sector BAM

2. Draw AP the bisector of the angle BAM
3. Draw PF $\perp$ AM
4. Bisect $\angle FPA$
5. Let PL be the bisect or
6. At L draw NL $\perp$ AM
7. With N as centre draw a circle with radius = NL

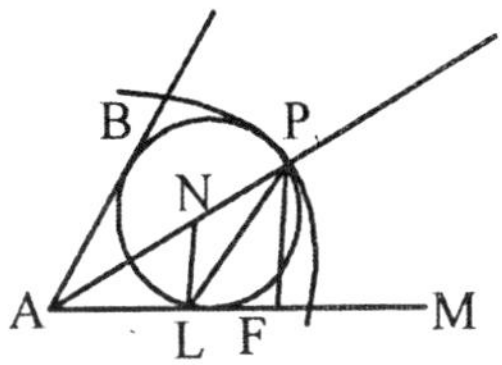

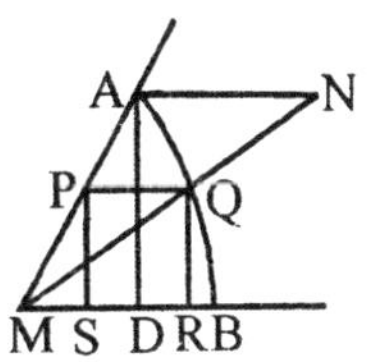

Ex. 40 1. Sector AMB is given

2. Draw AD $\perp$ MB
3. Draw AN || MB and make AN = AD
4. It meets the sector at point Q.
5. Q is one of the end points of the square
6. Complete the square PQRS.

Ex. 41 1. O is the circumcentre

2. DEF is pedal Δ.
3. AD, BE and CF are 3 perps.
4. AO makes an angle (90 - B) with AC or AE.
5. Similarly FE makes an angle B with AC.
6. Hence FE $\perp$ AO.
7. Take this other with make $\perp$ with side o of the Δ.

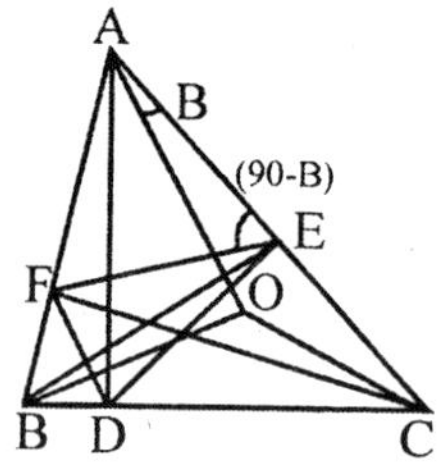

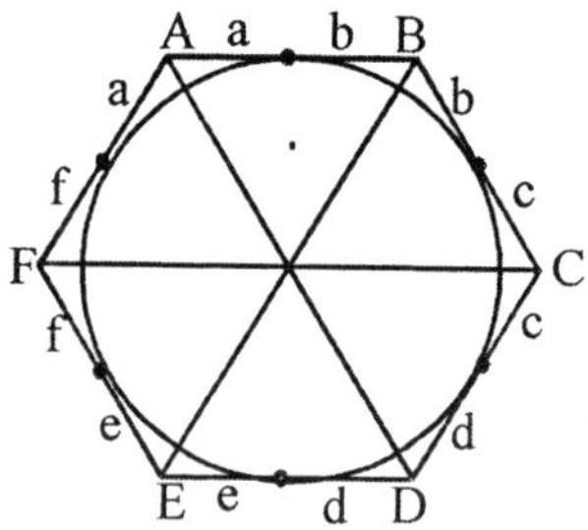

Ex. 42 (a) ABCDEF is a hexagon and is constructed round a circle touching all round it.

(b) AB + CD + EF = a + b+ c + d + e + f

(c) BC + DE + FA = b + c + a + e + f + a

$\therefore$ AB + CD + FE = BC + DF + FA

(d) Bisectors of $\angle s$ of A, B, C, D, E, F are concurrent with the centres.

(e) What is a true for a hexagon is also true of any polygon of even side.

Ex. 43 1. ABCDEF is a polygon of 6 sides inside a circle.

2. All the radii are equal. Hence all the angles are equal as shown
3. $\therefore \angle A + \angle C + \angle E = \angle B + \angle D + \angle F$
4. The $\perp$ bisectors of the sides are concurrent at the centre
5. In the above as well as the preceding examples the polygons are not regular.

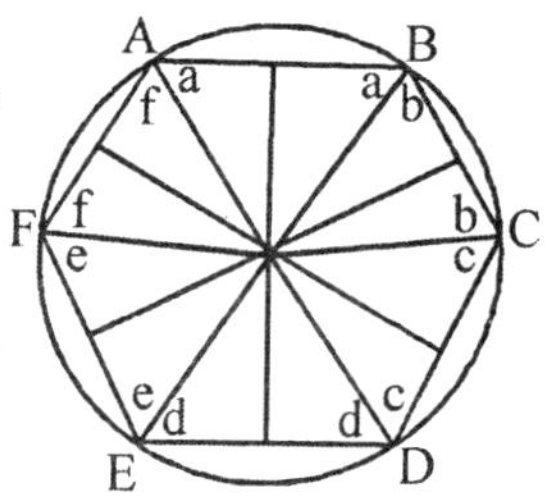

6. What is true of polygon of 6 sides is also true of polygon of any no. of even sides.

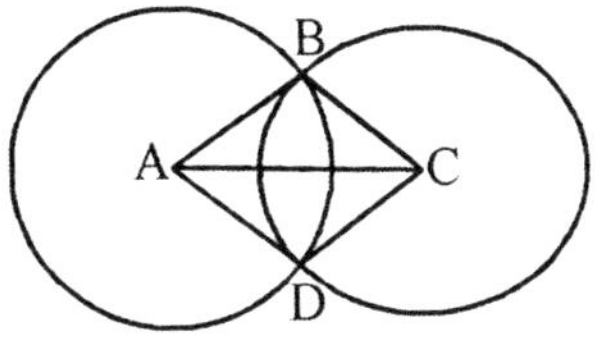

Ex. 45 2 quadrants = r^2 + Area common to both the ⊙ s

$\frac{\pi r^2}{4} + \frac{\pi r^2}{4} = r^2$ + Area Common to both OS.

$\frac{\pi r^2}{2} = r^2$ + Area common to both the circles

Ex. 47 $\angle RFP = \angle BRF + \angle RBF$

$= \frac{1}{2}(\angle BRP + \angle ROP)$

$= \frac{1}{2} \times 90° = 45°$

One circle is drawn with centre at C. Another circle is drawn.

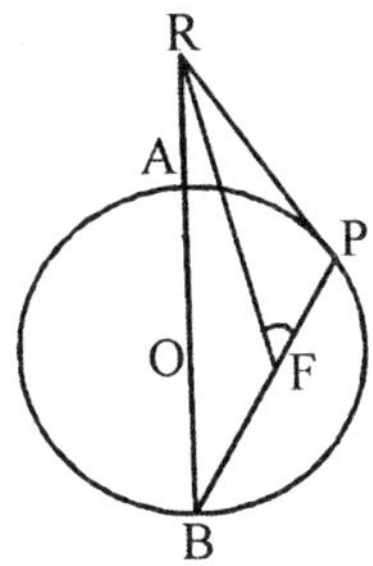

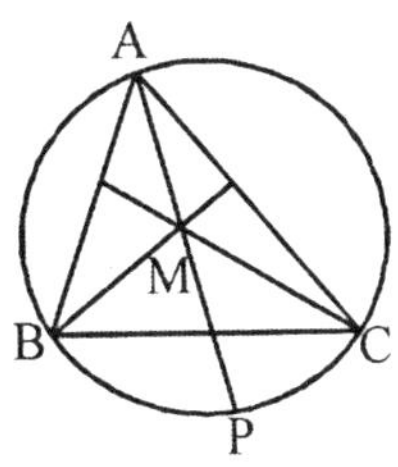

Ex. 48 $\angle BMP = \angle ABM + \angle BAM$

∴ $\angle BMP = \angle MBP$

∴ BP = PM

Similarly BP = PC

∴ BP = PM = PC

∴ P is the circum centre of Δ BMC.

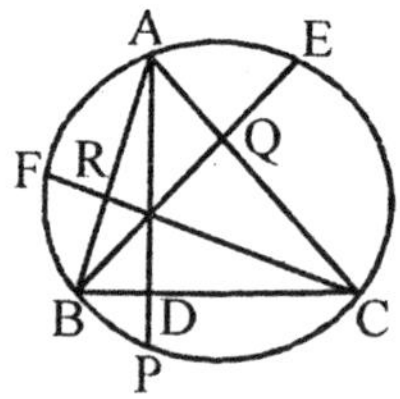

Ex. 49 ABC is the Δ. AD, BE and CF are 3 perps.

They meet the circumferences in P, Q, R respectively.

Arc PC = Arc QC as they subtend equal $\angle s$ at the circumference.

Thus C is the centre arc PQ.

Likewise Q and R are centres of Arc AC, Arc AB.

Ex. 50 Given: (1) OP and QR are 2 perp. raddi.

(2) From P and Q are drawn 2 perp chords PS and QR.

To Prove: PR || QS

Proof: $\angle PSQ = \frac{1}{2}\angle QOP = 45^\circ$

Similarly $\angle PRQ = 45^\circ$

$\therefore$ $\angle SPR = 45^\circ$

$\therefore$ $\angle SPR = \angle PSQ$

$\therefore$ $PR || QS$

Similarly If $PR || QS$

$PS \perp QR$

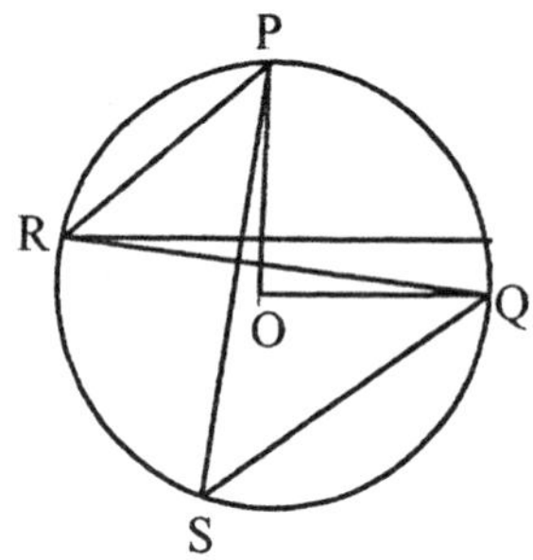

Ex. 51 $2\,CP^2 + 2DP^2$

$= 2(CM + MP)^2 + 2(CM - MP)^2$

$= 2\,(2CM^2 + 2MP^2)$

$= 4\,(DM^2 + MP^2)$

$\because CM = DM$

$= 4\,(DM^2 + MO^2)$

$\because MP = MO = 4DO^2$

$= 4AO^2 = AB^2$

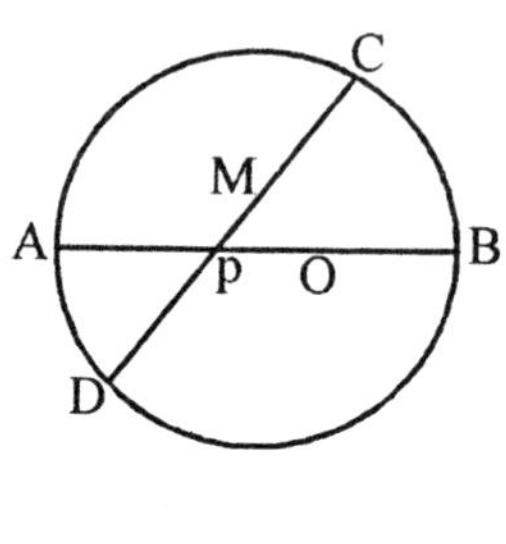

Ex. 52 $BX^2 = x^2 + x^2(1+\sqrt{2})^2$

$= x^2[1+1+2\sqrt{2}+2]$

$= x^2[4+2\sqrt{2}]$

$AC.AE = 2x[2+\sqrt{2}]^x$

$= x^2[4+2\sqrt{2}]$

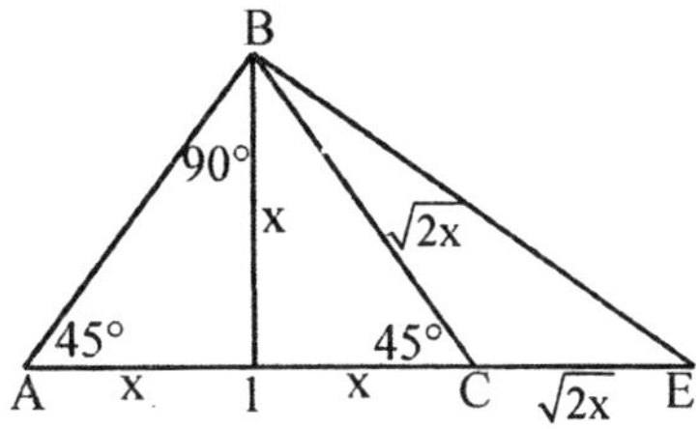

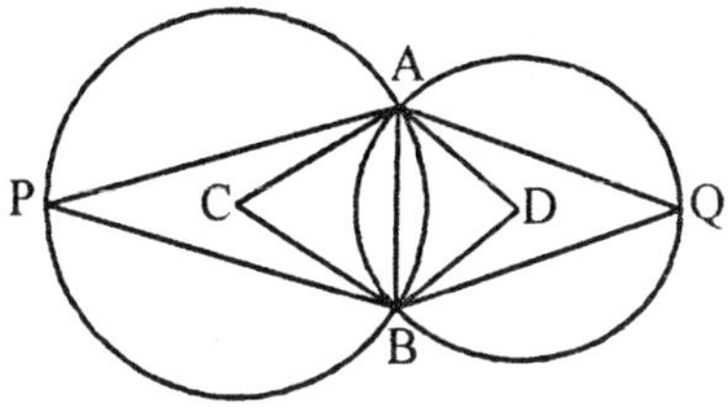

Ex. 54 $\angle CAD = \angle DBC = 90°$

$\therefore\ \angle ACB + \angle ADB = 180°$

$\therefore\ \angle APB + \angle AQB = 90°$

Ex. 56 BC = a. AD = h equating areas

$$x^2 + \frac{1}{2}x(a-x) + \frac{1}{2}x(h-a) = \frac{1}{2}ah$$

$$x^2 + \frac{xa}{2} - \frac{x^2}{2} + \frac{xh}{2} - \frac{x^2}{2} = \frac{1}{2}ah$$

$$\frac{ax}{2} + \frac{xh}{2} = \frac{1}{2}ah$$

$$\frac{x}{h} + \frac{x}{a} = 1$$

$$\frac{1}{h} + \frac{1}{a} = \frac{1}{x}$$

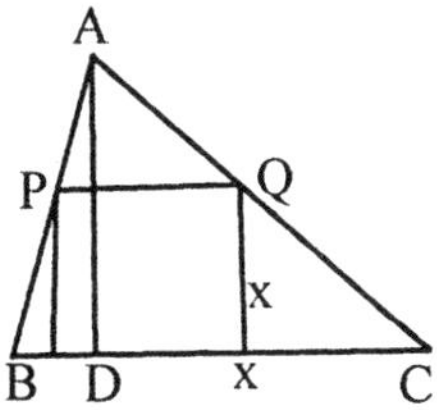

Ex. 57 1. PF is the direct common tangent to the circles with centres at A and B.

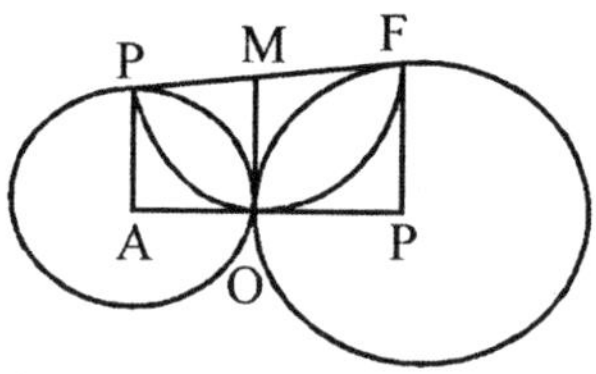

2. AP $\perp$ PF and BF $\perp$ PF
3. M is centre of PF
4. With M as centre a circle is drawn
5. Δs APM and AOM are congruent

   [AP=AO, PM=MO, AM is common]
6. Similarly Δs BFM and BOM are congruent
7. OM is at right angles PF
8. $\therefore$ Circle passes through.

Ex. 58 $\angle C$ is fixed

$\angle D$ is fixed

Hence ($\angle C - \angle D$) is also fixed.

Hence arc CD is fixed.

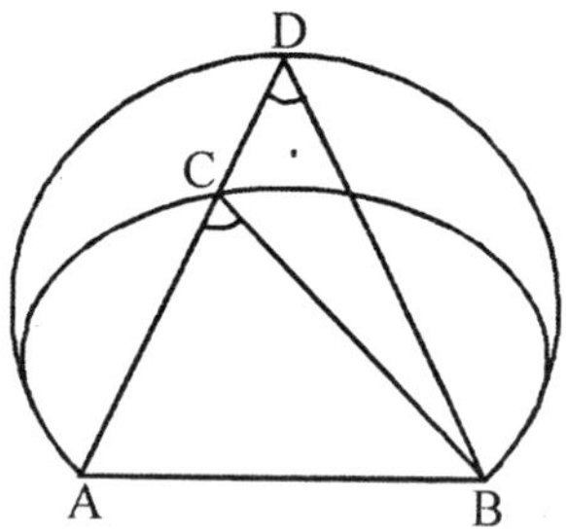

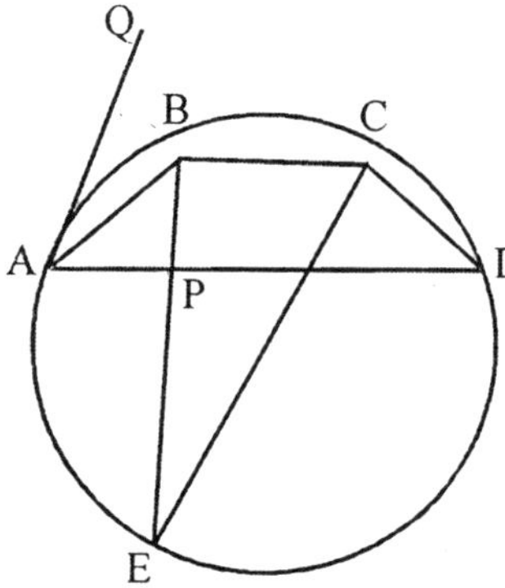

Ex. 60 $\angle A$=45°

$BP \perp AD$

$\angle ABP$=45°

$\angle ABC = 135°$

$\therefore$ $\angle EBC = 90°$

$\therefore$ CE is the diameter

$\angle QAB = \angle BEC = \angle BEC$

$\therefore$ $QA || EF$

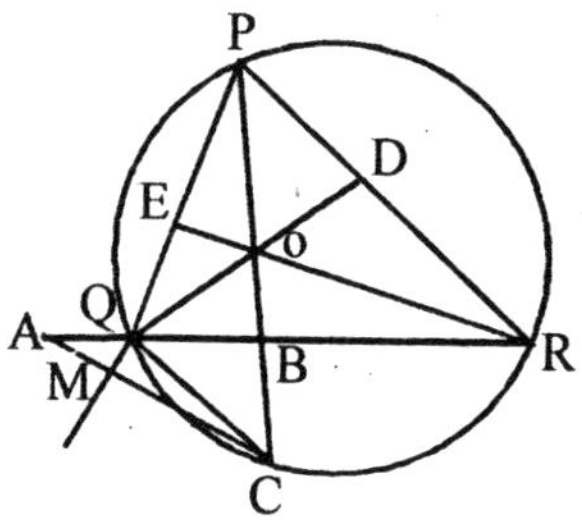

Ex. 61 1. O is the ortho centre.

2. $\angle CQR = \angle CPR$ as they are $\angle S$ in the same segment.

3. $\angle CPR = \angle DQR$ as $\angle ODR = \angle OBR = 90°$ and $\angle PRQ$ is common.

4. $\therefore \angle OQB = \angle CQB$, QB is common, $\angle QBO = \angle QBC = 90°$

5. $\therefore$ CB = BO

6. $\therefore$ AR is $\perp$ bisector of OC.

7. AR = CR

Ex. 62 (a) $\angle P + \angle C + \angle D = 180°$

$\angle Q + \angle D + \angle A = 180°$

$\angle P + \angle Q + \angle A + \angle C + 2\angle D = 360°$

$\angle P + \angle Q + 180° + 2\angle D = 360°$

$\angle P + \angle Q + 2\angle D = 180°$

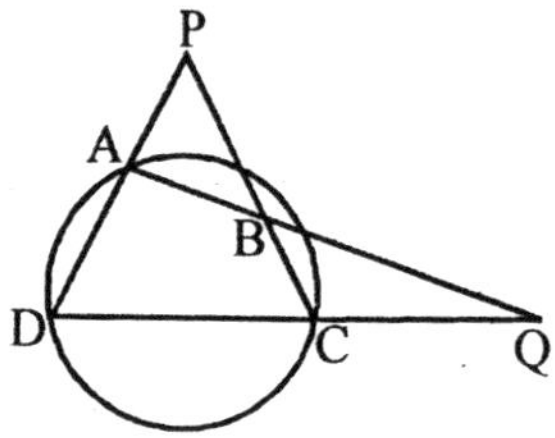

(b) $p = \angle 20°, \angle q = 40°$

$20 + 40 + 2\angle D = 180°$

$2\angle D = 180° - 60° = 120°$

$\angle D = 60°$ $\angle B = 120°$ $\angle C = 100°$ $\angle A = 80°$

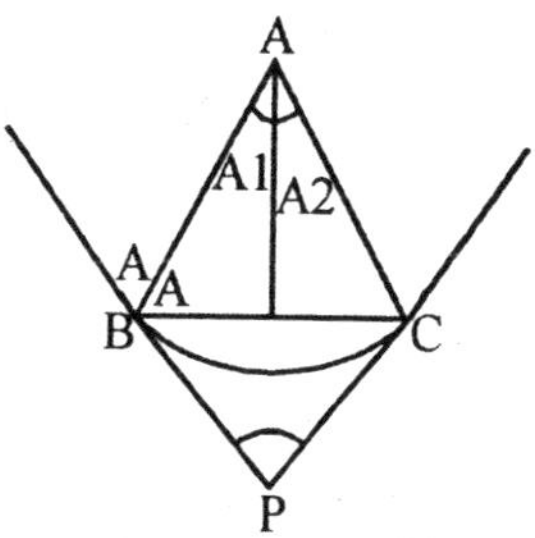

Ex. 63 1. Locus will be segment of a circle of (180 - 2A) on the lower side of BC that is opposite to A.

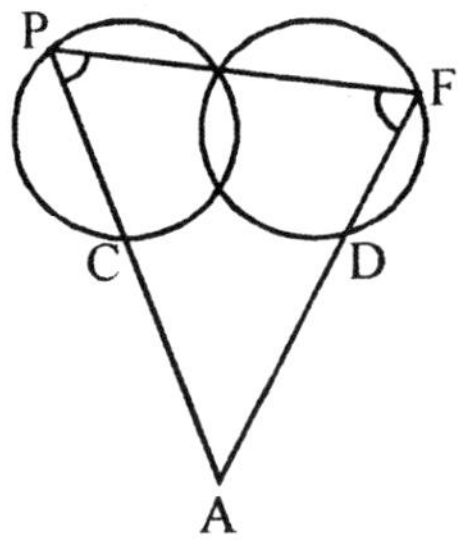

Ex. 64 C and D are points fixed on the 2 circles

$\therefore$ PC and FD will meet at an angle

$= 180 - \angle CPF - \angle DPF$.

Hence will be fixed. Hence it will be a segment of a circle subtending an angles

$= \angle CAD$ with CD.

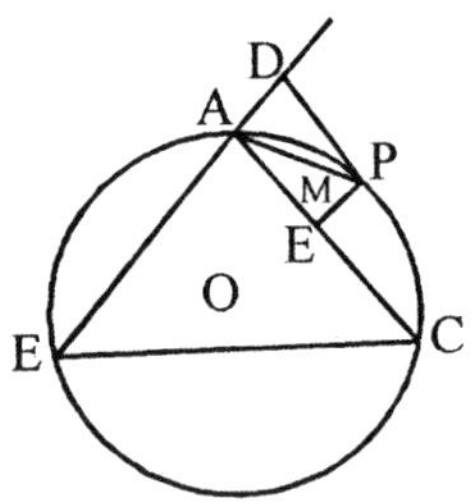

Ex. 65 ABC is a Δ in a circle. P is a point on the circumference. PD and PE are perpendiculars on AB and AC respectively. The circumcentre of Δ PDE will be at the point such as M, i.e. at a distance $\frac{PA}{2}$.

These will located on the semi circle of PO between P and C.

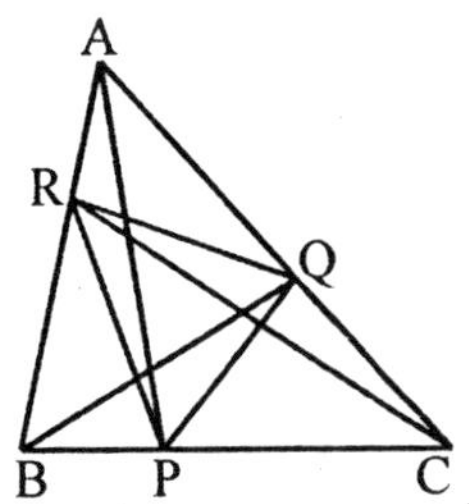

Ex. 66 Consider Δ BPR

$$BP = \frac{1}{3}BC, BR = \frac{2}{3}AB$$

$$\therefore \quad \Delta BPR = \frac{2}{9}\Delta ABC$$

Similarly $\Delta ARQ = \frac{2}{9}\Delta ABC$

and $\Delta RQC = \frac{2}{9}\Delta ABC$

$$\Delta ABC = \Delta PQR + \frac{2}{9}\,\Delta ABC + \frac{2}{9}\,\Delta ABC + \frac{2}{9}\,\Delta ABC$$

$$\Delta ABC = \Delta PQR + \frac{2}{3}\,\Delta ABC$$

$$\Delta PQR = \frac{1}{3}\,ABC$$

Exercise (H)

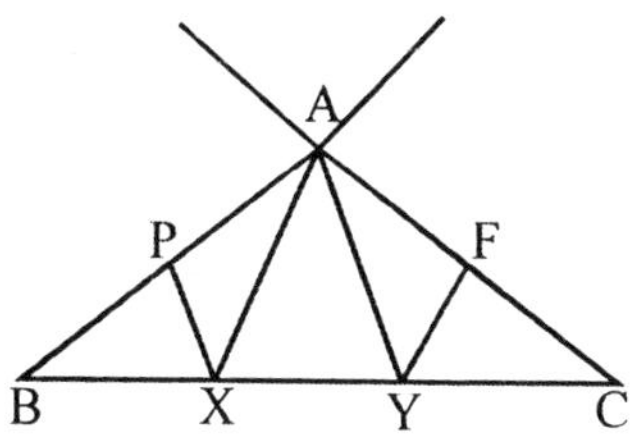

Ex. 1 $\angle BAC = 120°$

$\angle ABC = \angle ACB = 30°$

$\perp$ bisectors of AB and AC meet in X and Y.

AX = BX and AY = YC

AXY is equilateral

$\therefore$ BC = 3BX

Ex. 3 (a) $y + x + B = 180°$

$x + z + D = 180°$

$\therefore$ Adding

$2x + 72° + 180° = 360°$

$x = 54°$

(b) $2x + 90° + 180° = 360°$

$x = 45°$

(c) $2x + y + z = 180°$

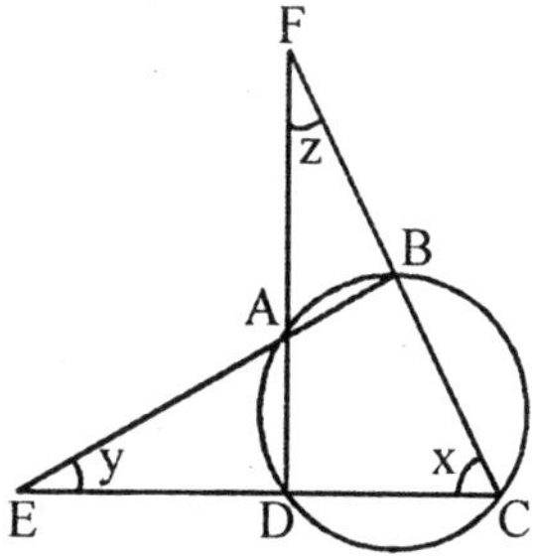

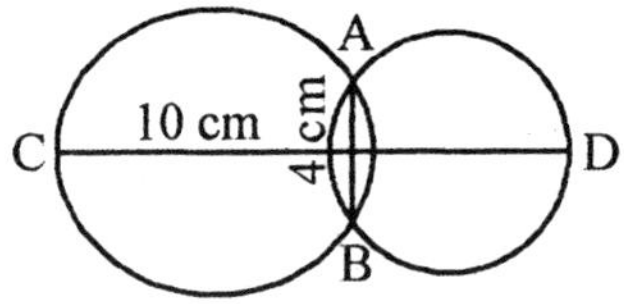

Ex. 4 CD = 10 cms

AB = 4 cms

$2\sqrt{r^2-4}+2r=10$

$\sqrt{r^2-4} = (5-r)$

$r^2-4 = 25-10r+r^2$

$29 = 10r$

$r = 2.9$

Ex. 5 $\angle ABF + \angle APF =$

$\angle FPD + \angle FCD = 180°$

Both ABFP and DCFP are cyclic

$\therefore \quad \angle PBC = \angle PAF$

and $\angle PCF = \angle PDF$

$\therefore$ Therefore

$\angle AFD = \angle BPC$

as $3\angle s$ of a Δ are 180°

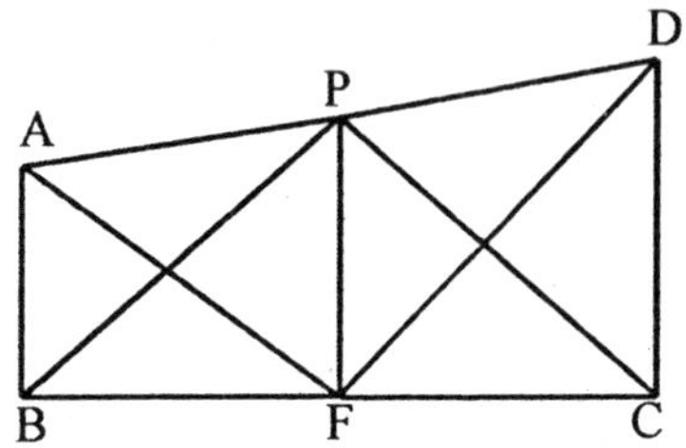

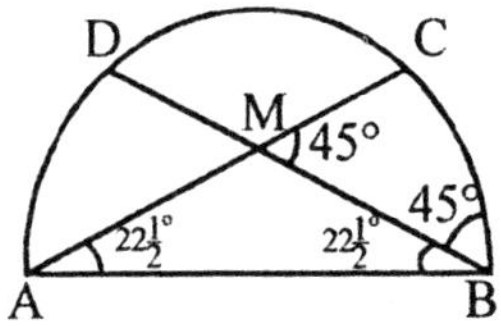

Ex. 6 $\angle ACB = 90°$

$\angle CMB = \angle CBM = 45°$

$MB^2 = CM^2 + CB^2$

$MB^2 = 2CM^2$

$MB = MA$

$MA^2 = 2CM^2$

Ex. 7 AT $= \sqrt{AP^2 - PT^2}$

$= \sqrt{\frac{13^2}{4}.5-121} = \sqrt{\frac{169\times5}{4}-121} = \sqrt{\frac{845-484}{4}}$

$$= \sqrt{\frac{361}{4}} = \frac{19}{2}$$

$$\frac{13}{2x} = \frac{19}{2y}$$

$$13y = 19x \qquad ...(1)$$

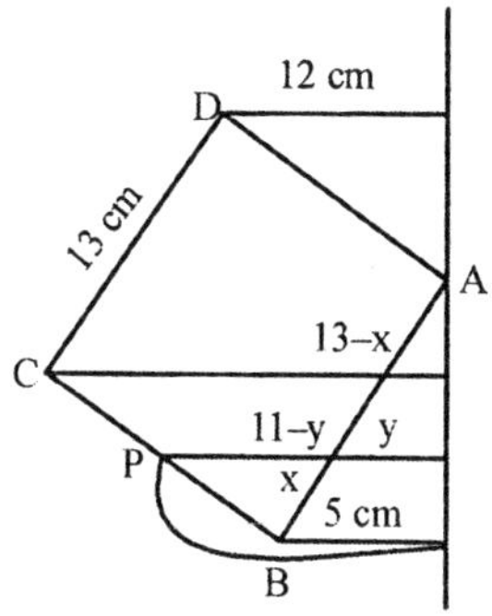

$$\frac{13-x}{\frac{19}{2}} = \frac{11-y}{\frac{13}{2}}$$

$$169 - 13x = 209 - 19y$$

$$19y - 13x = 40 \qquad ...(2)$$

From (1) + (2)

We get $x = \frac{65}{24}, y = \frac{5}{24}$

Distance of B = 5 cms

Distance of C = 7 cms

Distance of D = 12 cms

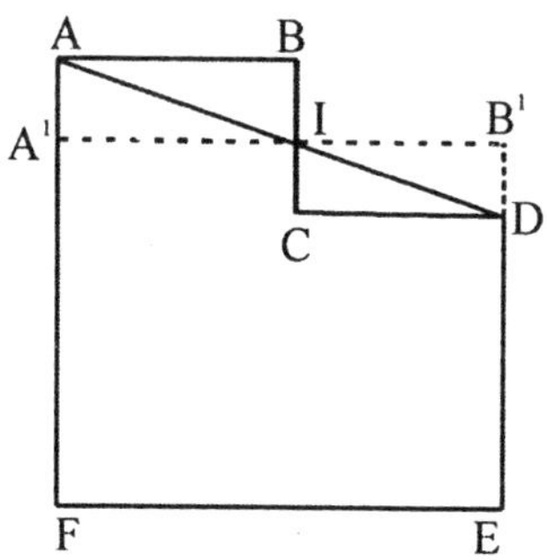

Ex. 8 ABCDEF is the original figure to be divided into 2 equal parts.

Area ABI = Area CID

Area AA^1IB = Area IB^1D

$\therefore$ Area A^1B^1EF is the new area of ABCDEF

Divide FA^1 into two equal parts at P.

Draw a line PQ || FE.

This is dividing line for the figure ABCDEF.

Ex. 9 Given : $BD = x^2, DC = y^2,$
$AD = xy$

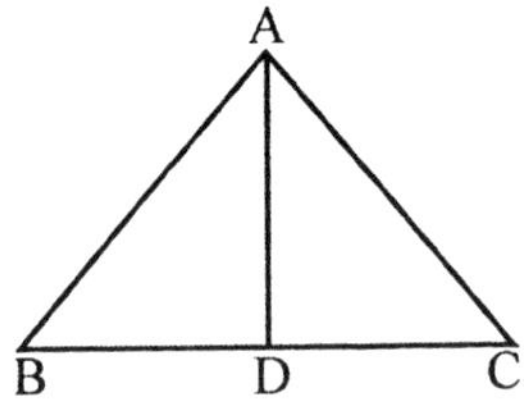

$$AB = \sqrt{x^4 + x^2y^2} = x\sqrt{x^2 + y^2}$$

$$AC = \sqrt{y^4 + x^2y^2} = y\sqrt{x^2 + y^2}$$

$$\sqrt{AB^2 + AC^2} = \sqrt{x^2\left(x^2 + y^2\right) + y^2\left(x^2 + y^2\right)}$$

$$= (x^2 + y^2) = (BD + DC) = BC$$

$\therefore \quad \angle A = 90°.$

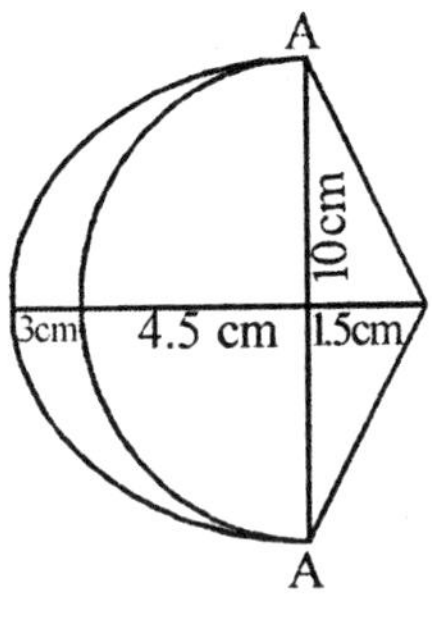

Ex. 10 $r^2 = 5^2 + 1.5^2$

$= 25 + 2.25$

$= 27.25$

$r = 5.25$

Ex. 11 (1) $r_1^2 = 16.00 + (r_1 - 1)^2$

$= 16.00 + r_1^2 - 2r_1 + 1$

$2r_1 = 17.00$

$$r_1 = \frac{17}{2} = 8.5 \text{ cm}$$

(2) $r^2_2 = (r_2 - 2)^2 + 16$

$r^2_2 = r^2_2 - 4r^2_2 + 4 + 4 + 16$

$4r_2 = 20$

$r_2 = 5.0$

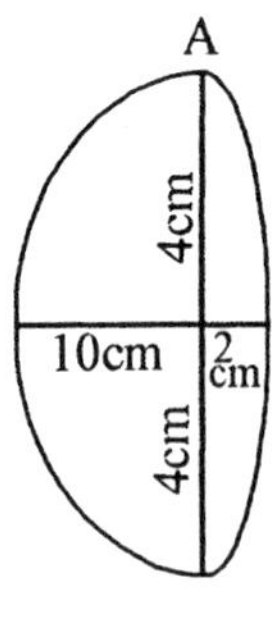

Ex. 12

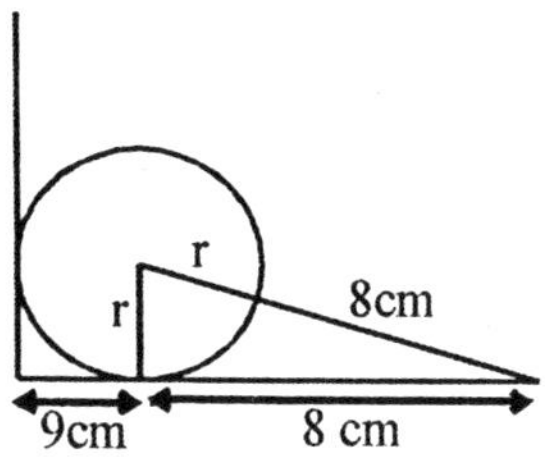

$$(r+8)^2 = r^2 + (4.5+8.0)^2$$

$$r^2 + 16r + 64 = r^2 + \frac{625}{4}$$

$$16r = \frac{625}{4} - 64$$

$$= \frac{625-256}{4} = \frac{369}{4}$$

$$r = \frac{369}{64} = 5.675 \text{ cm}$$

Ex. 13 Given: (1) ABCD is a rectangle.

(2) AE = 2BF

(3) CD = 2AD

(4) Quad DEFC = 29cm^2

Fine the value of AE

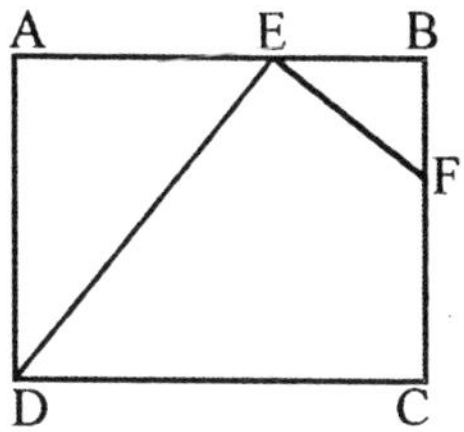

$$50 = 29 + \frac{10x}{2} + \frac{10x - 2x^2 2}{2}$$

$$21 = 5x + 5x - x^2$$

$\therefore$ $$x^2 - 10x + 21 = 0$$

$$x = 7 \text{ or } 3$$

$$AE = 2x = 6$$

Ex. 14 $\frac{x}{4} = \frac{x\sqrt{2}}{x\sqrt{2}+x4}$

$$\frac{x_1}{4} \pm \frac{x_1 + 4 + 4\sqrt{2}}{4\sqrt{2}}$$

$$x(x\sqrt{2} + x + 4) = 4 - x\sqrt{2}$$

$$\sqrt{2}\, x_1 = x_1 4$$

$$(\sqrt{2} + 1)$$

$$x^2(\sqrt{2}+1) = 4\sqrt{2} - 4x$$

$$\left(\sqrt{2}-1\right)x_1 = 4\left(\sqrt{2}+1\right)$$

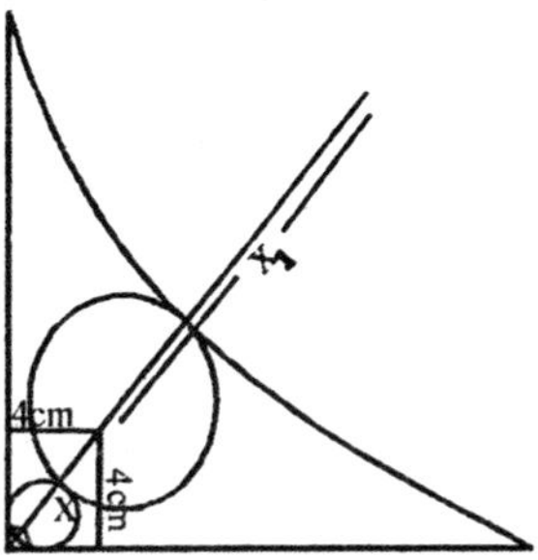

$$x = \frac{4\left(\sqrt{2}-1\right)}{\left(\sqrt{2}+1\right)}$$

$$x_1 = 4\left(\sqrt{2}+1\right)^2$$

$$=4\left(\sqrt{2}+1\right)^2$$

$$=4\,(2+1-2\sqrt{2})$$

$$=12-8\sqrt{2}$$

$$=4\,(3+2\sqrt{2})$$

$$=12+8\sqrt{2}$$

Ex. 15 (1) Draw AP || CD

∴ Arc AC = Arc PD

Arc AC + Arc BD = Arc PD + Arc BD

= Arc BP

Arc BP is a semi ⊙ because it subtends a right ∠ at the circumference

∴ Arc AC + Arc BD is semi ⊙

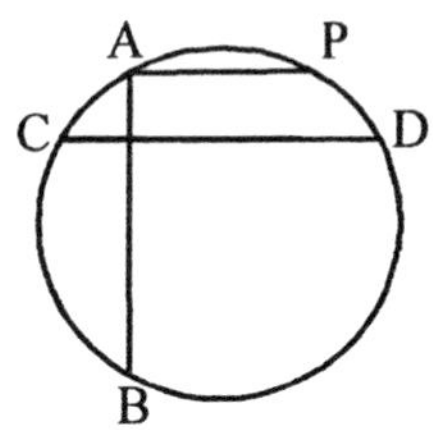

Similarly

If Arc AC + Arc BD is semi-circle

Arc AC + Arc BD = Arc PD + Arc BD

= Arc BP

∴ It subtends a right

$\angle BAP = 90^\circ$

Arc $AB \perp CD$.

Ex. 16 In a right triangle the circum-radius = $\frac{c}{2}$

The radius of the in-circle = $\frac{\Delta}{S}$

$$\therefore \quad \frac{c}{2}+\frac{\Delta}{S}$$

$$=\frac{c}{2}+\frac{ab}{2s}=\frac{c}{2}+\frac{ab}{(a+b+c)} \qquad =\frac{c(c+a+b)+2ab}{2(a+b+c)}$$

$$=\frac{c^2+c(a+b)+2ab}{2(a+b+c)}=\frac{(a^2+b^2)+c(a+b)+2ab}{2(a+b+c)}$$

$$=\frac{(a+b)^2+c(a+b)}{2(a+b+c)}=\frac{(a+b)(a+b+c)}{2(a+b+c)}=\frac{a+b}{2}$$

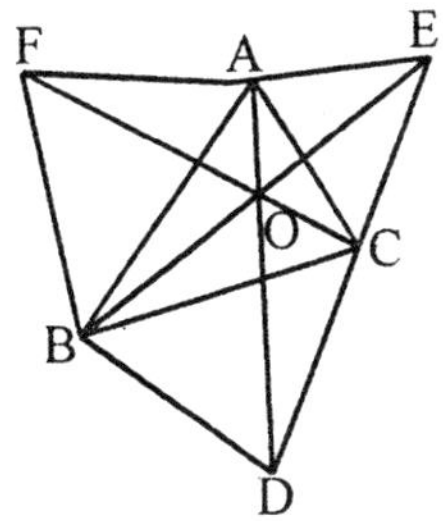

Ex. 18 ABC is a Δ on AB, BC and CA. 3 equilaterals are constructed. They are DBC, ACE and AFB.

We have to prove that AD and BE and CF are concurrent. Let AD and BE meet in O. We have to show that F, O, C are in the same st. line.

Now $\qquad \Delta BCE \equiv \Delta ADC$

$\therefore \qquad \angle OBC = \angle ODC$

$\therefore$ OBDC is a cyclic quadrilateral

$\therefore \qquad \angle BOC + \angle BDC = 180°$

Since $\qquad \angle BDC = 60°$

$\qquad \angle BOC = 120°$

Similarly O AEC is cyclic

$\therefore$ $\angle AEC + \angle AOC = 180°$

Sin $\angle AEC = 60°$

$\angle AOC = 120°$

Because $\angle AOC = \angle BOC = 120°$

$\therefore$ $\angle AOB = 120°$

$\therefore$ AFBO is cyclic

$\angle AOF = \angle ABF$

$\therefore$ $\angle AOF = 60°$

$\therefore$ $\angle AOF + \angle AOC = 180°$.

$\therefore$ C, O, F are in one st. line.

Ex. 19 1. A, B, C, D are the centres of the circles.

2. AB, BC, CD and DA are straight lines.

3. P, Q, R, S is a quadrilateral

4. The angles of P, Q, R, S are

$$\frac{A+B}{2}, \frac{B+C}{2}, \frac{C+D}{2}, \frac{D+A}{2}$$

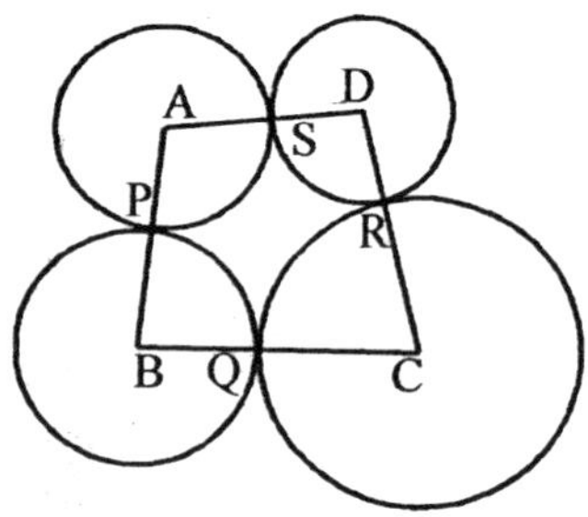

5. $\therefore$ PQRS is cyclic.

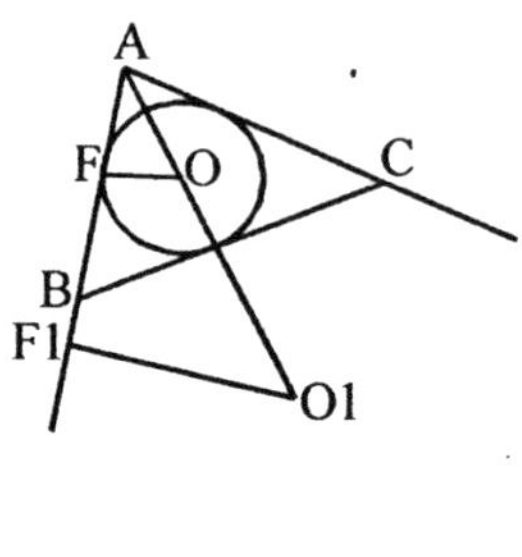

Ex. 20 BAC = 90°

OF = radius of inscribed circle.

O_1F_1 = radius of escribed circle.

OF = AF

$O_1F_1 = AF_1$

AOO_1 bisects $\angle A$ which is 90°

$O_1F_1 - OF = AF_1 - AF$

$= FF_1 = BC$

Ex. 21 $\angle APB + \angle CPD = 180^\circ$

Draw $DE \parallel AP$

and $CE \parallel BP$

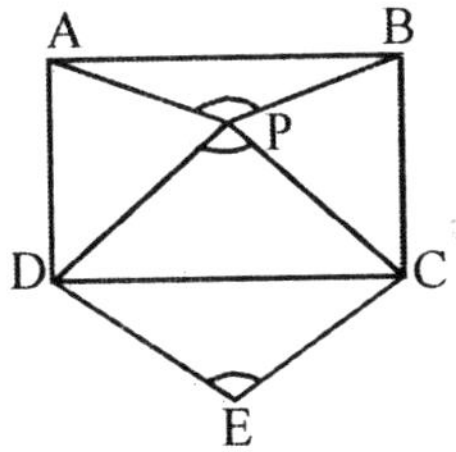

$\angle APB = \angle DEC$

$\angle DEC + \angle CPD = 180^\circ$

$\therefore$ PCED is cyclic

$\therefore$ $\angle PDC = \angle PEC$

But $\angle PEC = \angle PBC$

$\angle PDC = \angle PBC$

Ex. 22 Similar to Ex. 21

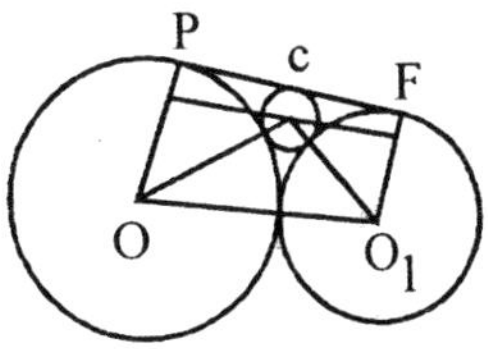

Ex. 25 $PF = \sqrt{(a+b)^2 - (a-b)^2} = 2\sqrt{ab}$

$OP = \sqrt{(b+c)^2 - (b-c)^2} = 2\sqrt{bc}$

$OF_1 = \sqrt{(a+c)^2 - (a-c)^2} = 2\sqrt{ac}$

$\sqrt{ab} = \sqrt{bc} + \sqrt{ac}$

$\frac{1}{\sqrt{c}} = \frac{1}{\sqrt{a}} + \frac{1}{\sqrt{b}}$

Ex. 27 $\angle FCP = \angle CPA - \angle CFP$

$= \angle TAP - \angle CFP$

$= \angle FDA - \angle CFP$

$= \angle FDA - \angle CDA$

$= \angle FDC$

$\therefore$ FC = FD

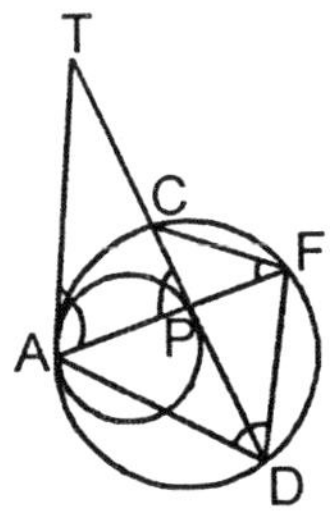

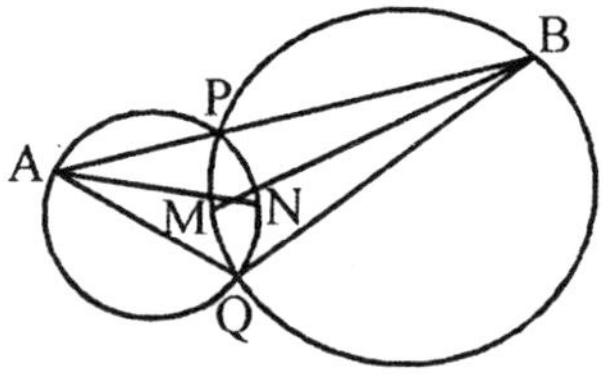

Ex. 28 The bisectors of A and B pass through the mid pts of 2 arc PF one in each of the two circles. They meet at an angle = 90 + $\frac{\angle AQB}{2}$ Therefore the locus is a circle making an angle = 90 + $\frac{x}{2}$ over MN, M and N being the centre of arcs in each circle and x = $\angle AQB$.

Ex. 29 1. ABCD is a rectangle side AB = side CD

2. Opp. sides of a rhombus are parallel
3. Therefore AP || CR, BP || DR
4. $\therefore \Delta ABP \equiv \Delta CDR$
5. AP = CR and PB = RD
6. $\therefore$ AS = CQ

DS = BQ

In rhombus, All the sides are equal

$\therefore$ BQ = DR and AS = CQ

Hence segment APB = segment CQD.

Also segment ADS = segment BCQ.

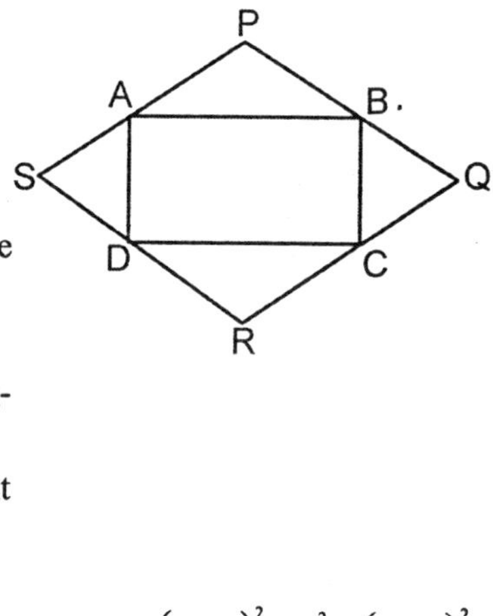

Ex. 30 $\left(r+\frac{x}{4}\right)^2 = \frac{x^2}{16} + \left(\frac{x}{2}-r\right)^2$

$\frac{rx}{2} + xr = \frac{x^2}{4}$

$\frac{3rx}{2} = \frac{x^2}{4}$

6r = x

r = $\frac{x}{6} = \frac{AB}{6}$

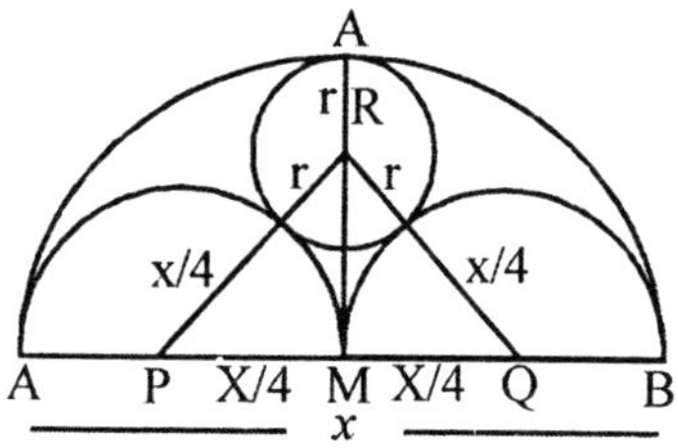

Ex. 31 $RS^2 - AB^2$

$= (RS + AB)(RS - AB)$

$= (2RA + AB + AB)(2RA)$

$= 2(RA + AB)(2RA)$

$= 4RB.RA$

$= PP_1^2$

$\therefore RS^2 = PP_1^2 + AB^2$

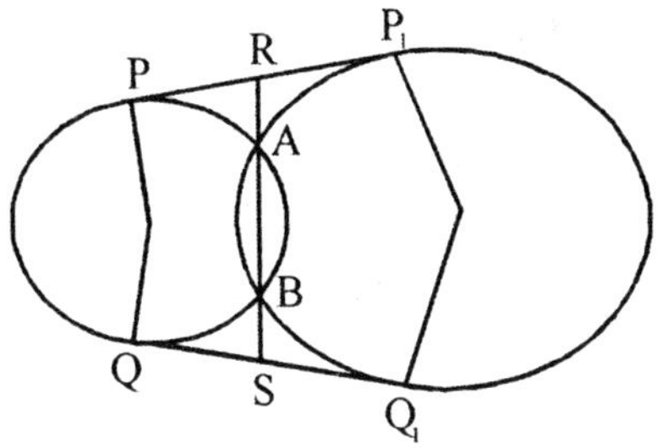

Ex. 32 Make AD = AC

$AB + AC = BD$

$\angle BDC = \frac{A}{2}$

$\angle P = \angle A$

$\angle PFC = \angle BDC = \frac{A}{2}$

BDFC is cyclic

BD is diameter

$BD > BF > CP + PC$

$BD + BC > BP + PC + BC$

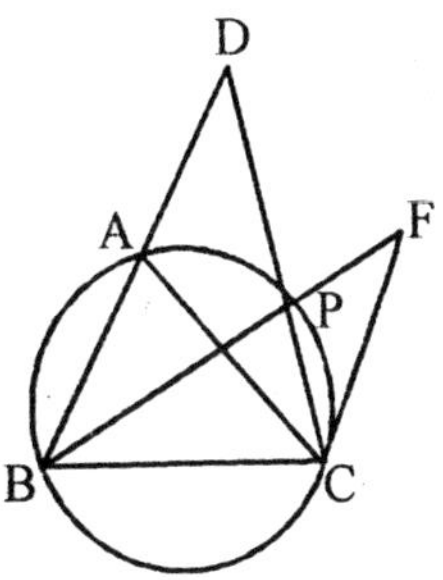

Ex. 35 $\theta = \alpha + \gamma$

But $\gamma = \beta + D$

$\therefore \theta = \alpha + \beta + D$ ...(1)

$m = \theta + \beta$

$\theta = m - \beta$

$m = B - \alpha$

$= B - \alpha - \beta$ ...(2)

Add (1) and (2) $2\theta = B + D$

$= \frac{180}{2} = 90°$

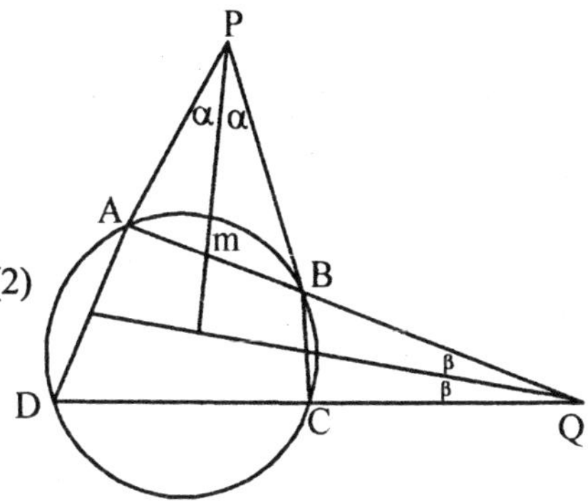

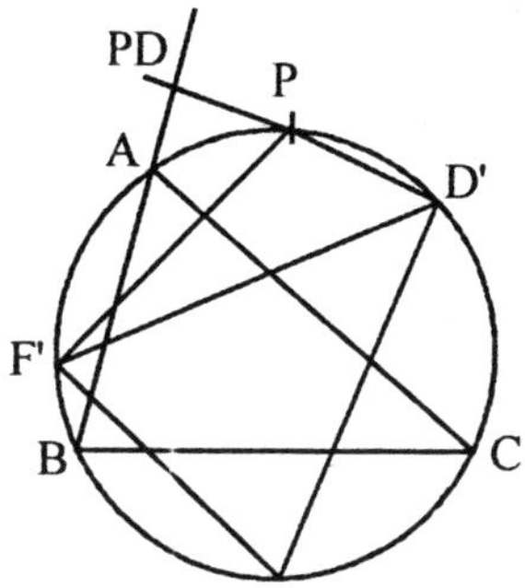

Ex. 36 (1) $\angle ABC = \angle E'PD' =$

$\angle E'F'D' = \angle B$

$\angle BAC = \angle DPF' \div \angle C = \angle D'E'F' = \angle A$

The third angle = $\angle C$ automatically.

Ex. 37 Distance between centres

$= \sqrt{r_1^2 + r_2^2}$

(Direct or common tangent)^2

$= (r_1^2 + r_2^2) - (r_1^2 - 2r_1r_2 + r_2^2) = 2r_1r_2$

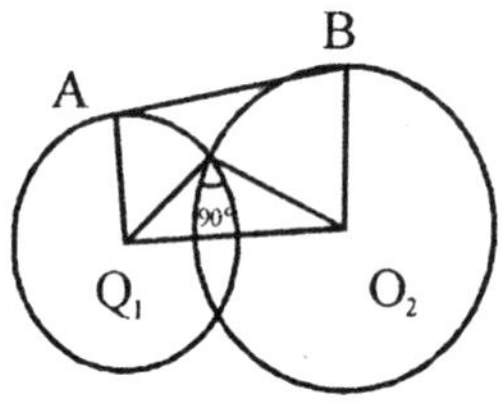

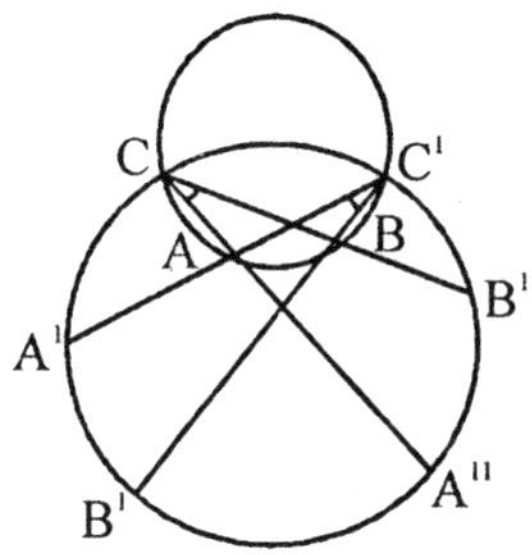

Ex. 38 Circle is given and the two points through which it passes are A and B. On AB draw a circle which will enclose the given angle. It cuts the given circle in 2 points C and C^1. The 2Δs are CAB and C^{1}AB.

Ex. 40 (1) XYZ is pedal Δ of DEF which is the incircle of ΔABC.

(2) AB is tangential to DEF

(3) $\therefore \angle AFE = \angle EDF$

(4) $DX \perp FE, EY \perp DF$

(5) $\therefore$ XYDF is cyclic

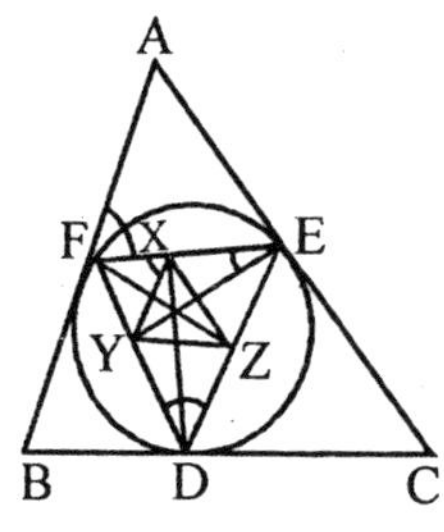

(6) ∠FXY = ∠EDF

(7) ∴ from (3) + (6) ∠FXY = ∠EDF

Ex. 41 First find out the locus of the centres of these circles. It will be two perpendicular lines AO and BO. O is at distance equal to radius of these circles. From O draw a circle $\frac{1}{4}$ radius of these circles. That will be the required locus.

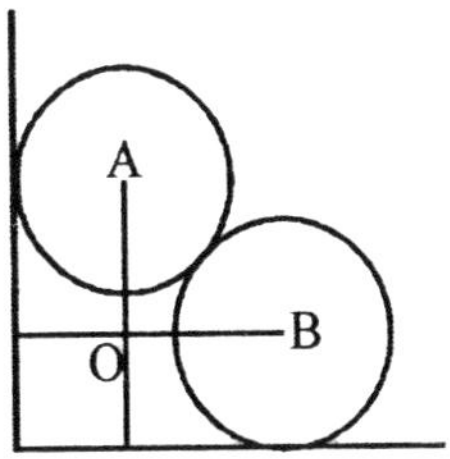

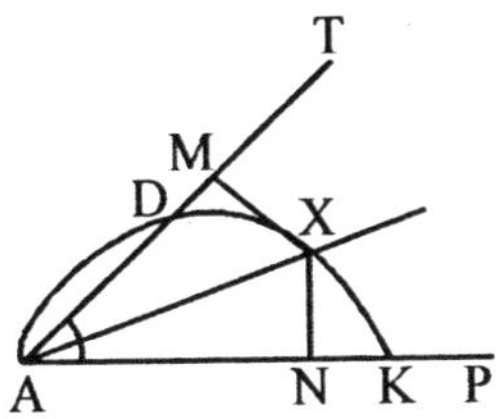

Ex. 42 (1) TAP is any fixed angle. X is a point on the bisector of ∠TAP. Through A and X any circle is drawn cutting AT and AP in D and K. We have to prove that either (AD + AK) or (AD-AK) is fixed.

Proof: From X drop XM and XN perpendiculars to AT and AP respectively.

Then ΔDMX ≡ ΔKMN

∴ DM = NK

AD + AK = AM - MD + AN - NK

But MD = NK and AM = AN

∴ AD + AK = AN + AN = 2AN

This is fixed.

(2) In the second case D and K is on lower side of AX. ∠ADX = ∠AKX

ΔDMX = ΔKNX

∴ DM = NK

∴ AK = AN + NK

= AN + MD

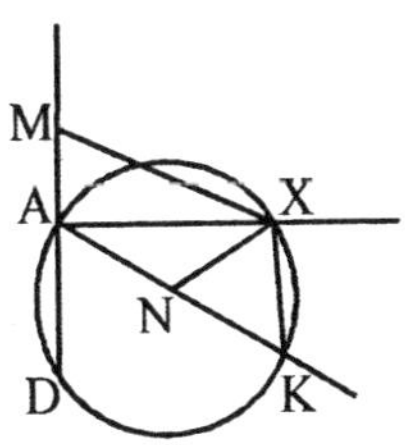

$$= AM + AM + AD$$

$$\therefore \quad AK - AD = 2AM$$

Ex. 43 M and N are centres of 2 circles. N is located on the circle whose centre is M. RF and TS are 2 direct common tangents to these 2 circles. They meet at point A. M and N are on the birector of $\angle FAS$ or $\angle RAT$. Join MT.

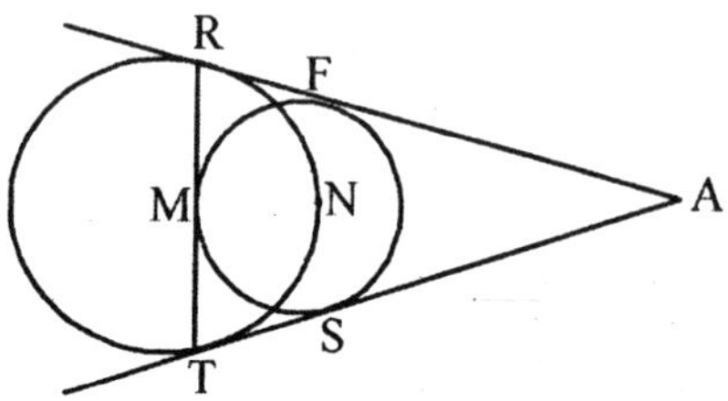

$\angle NTS = \angle NRT$ (Th 59).

Because $RN = TN$ $\angle NRT = \angle NTR$

$\therefore$ $\angle NTS = \angle NTR$

$\therefore$ NT is the birector of $\angle RTA$

$\therefore$ N is the in centre of ΔRTA.

$\therefore$ RT is a tangent to the circle with centre N.

Ex. 44 (1) On the left AE is drawn so that $\angle DAE = \angle ADC$ it will be parallel.

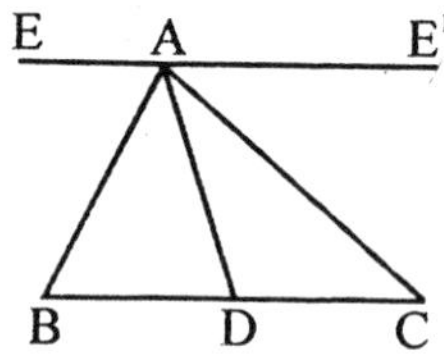

(2) On the right side $\angle CAE' = \angle ABC$ because $\angle ADC = \angle DAE$ then cancelling $\angle BAD$

$$\angle ABC = \angle CAE'$$

$\therefore$ It is tangent.

Ex. 45 From 24 of Exercise N

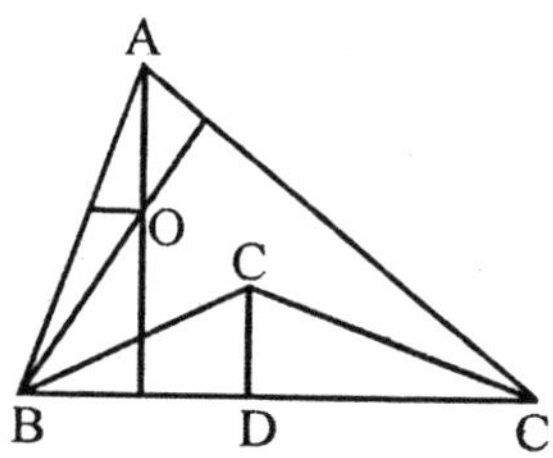

$OA = 2CD$

$a = 2BD$

$BD^2 + CD^2 = r^2$

$$\frac{a^2}{4} + \frac{OA^2}{4} = r^2$$

$$a^2 + OA^2 = 4r^2$$

$$BC^2 + OA^2 = 4r^2$$

Ex. 46 $\frac{x}{5} = \frac{2x}{\sqrt{4x^2 - 576}}$

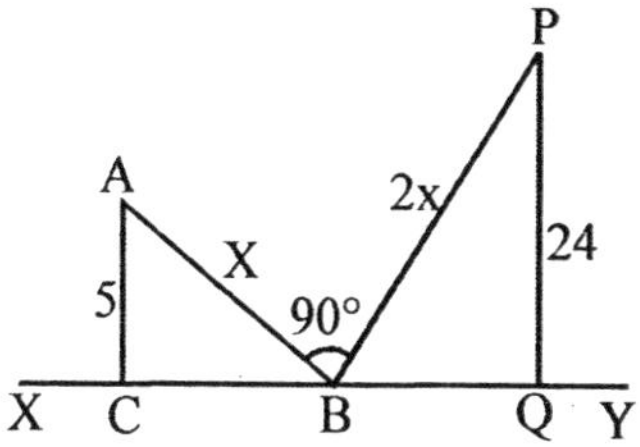

$X\sqrt{4x^2-576} = 10x$

$4x^2 - 576 = 100$

$4x^2 = 676$

$x^2 = 169$

$x = 13$ cm

BC = 12 cm,

AP = $13\sqrt{5}$cm, QC = 22 cms.

Ex. 47 (a) Circumcentre is fixed.

(b) For centroid take a distance = $\frac{1}{3}$ ht and draw a circle about the centre of the base and terminated by the base.

Ex. 48 Bisect $\angle A$ and $\angle B$

Let the bisectors meet in M. MP, ME and MF are perpendiculars to AB, AD and BC respectively

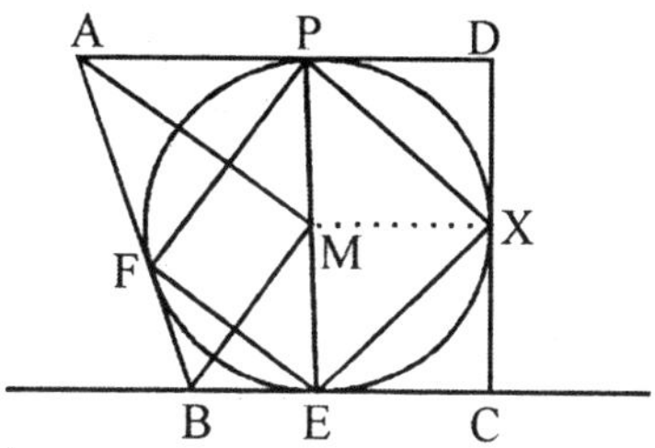

AP = AF

BF = BE

AP + BE = AF + BF

But we are given that

AD + BC = AB + DC

$\therefore$ PD + EC = DC

From DC take CE = CX

$\therefore$ PD = DX

Join MX

$$\angle PFE = \frac{A+B}{2}$$

$$\angle PXE = \frac{C+D}{2}$$

$$\angle PFE + \angle PXE = \frac{A+B+C+D}{2} = 180°s$$

$\therefore$ PFEX is cyclic

Δ PFE has radius = MF

MP = MX

PD = DX

$\therefore$ Δ MPD = Δ MXD

$\angle MPD = \angle MXD = 90°$

$\therefore$ CD is a tangent to PFEX.

Ex. 49 Draw AC common tangent to the 2 circles

$\angle PCF = \angle PCA + \angle FCA$

O_1P and O_1C are both $\perp$ to

$\angle PCA = 90 - x$

$\angle FCA = 90 - y$

$\angle PCF = 180 - (\angle x + \angle y)$

$= 180 - \dfrac{\angle PAF}{2}$

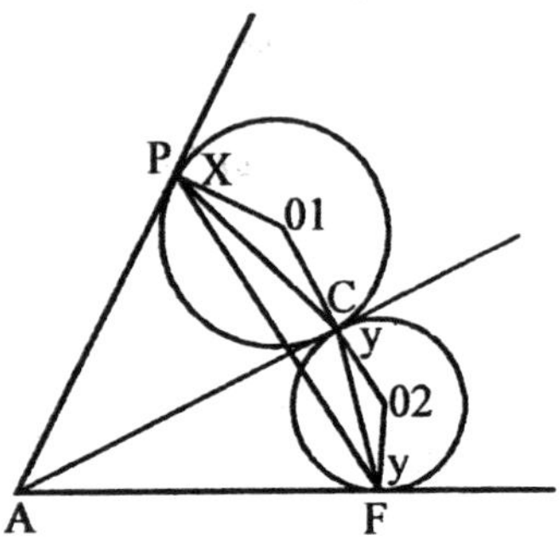

Ex. 52 S and T are mid-points of AB and CD

P and F are mid-points of AC and BD

$SP = \frac{1}{2}$ BC and $FT = \frac{1}{2}$ BC

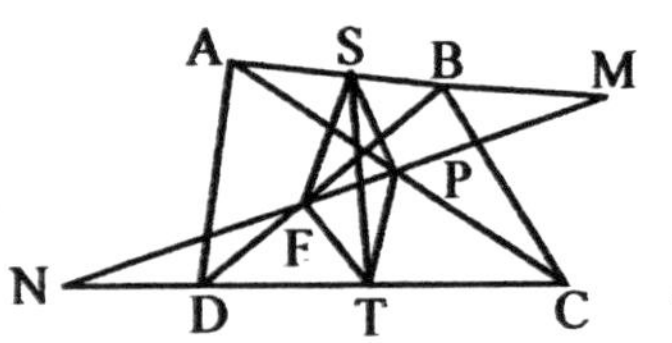

$\therefore$ SP = FT

Similarly, $PT = \frac{1}{2}$ AD

and $SF = \frac{1}{2}$ BC

$\therefore$ PT = SF

But AD = BC

$\therefore$ SP = PT = TF = FS

$\therefore$ SPTF is a rhombus

We have to prove $\angle AMP = \angle FNT$

$$\angle AMP = \angle ASP - \angle SPM$$

$$\angle FNT = \angle FTC - \angle TFN$$

But $\angle ASP = \angle FTC$

$$\angle SPF = \angle TFN$$

$\therefore$ $\angle AMP = \angle FNT$

Ex. 54 $AB = x$

$BC = 2x$

$AC = \sqrt{5}\,x$

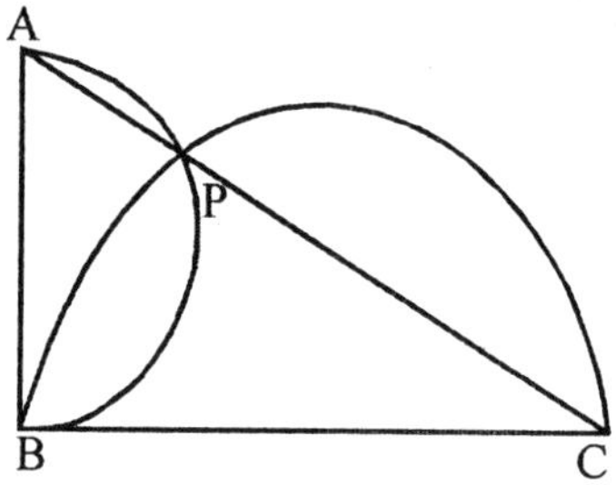

Area ABC

$$\cdot = \frac{1}{2} \odot x \times 2x = x^2$$

$$x^2 = \frac{1}{2}\sqrt{5}x \times BD$$

$$BD = \frac{2x}{\sqrt{5}} = \frac{2}{5}\sqrt{5}x = \frac{2}{5} \times \text{diagonal}$$

Ex. 55 (1) PQRS is drawn in a circle

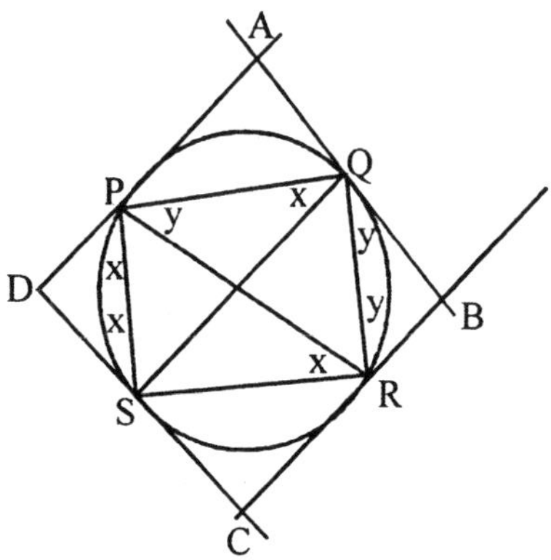

(2) Tangents are drawn to the circle at P, Q, R, S

(3) They meet in A, B, C, D.

(4) $\angle DPS = \angle PRS = x$

$\angle DSP = \angle PQS = x$

(5) Similarly

$\angle BQR = \angle BRQ = \angle 2PR$

$= \angle QSR = y$ in fig.

(6) Because $\angle D \div \angle B = 180°$

$2x + 2y = 180°$

$x + y = 90°$

Hence diagonals are at 90°

Ex. 56

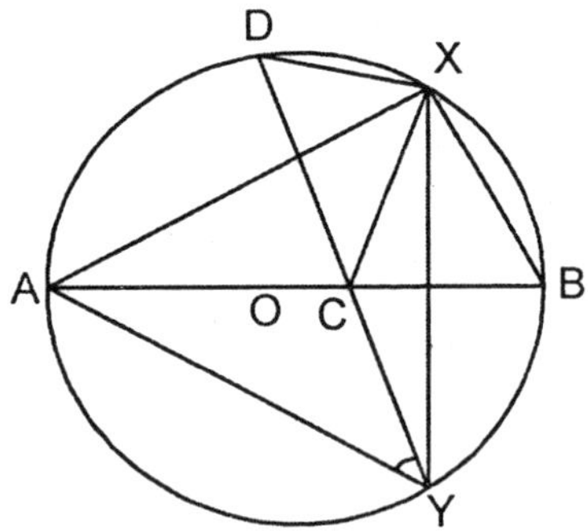

$\angle DXA = \angle DYA$

$\angle CXA = \angle DYA$

$\therefore$ $\angle DXA = \angle CXA$

XB is at right $\angle s$ to AX

$\therefore$ XB is the exter not bisector

Ex. 57 (1) Let equal circles have centres at M and N respectively.

(2) Let T be the centre of double the circle.

(3) BT = 2R

(4) M, A, N are in one st. line.

(5) BA meets the circle in F.

(6) TF = 2R = 2MA.

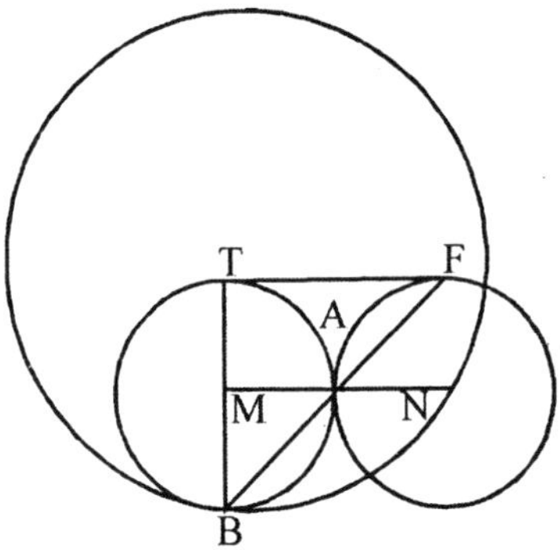

Ex. 58 $\angle MOA = 90 - \angle ABD$

$\angle ABD = \angle ACD$

$\therefore$ $\angle MOA = 90 - \angle ABD$

$\therefore$ $\angle MOA + \angle ACD = 90°$

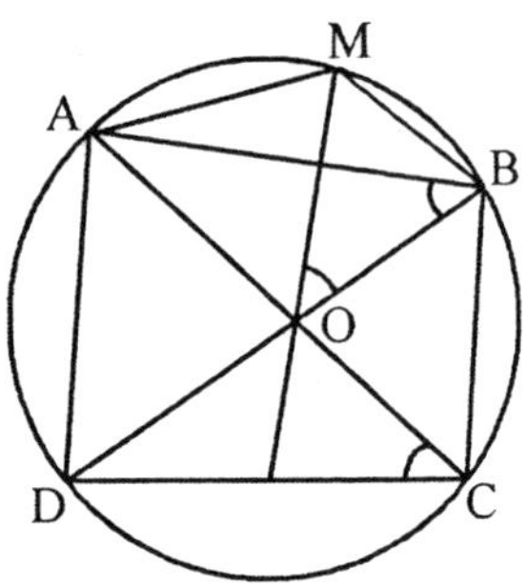

Ex. 59 Let ABCD be a square inside a circle. And let PQRS be the square inside the smaller segment of AB. be = 2x. Let the side of the small square = 2x. Let the radius of the circle = 2r. Then AB = $\frac{2r}{\sqrt{2}}=\sqrt{2}r$. The sides of the Δ OSU will be OS = r, SU = x, OU= $\left(\frac{r}{\sqrt{2}}+2x\right)$

Then by Pythagoras theorem,

$$r^2=\left(\frac{r}{\sqrt{2}}+2x\right)^2+x^2$$

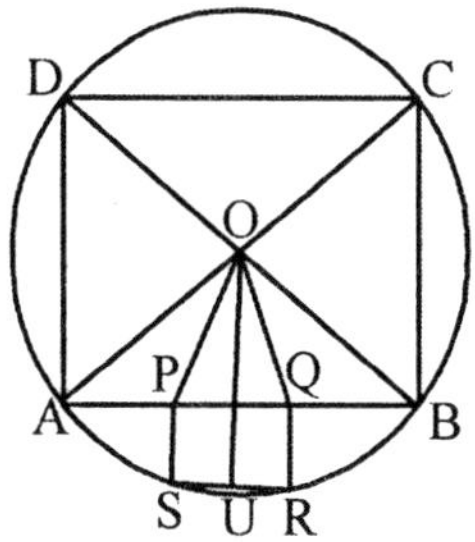

$$r^2=\frac{r^2}{2}+4x^2+4x.\frac{r}{\sqrt{2}}+x^2$$

$$0=\frac{r^2}{2}+2\sqrt{2}xr+5x^2$$

$$0=10x^2+4\sqrt{2}xr-r^2$$

$$0=\left(10x-\sqrt{2}r\right)\left(x+\frac{r}{\sqrt{2}}\right)$$

$$10x=\sqrt{2}r$$

$$2x=\frac{\sqrt{2}r}{5}=\frac{AB}{5}$$

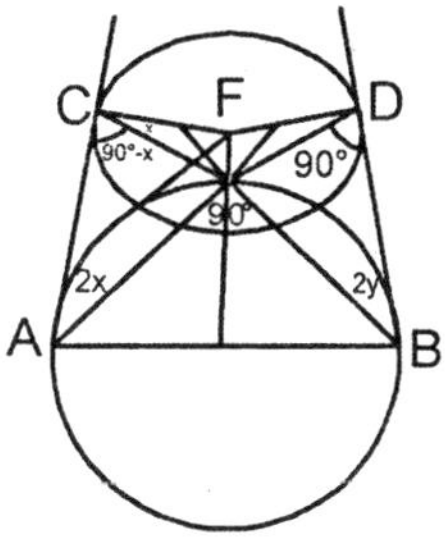

Ex. 63 To prove that

$$AC^2+BD^2=AB^2$$

$$AC=AE \text{ and } BD=BE$$

BE is produced to EF

$$\angle DEF=90+y$$

$$\angle FEC=x$$

$\therefore$ CED $=90+x+y$

$$\angle CEA+\angle FED+\angle DEF=270^\circ$$

$\therefore$ $\angle AEB=90^\circ$

$\therefore$ E is on the circle

$\therefore$ $AB^2 = AE^2 + EB^2$

$= AC^2 + DE^2$

Exercise I

Ex. 11 $AB^2 + BC^2 = 2AM^2 + 2BM^2$

If M is point where AM = MB

$AD^2 + CD^2 = 2AM^2 + 2DM^2$

Adding,

$AB^2 + BC^2 + CD^2 + AD^2 = 4AM^2 + (2BM^2 + DM)$

If N is point of bisector of BD

$= AC^2 + 2(2NM^2 + 2DN^2)$

$= AC^2 + 4DN^2 + 4NM^2$

$= AC^2 + BD^2 + 4MN^2$

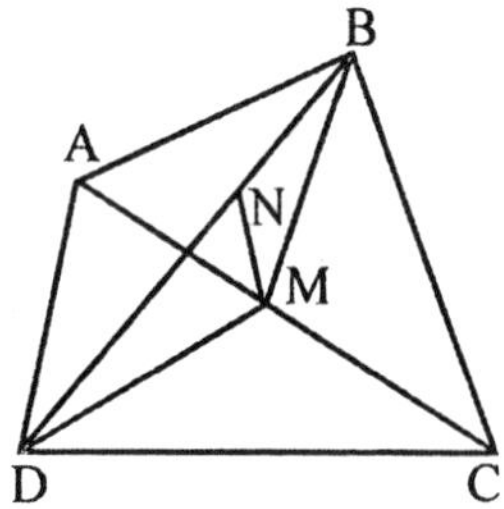

Ex. 13 $mBX = nCX$

$AB^2 = AX^2 + BX^2 - 2BX.XD$

$AC^2 = AX^2 + CX^2 + 2XC.XD$

$mAB^2 = mAX^2 + mBX^2 - 2mBX.XD$

$nAC^2 = nAX^2 + nCX^2 + 2nCX.XD$

$m.AB^2 + n.AC^2 = (m + n)AX^2 + mBX^2 + nCX^2$

The last term cancels as mBx = nCX.

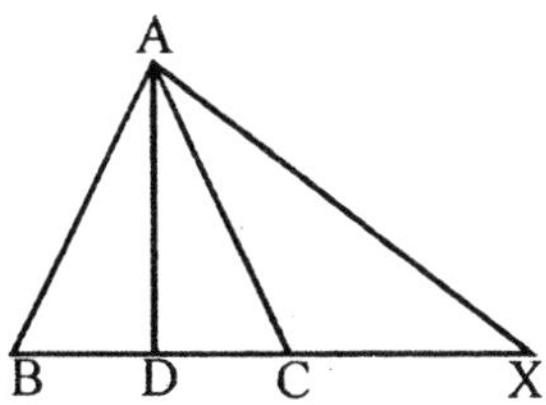

Ex. 14 $mBX = nCX$

$AB^2 = AX^2 + BX^2 - 2BX.XD$

$AC^2 = AX^2 + CX^2 + 2XC.XD$

$mAB^2 = mAX^2 + mBX^2$

$-2m. BX.XD$

$nAC^2 = nAX^2 + nCX^2$

$+ 2n. CX.XD$

$m(AB^2) - nAC^2 = (m-n)AX^2 + nBX^2 - nAX^2$

Ex. 15 $AB^2 + AC^2 = 2AD^2 + 2BD^2$

$AC^2 + BC^2 = 2CF^2 + 2AF^2$

$BC^2 + AB^2 = 2BE^2 + 2EC^2$

Adding the 3 equations

$2(AB^2 + BC^2 + CA^2) = 2(AD^2 + BE^2 + CF^2) + 2(BD^2 + AF^2 + EC^2)$

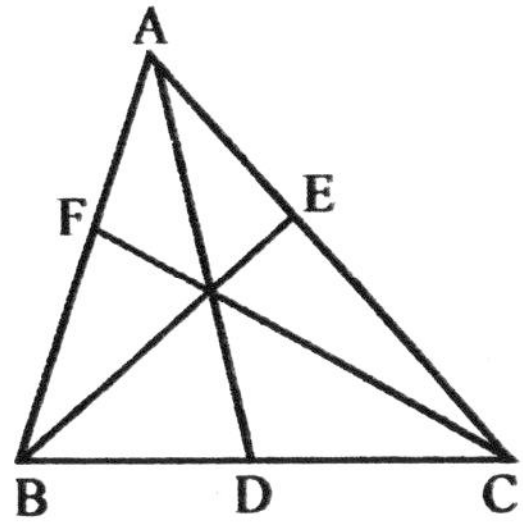

$$2(AB^2 + BC^2 + CA^2) = 2(AD^2 + BE^2 + CF^2) + 2\left(\frac{BC^2}{4} + \frac{AB^2}{4} + \frac{CA^2}{4}\right)$$

$$2(AB^2 + BC^2 + CA^2) - \left(\frac{BC^2 + AB^2 + CA^2}{2}\right) = 2(AD^2 + BE^2 + CF^2)$$

$3(AB^2 + BC^2 + CA^2) = 4(AD^2 + BE^2 + CA^2)$

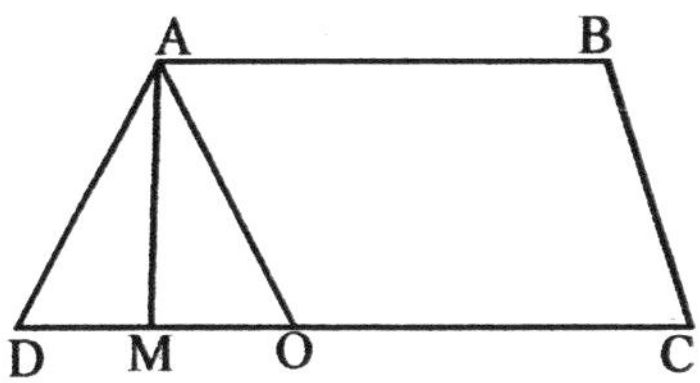

Ex. 16 $AC^2 = AM^2 + MC^2$

$AD^2 = AM^2 + MD^2$

$AC^2 - AD^2 = MC^2 - MD^2$

$= (MC + MD)(MC - MD)$

$= DC.D'C$

$= DC.AB$

Ex. 17 $AF^2 + FB^2 - 2AF.FB$

$= (AF - FB)^2$

$= [(AP + PF) - (PB - PF)]^2$

$AP = PB$

$= 2PF^2 = 4PF^2$

$AF^2 + FB^2 = 2FA.FB + 4PF^2$

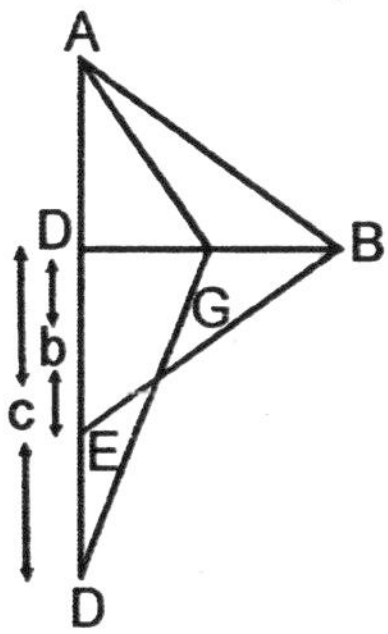

Ex. 19 $c^2 = a^2 + b^2 + 2a.Dc$

$$DC = \frac{c^2 - a^2 - b^2}{2a}$$

$$CF = \sqrt{\left(\frac{c^2 - a^2 - b^2}{2a}\right)^2 + c^2}$$

$$CF = \sqrt{\frac{c^4+a^4+b^4-2c^2a^2+2a^2b^2-2b^2c^2+4a^2c^2}{4a^2}}$$

$$BE = \sqrt{DE^2+DB^2}$$

$$= \sqrt{b^2+\left(\frac{c^2-a^2-b^2}{2a}+a\right)^2}$$

$$= \sqrt{b^2+\frac{c^4+a^4+b^4+2c^2a^2-2b^2c^2-2b^2a^2}{4a^2}}$$

$$= \sqrt{\frac{4a^2b^2+c^4+a^4+b^4+2c^2a^2-2b^2c^2-2a^2b^2}{4a^2}}$$

$$= \sqrt{\frac{c^4+a^4+b^4+2c^2a^2-2b^2c^2+2a^2b^2}{4a^2}}$$

$\therefore$ $BE = CF$

Ex. 20 $AC^2 = AD^2 + DC^2 - 2DC.DF$

$BD^2 = BC^2 + DC^2 - 2DC.GC$

$AC^2 + BD^2 = AD^2 + BC^2 + 2DC^2 -$
$2DC.DF - 2DC.GC$

$= AD^2 + BC^2 + 2DC(DC - DF - GC)$

$= AD^2 + BC^2 + 2DC.FG$

$= AD^2 + BC^2 + 2DC.FG$

$= AD^2 + BC^2 + OC.AB$

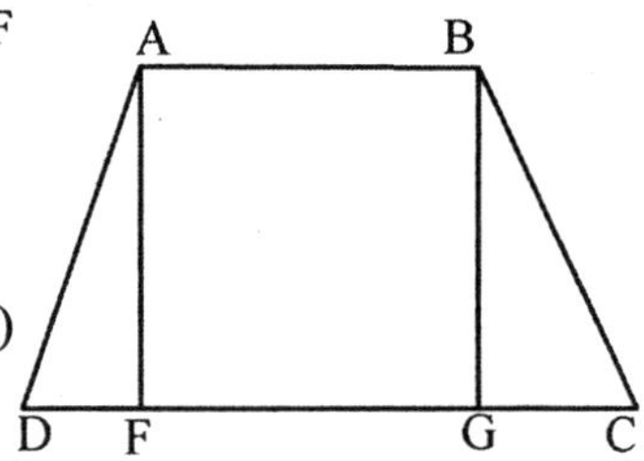

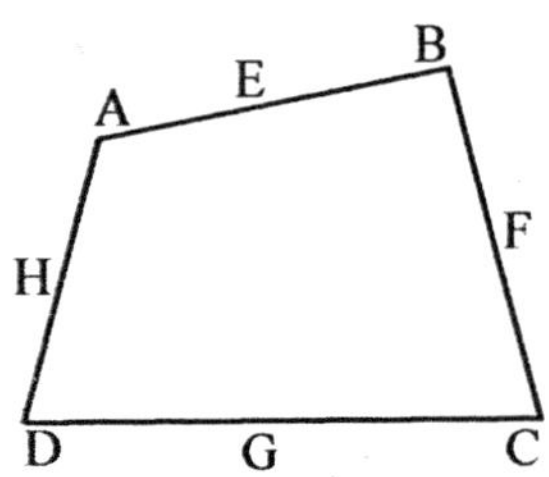

Ex. 21 $AC^2 + BD^2 = 4HG^2 + 4GF^2$

$= 2HG^2 + 4GF^2 + 2EF^2$

$= 2HG^2 + 2GF^2 + 2GF^2 + 2EF^2$

Ex. 22 $DB^2 + DC^2 = 2BX^2 + 2DX^2$

$DB^2 + DC^2 = 2BX^2 + 8XG^2$

$AB^2 + AC^2 = 2BX^2 + 2AX^2$

$= 2BX^2 + 18XG^2$

$DB^2 + DC^2 - 8XG^2 = 2BX^2$

$AB^2 + AC^2 - 18XG^2 = 2BX^2$

$AB^2 + AC^2 - 10XG^2 = DB^2 + DC^2$

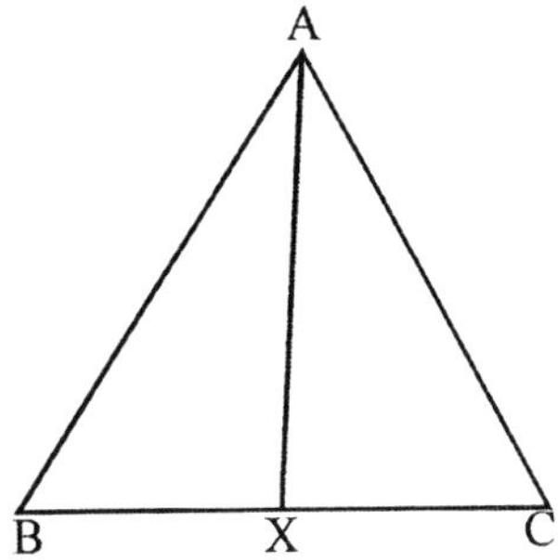

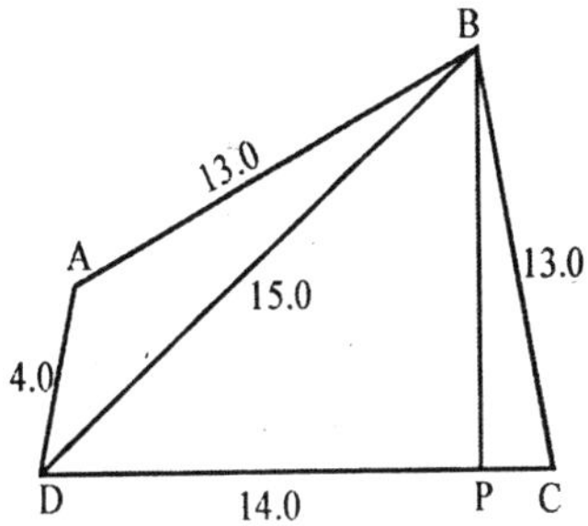

Ex. 24 $15^2 = 13^2 + 14^2 - 2.14.PC$

$-140 = -28.PC$

$5 = PC$

$12 = BP$

$\text{Cos } BCP = \frac{5}{13}$

$15^2 = 13^2 + 14^2 + 2.4.13$ on DAR

Cos DAQ = Cos BCP

$15^2 = 13^2 + 14^2 + 2.4.AQ$

$\frac{40}{104}$

Ex. 27 $AY^2 = AD^2 + DY^2$

$= AD^2 + (BD - BY)^2$

$= AD^2 + BD^2 - 2BD.BY + BY^2$

$= AB^2 - BY(2BD - BY)$

$= AB^2 - BYYC$

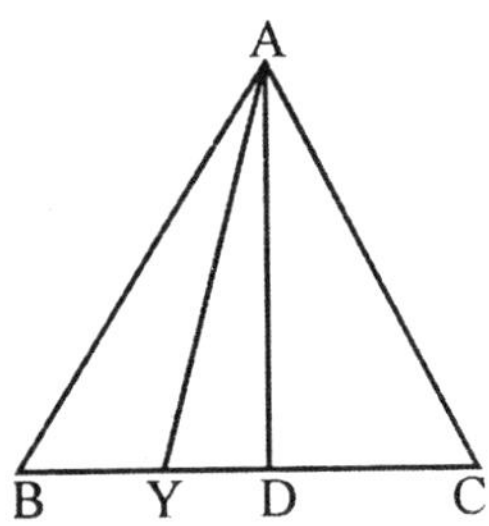

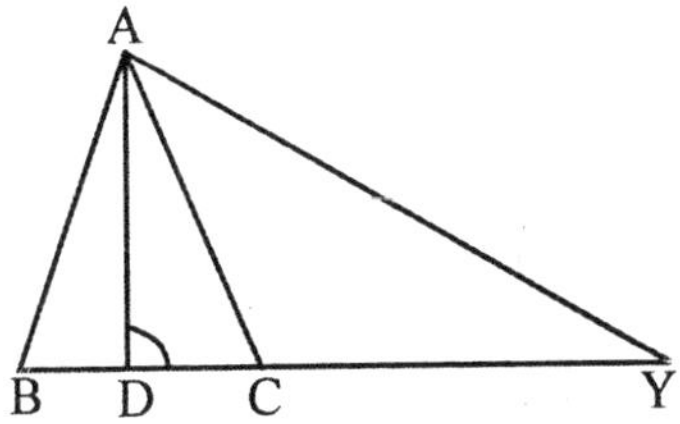

$AY^2 = AD^2 + DY^2$

$= AD^2 + DC^2 + CY^2 + 2DC.CY$

$= AC^2 + CY(CY + 2DC)$

$= AC^2 + CY.BY$

Ex. 26 $AB^2 = AY^2 + YB^2 + 2AY.YB$

$AY^2 + YB^2 = AB^2 - 2AY.YB$

$= AB^2 - 2(AX + XY)(AX - AY)$

$= 4AX^2 - 2(AX^2 - XY^2)$

$= 2AX^2 + 2X^2Y^2$

$= 2(AX^2 + XY^2)$

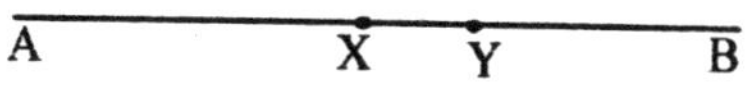

Exercise J

Ex. 8 (a) If it is acute angled Δ then AE, BD and CF are 3 perps.

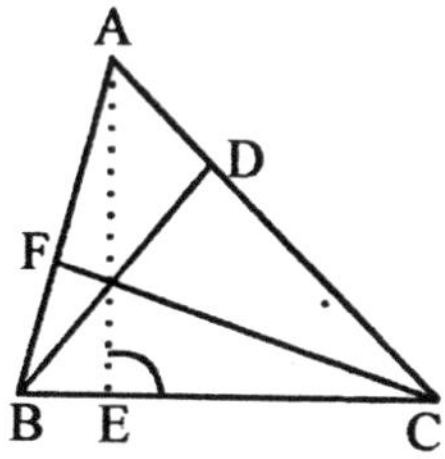

BF.BA = BE.BC

CA.CD = CE.BC

Adding we get,

BF.BA + CA.CA = BC (BE+EC)

= BC.BC

$= BC^2$

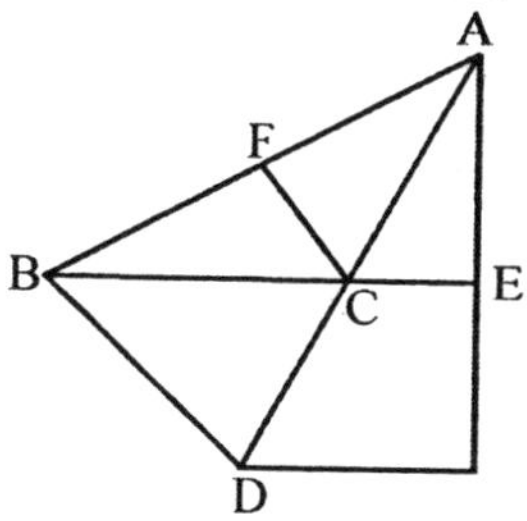

(b) If $\angle C$ is obtuse then $\angle A$ and $\angle B$ are acute then

BF.BA = BC. BE and AC.CD = CE. BC

$\therefore$ BF.BA – AC.CD = BC.(BC) = BC^2

Ex. 10 (A) Problem : M is the centre of a given circle and P is the point from which we have to draw a secant to the circle which will be bisected at the point meeting the circle.

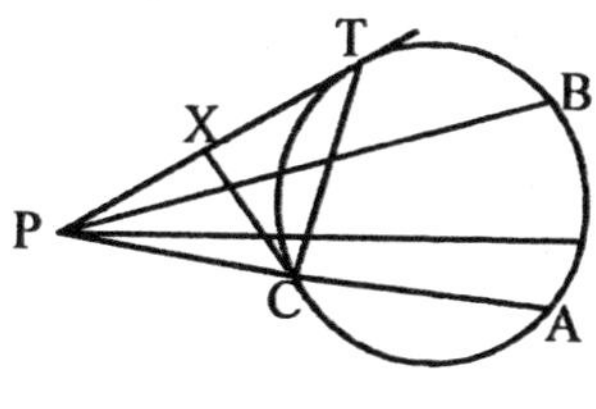

Method : From P draw a tangent PT to the circle On PT draw a semi cricle. Let it meet the Circle in point C. With PC draw an arc of a circle cutting at pt. A. Draw PA to meet the Circle at B. PB is the required secant.

Proof : $PC^2 + PC^2 = PT_2$

$2PC^2 = PT^2$

$2PA.PB = PT^2$

$\therefore$ A is the point where secant is divided in 2 equal parts.

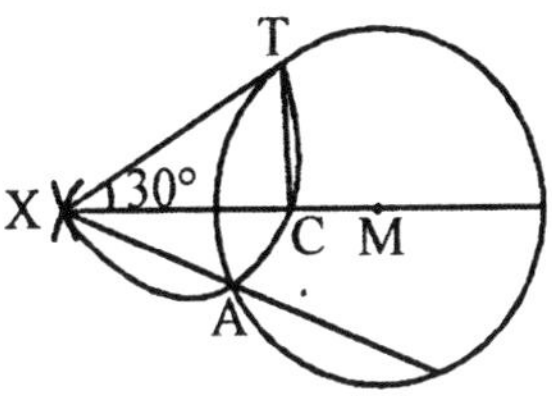

(B) **Problem:** M is the centre of the circle. X is the point from which secant is to be drawn to the circle which will be trisected at its point of meeting the circle From X draw XT tangent to the Circle. We have to draw a rt. angled Δ. with angle = 30°. Let XC T be the Δ. With angle TXC = 30°. With TC cut off the circle at A. A is the point when the secant is all off at 1/3 of its distance.

Construction: From X draw XT a tangent to the circle with M as centre. Construct right tangent. XLP = 30°. With X as centre and TL as radius cut off the circle at A. XAB is the secant required.

Proof: In right angled Δ with one angle = 30° and the other 60° we get a ratio of the side as $1:\sqrt{3}:2$

$\therefore$ The hypotenuse is twice the side of the Δ.

Ex. 11 M is the mid pt. of AB. G and H are the mid points of ED and FC respectively.

$GH^2 = 4GM^2$

$ED^2 = 4EG^2$

$AB^2 = 4AM^2$

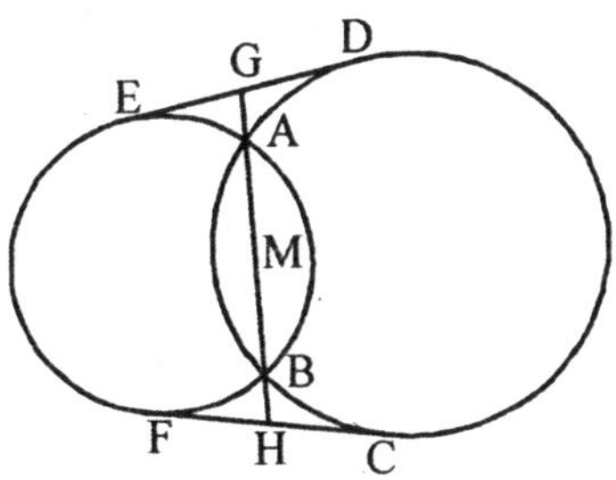

$$EG^2 = GB.GA$$
$$= (GM + AM)(GM - AM)$$
$$= GM^2 - AM^2$$
$$GM^2 = AM^2 + EG^2$$

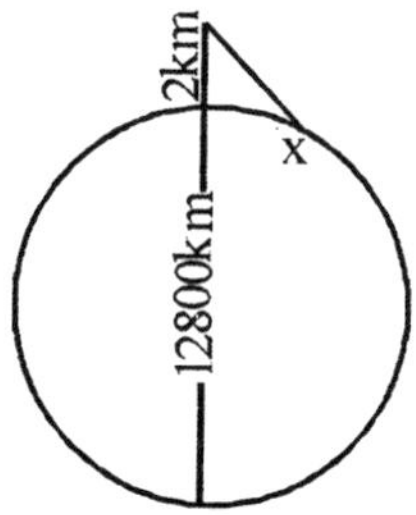

Ex. 12 Height of man = 2 km

Diameter of earth = 12800 km

The distance which he can observe

$$\sqrt{12800002 \times 2}$$
$$= 160.012 \text{ km. appr.}$$

Ex. 13 AB is the gate

It is Circular

$\therefore$ AC is chord of half the arc which is equal to

$$AC = \sqrt{64+16}$$
$$= \sqrt{128}$$
$$= 8\sqrt{2} \text{ meter.}$$

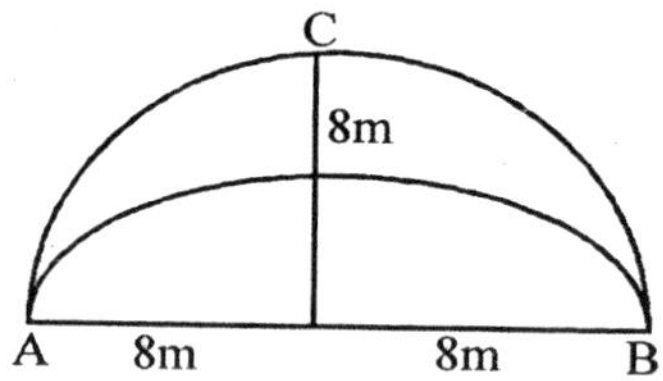

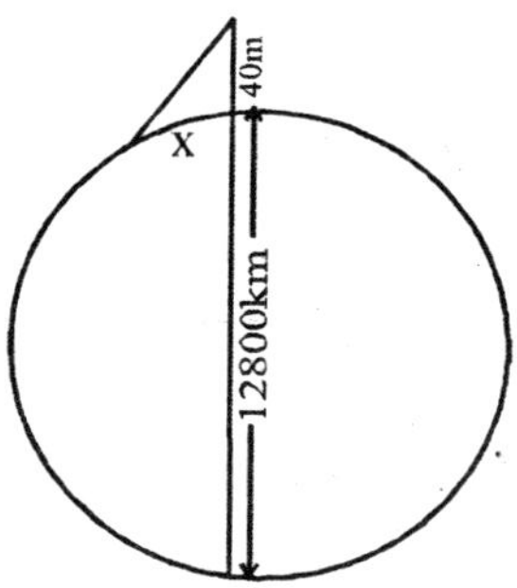

Ex. 14 Actual Ht. of Chimney = 40 m

Dia. of the earth= 1280000 m.

$$\therefore x = \sqrt{(12800040) \times 40}$$
$$= \sqrt{12800040 \times 40}$$
$$= \sqrt{51201600}$$
$$= 22.626 \text{ km}$$

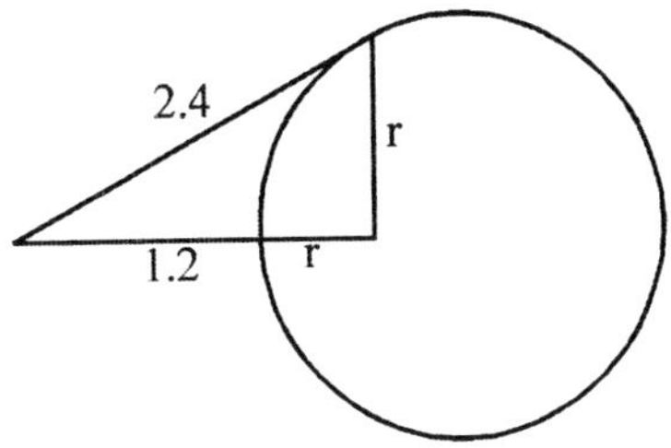

Ex. 15 $(r+1.2)^2 = r^2 + 2.4^2$

$2.4r = 5.76 - 1.44 = 4.32$

$$r = \frac{4.32}{2.4} = 1.8 \text{ km}$$

Exercise M

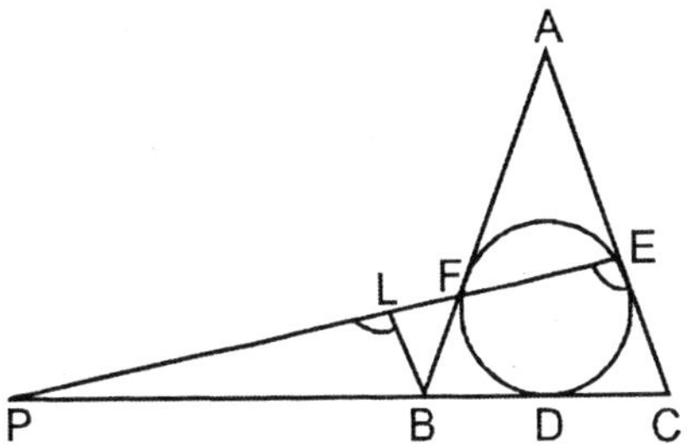

Ex. 3 Draw $BL \parallel CE$

$\angle BLP = \angle CEP$

Because $AF = AE$

$\angle BFL = \angle AFE$

$= \angle AEP$

$\therefore \angle BLF = \angle AEP$

$= \angle BFL$

$\therefore BL = BF$

$$\frac{BD}{CD} = \frac{BF}{CD} = \frac{BL}{CD} = \frac{BL}{CE} = \frac{PB}{PC}$$

$\therefore$ P and D divide BC harmonically.

Ex. 7 Produce CA to Q so that $AB = AQ$

CAP and CQB are 2 similar Δ

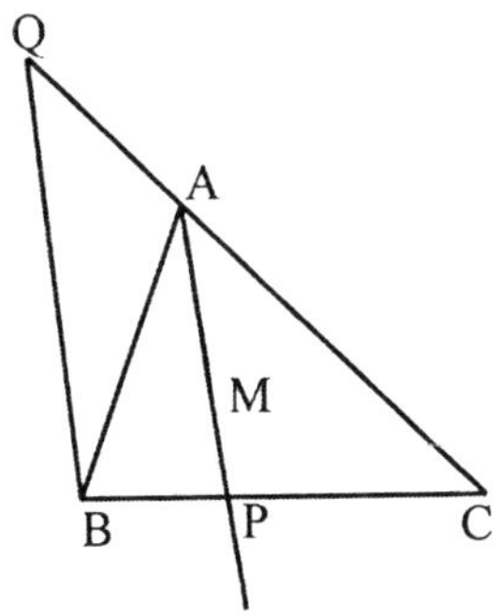

$$\frac{PC}{BC} = \frac{AC}{CQ}$$

$$\frac{PC}{a} = \frac{b}{b+c}$$

$$PC = \frac{ab}{b+c}$$

$$\therefore \quad \frac{AC}{PC} = \frac{AM}{MP}$$

$$\frac{b}{\frac{ab}{b+c}} = \frac{AM}{MP}$$

$$\frac{b+c}{a} = \frac{AM}{MP}$$

Ex. 9 (1) PX and PY are 2 tangents from any point P.

(2) XY is the chord of contact.

(3) AM, AZ and AN are the three perps. from A on PX, XY and PN respectively.

(4) $\angle AYN = \angle AZN$

$\angle AYN = \angle AXZ$

$\angle AXZ = \angle AMZ$

$\angle AZN = \angle AMZ$

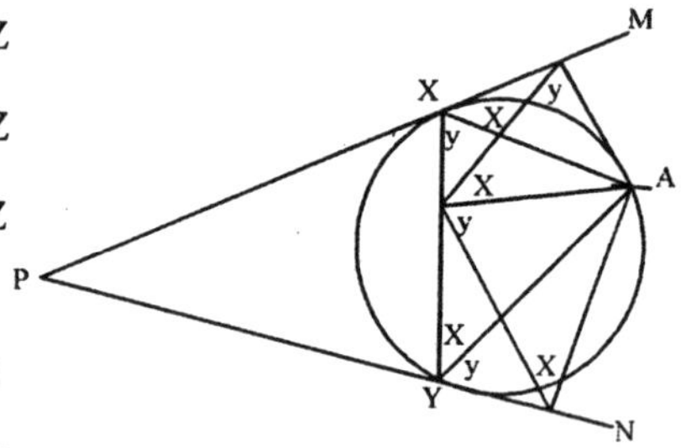

(5) Similarly

$\angle AYZ = \angle ANZ$

$\angle ANZ = \angle AXM$

$\angle AXM = \angle AZM$

So $\angle ANZ = \angle AZM$

(6) $\therefore$ AMZ and AZN are equiangular Δ s

(7) $\therefore$ $\frac{AN}{AZ} = \frac{AZ}{AM}$

$\therefore$ $AZ^2 = AN.AM$

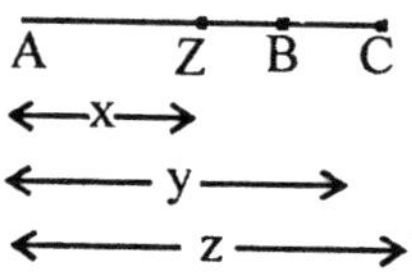

Ex. 11 $\frac{x}{y-x} = \frac{z}{z-y}$

$xz - xy = zy - xz$

$2xz = xy + zy$

$$\frac{2}{y}=\frac{1}{x}+\frac{1}{z}$$

$$\frac{2}{AB}=\frac{1}{Ax}+\frac{1}{Ay}$$

Ex. 13 $\frac{AX}{XB}=\frac{a}{b}$ Also $\frac{AY}{YB}=\frac{a}{b}$

$XB=\frac{b}{a}AX$ $YB=\frac{b}{a}AY$

$\frac{XB}{YB}=\frac{AX}{XY}=\frac{a}{b}$ $\frac{YB}{AY}=\frac{b}{a}$

Ex. 14 Construction : Draw $AD \perp BC$

Draw $AE \| BC$

Make $AE = AD$

Join BE

It meets AC in F

F is one corner of the square

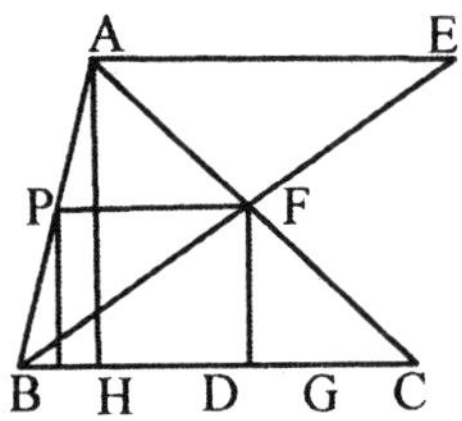

Proof: $\frac{AE}{PF}=\frac{AB}{BP}=\frac{AD}{PH}$

But $AE = AD$

$\therefore$ $PF = PH$

Like this 3 squares are possible corresponding to three sides.

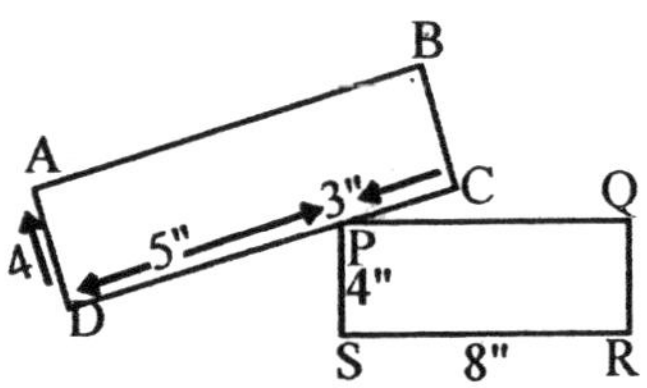

Ex. 15 Ht. of $P = 4^{11}$

$DS = 3^{11}$

Ht of $C = 8\times\frac{4}{5}=\frac{32}{5}=6.4$

$$\text{Ht. of B} = 6.4 + \frac{4 \times 3}{5}$$

$$= 6.4 + 2.4 = 8.8$$

$$\text{Ht. of A} = 4 \times \frac{4}{5} = \frac{16}{5}$$

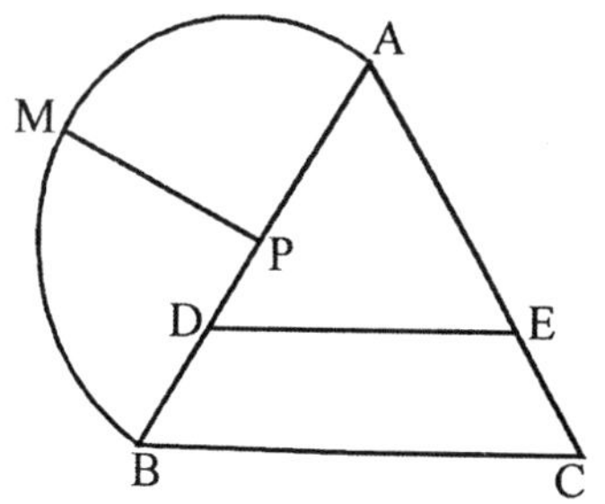

Ex. 27 Construction: Draw a semicircle on AB. Draw perp. bisector of AB. It meets the semicircle in point M. With AM as radius cut AB at D from A. Draw DE parallel line to BC.

Proof: $$\frac{\text{Area of } \Delta ABC}{\text{Area of } \Delta ADE} = \frac{AB^2}{AD^2}$$

$$= \frac{2AD^2}{AD^2} = \frac{2}{1}$$

Ex. 28 Construction: (1) Draw a right angled Δ with $\angle ABD = 30°$ and $\angle DAB = 90°$

(2) From AB cut off AP = AD.

(3) With A as centre and radius = PD cut off AQ

(4) From P and Q draw $\parallel$ lines to BC.

These will trisect the Δ.

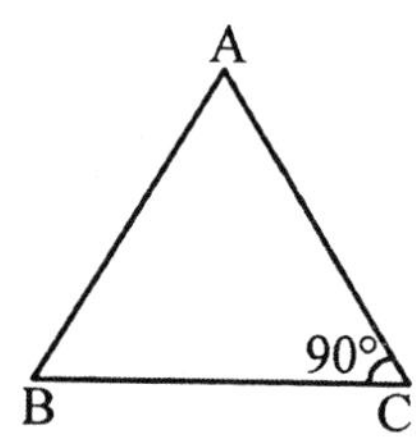

Proof: $\angle ADB = 60^\circ$ and $\angle ABD = 30^\circ$

$$\frac{AB}{DB} = \frac{1}{2}, \frac{AD}{AB} = \frac{1}{\sqrt{3}}, \frac{AP^2}{AB^2} = \frac{1}{3}$$

$$DP = \sqrt{2}AD$$

$$AQ = \sqrt{2}\,AD$$

$$AQ^2 = 2AD^2$$

$$\Delta AQ4 = 2\Delta APX$$

$$\Delta ABC : \Delta AQY : \Delta ABC = 3:2:1$$

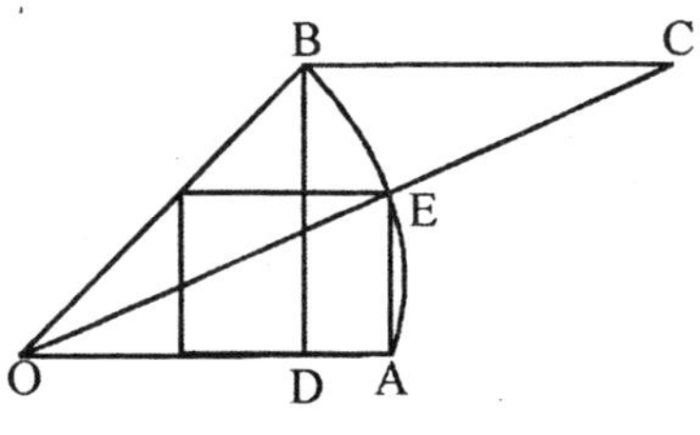

Ex. 29 OAB is a sector. From B draw $BD \perp OA$. Take BC at right $\angle$s to BD and equal to BD. Join OC. In meets the circle in E. E is the corner of the square.

If $\angle BOA = 60^\circ$ the side will be

$$= \frac{\sqrt{7+210}}{\sqrt{3}}$$

If $\angle BOA = 40^\circ$ the side will be $= \frac{\sqrt{5}}{1}$

Ex. 30 Given : $AB = AC = 2BC$

ΔABC and ΔDBC are similar Δ s

$$\frac{\text{Area of } \Delta ABC}{\text{Area of } \Delta DBC} = \frac{AB^2}{BC^2} = 4$$

$$\frac{\text{Area of } \Delta BED}{\text{Area of } \Delta AED} = \frac{BE}{AE} = \frac{3}{1}$$

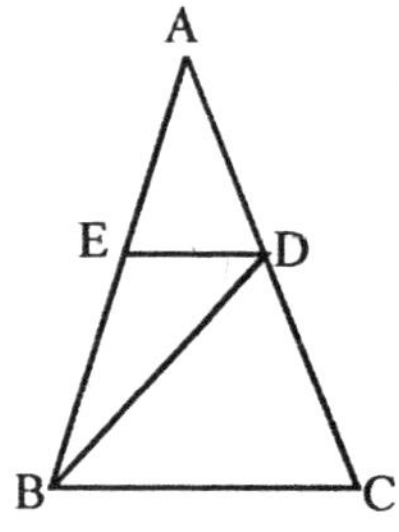

Exercise N

Ex. 2 The points diametrically opposite to A, B and C are three points for which the Simpson's line coincides with the side of the triangle.

Ex. 4 Suppose you are given a quadrilateral ABCD. Produce $\overline{DA}$ and $\overline{CB}$ to meet in P. Draw a circle about PDC. On PD, CB and DC any point on the circle are such that the feet of perp. are in one line.

Similarly produce AB and DC to meet in F. Again draw a circle about ADF. Hence any point of the circle ADF is such that the feet of perp. are in one line. These 2 circles meet in L. Hence L is such a point that all the perpendiculars to this would be in one line.

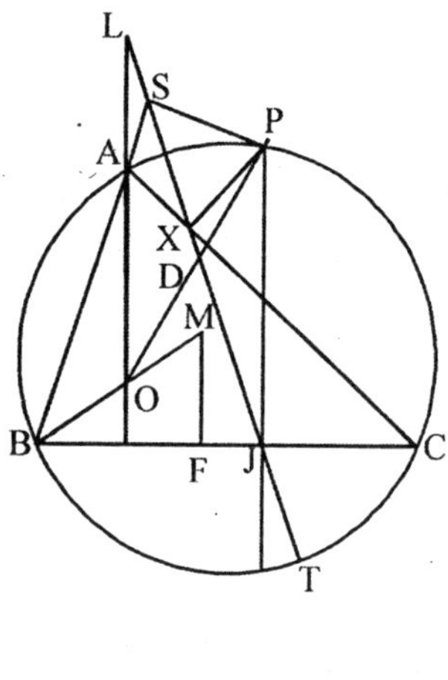

Ex. 11 (1) O is Ortho centre.

(2) M is the circum-centre.

(3) $MF \perp BC$

(4) Simpson line SXJ produced meets OA in L

(5) $2MF = OA$

(6) $2MF = PJ - JT$

or $PJ = 2MF + JT$

$= OA + JT$

$= OA + LA$

$= OL$

(7) PJOL is a parallelogram

(8) Midpoint D of OP is on LJ (Simpson)

Exercise P

Ex. 1 Three points A, B, C are given. We have to find a point P such that $PA.PC = PB^2$

On AC draw a semi-circle. At B make $\angle DBC = 45°$. D is on circle. Draw $DP \perp AC$. P is the required point.

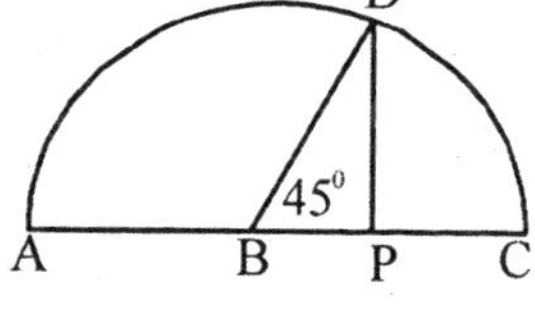

Ex. 2 Given: (1) ABC is a Δ. P is any point on BC.

(2) $PF \parallel AC, PD \parallel AB$.

To Prove: (Area of AFD)^2 = (Area ΔBFP) x (Area ΔPDC)

Proof: $\frac{\Delta_1}{\Delta_2} = \frac{AF}{FB}$ as $PD \parallel AB$

$\frac{\Delta_3}{\Delta_1} = \frac{DC}{AD}$ as $PF \parallel AC$

But ΔBFP is equiangular to ΔPDC

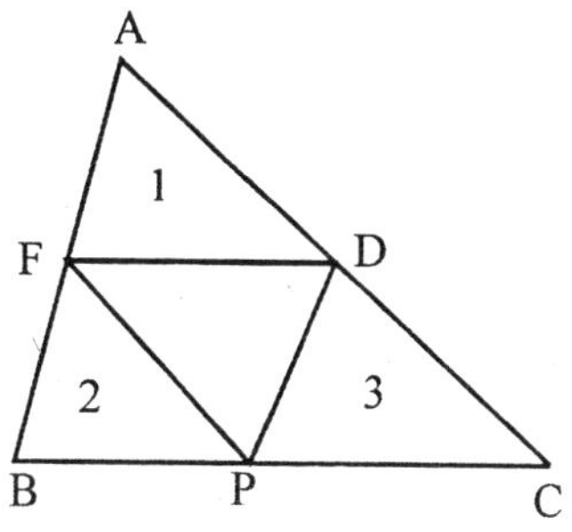

$\therefore \quad \frac{FB}{FP} = \frac{PD}{DC}$

$\frac{FB}{AD} = \frac{AF}{DC}$

By alternendo

$\frac{FB}{AF} = \frac{AD}{DC}$

$\frac{AF}{FB} = \frac{DC}{AD}$

$\therefore \quad \frac{\Delta_1}{\Delta_3} = \frac{\Delta_3}{\Delta_1}$

Ex. 4 (1) AB is a chord of the circle.

(2) P is a point.

(3) PA and PB are joined.

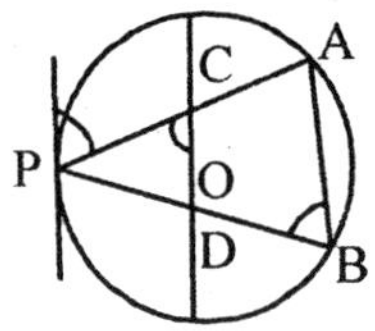

(4) A line CD is drawn which is $\parallel$ to tangent at P.

(5) Angles are as shown.

(6) Δs PBA and PCD are similar.

(7) $\dfrac{PC}{PB} = \dfrac{PD}{PA}$

(8) $\therefore$ PC. PA = PB. PD

Ex. 7 Divide the points AC internally as well as externally in the ratio of AP:PB. On these two points draw a semi-circle.

Ex. 8 (1) Let O be the centre of the circle. And y be the point where the chords meet at right angles. We have to find the locus of mid points such as M and M^1

(2) In ΔOMY, $MO^2 + MY^2 = 2YX^2 + 2MX^2$

(3) YX is fixed as it is $\dfrac{OY}{2}$.

(4) Locus of M is a circle with X centre and MX as radius.

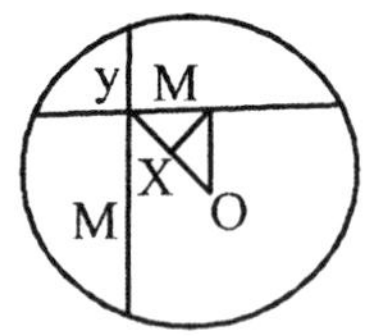

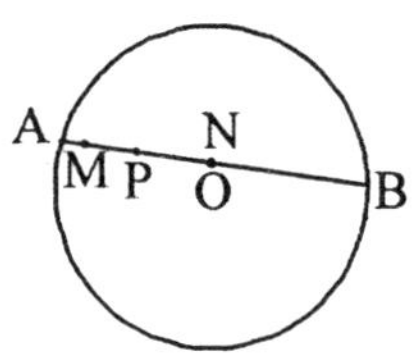

Ex. 9 P is a point inside a circle. From this point draw a diameter to the circle PA and PB are lines to the circle. Divide PA and PB in the given ratios at M and N and MN as diameter draw a circle. That is the locus. Same thing holds good for points outside the circle.

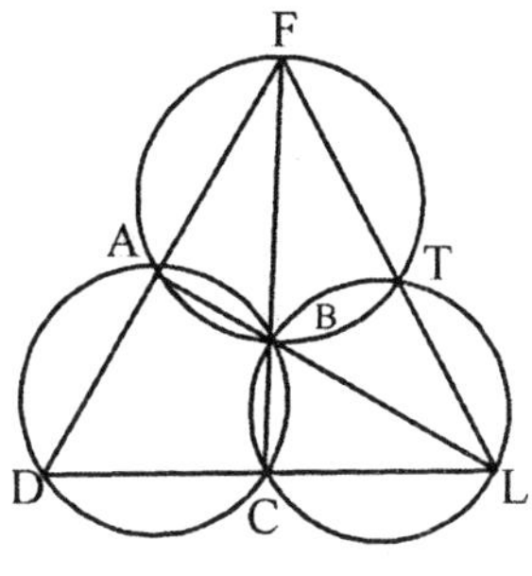

Ex. 10 (1) AB and CD are 2 sets of lines. They form a quadrilateral.

(2) $\overrightarrow{DA}$ and $\overrightarrow{CB}$ meet in F.

(3) $\overrightarrow{AB}$ and $\overrightarrow{DC}$ meet in L.

(4) $4\,\Delta s$ are formed ABF, BCL, DCF, ALD.

(5) Their circumferences pass through a common point i.e. T.

Ex. 11 If T is the common meeting point of the circumcircles of the $4\,\Delta s$ mentioned in the last example and if TX, TY, TZ and TM are joined to the respective orthocentres then the midpoints of TX, TY, TZ and TW are in one line that is the common Simpson line of the point T (In this connection reference is inverted to example no. 11 of exercise N).

Ex. 12 Join these points and divide the line so obtained both internally and externally in the ratio of the 2 perps.Through these points the line shall pass.

Ex. 13 (1) The in-circle always touches the base between the foot of the perpendicular and the mid-point of the base. Hence the in-circle is always inside the nine point circle. Hence its diameter is always less than the circumradius.

(2) In an equilateral Δ only diameter of the incircle in equal to the circum-radius.

Ex. 14 ABC is a Δ in which $\angle A = 36^\circ$ and $\angle B = \angle C = 72^\circ$. A point X is taken on AB so that AX = BC. The configuration of Δ is such that AX = BC = CX. Draw $CM \perp AB$. M is the midpoint of BX. Draw $XN \perp AC$ and AX = CX. XN is the perpendicular bisector of AC.

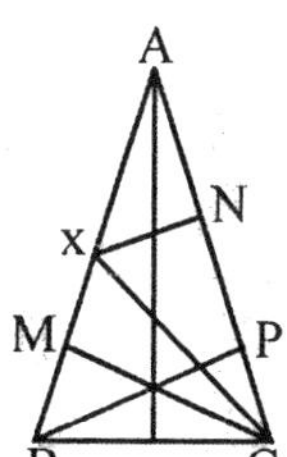

Draw $BP \perp AC$

$NC - PC =$ projection of BX on AC

$\Delta BCM \equiv \Delta BPC$

$NC - PC = NC - BM$

$$=\frac{AC}{2}-BM$$

$$=\frac{AC-2BM}{2}=\frac{AX}{2}=\frac{BC}{2}$$

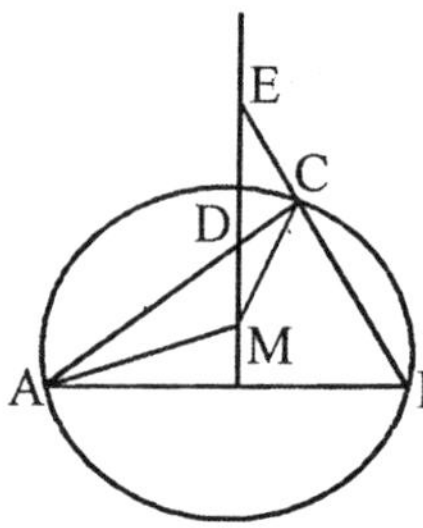

Ex. 15 $\angle AMC = 2\angle B$

$\angle MCA = 90 - \angle B$

$= \angle CED$

MC is a tangent to the circle CDE

$\therefore$ $MC^2 = MD.ME$

Ex. 16 Draw a line AB through C that is through one intersection. Divide it in the given ratio at P. Join PD. Through C draw $EF \| PD$.

EF is the required line.

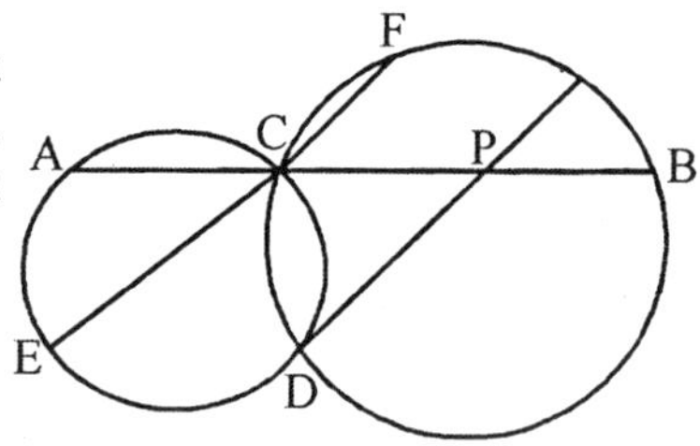

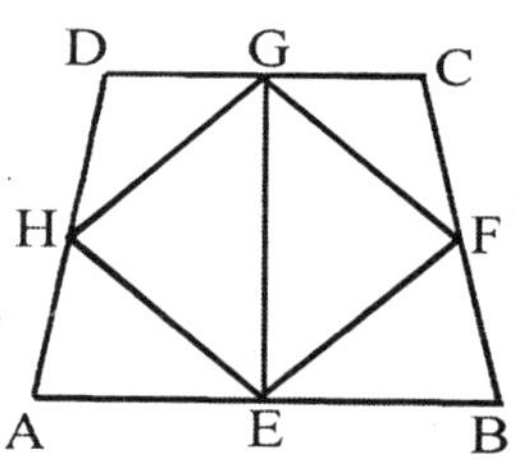

Ex. 17 The area of ABCD is given.

EFGH is a parallelogram.

The point E and G are given. Area of EFGH $= \frac{1}{2}$ Area of the main quadrilateral ABCD = given

$\therefore$ CBEG is such that $CB \| GE$ ' G is fixed.

Hence C moves so that it is parallel to GE.

Same is true for D.

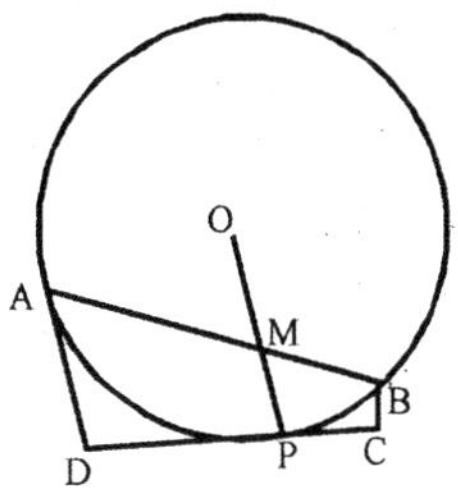

Ex. 20 (1) Join A to B

(2) Divide AB in the given ratio at M

(3) $\frac{AM}{MB}=a$

(4) Join OM produce to meet the circle in P

(5) At P draw a tangent

(6) Draw $AD \perp DC$.

$BC \perp DC$

(7) DC is required line.

Ex. 22 $\frac{AD}{DP}=\frac{AB}{PM}$

$\frac{AP}{AD}=\frac{PM}{DC}$

Multiplying the two equations

$\frac{AP}{DP}=\frac{AB}{DC}$

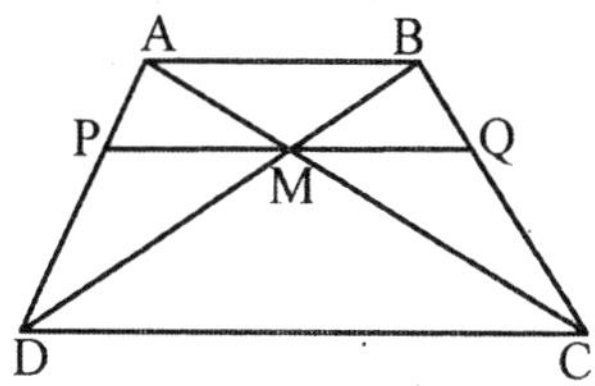

Ex. 23 Given : (i) AB is a diameter of the circle.

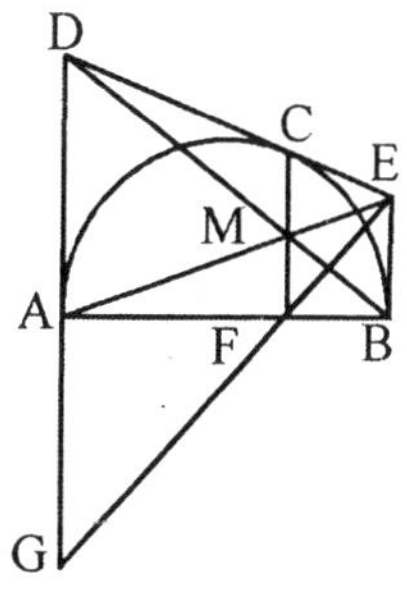

(ii) DE is a tangent at C, a point on the C.

(iii) AD and BE are tangents at A and B.

(iv) CF is $\perp$ AB.

To Prove: AE and BD meet at Mid point of CF.

Proof: Join EF and let it meet DA produced in G. Let AE meet CF in M.

ΔFEB and ΔAFG are similar.

$\frac{EB}{AG}=\frac{BF}{AF}$

$$\therefore \qquad \frac{EB}{AG}=\frac{CE}{CD}$$

$$\therefore \qquad \frac{EB}{AG}=\frac{EB}{AD}$$

$$\therefore \qquad AG=AD$$

$$\frac{AG}{MF}=\frac{AE}{ME}$$

$$\frac{AD}{CM}=\frac{AE}{ME}$$

$$\therefore \qquad MF=CM$$

Ex. 24 (1) A, B, C, D are 4 points in one line.

(2) P is a point such that circles PAB and PCD touch each other at P.

(3) If PT is the common tangent then

$$\angle BPT=\angle PAB \qquad \text{(Th 59)} \qquad ....(1)$$

and $$\angle CPT=\angle PDC \qquad \text{(Th 59)} \qquad ....(2)$$

Adding, we get

$$\angle BPT+\angle CPT=\angle PAB+\angle PDC$$

To this we add $\angle APD$ is both sides so that the R.H.S. becomes 180°

$$\angle BPT+\angle CPT+\angle APD=\angle PAB+\angle PDC+\angle APD$$

$$\angle BPT+\angle CPT+\angle APT+\angle DPT=180°$$

But $\angle BPT+\angle DPT=\angle BPD$

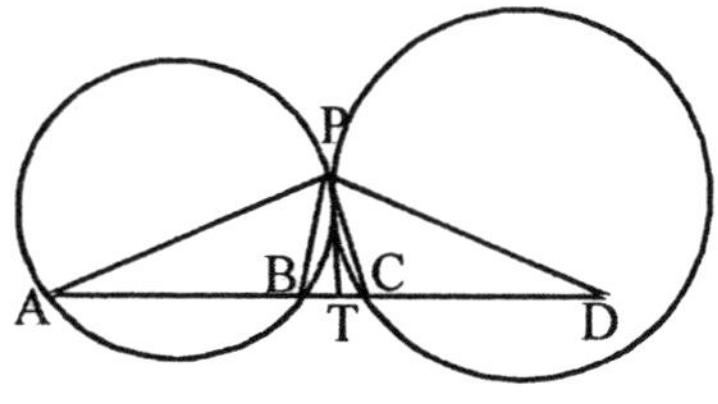

$$\angle CPT+\angle APT=\angle APC$$

$$\therefore \qquad \angle APC+\angle BPD=180°$$

Ex. 25 $\frac{1}{2}DE.AC+\frac{1}{2}DF.BC=\frac{1}{2}BC.AC$

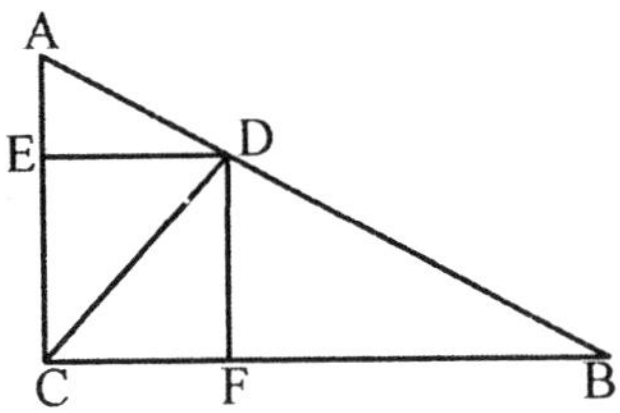

$DE.AC+DF.BC=AC.BC$

$DE.AC+DE.BC=AC.\ BC$

$\because$ $DE=DF$

$DE\,(AC+BC)=AC.BC$

Ex. 26 (1) Centre of AO is the centre of the circle AFOE

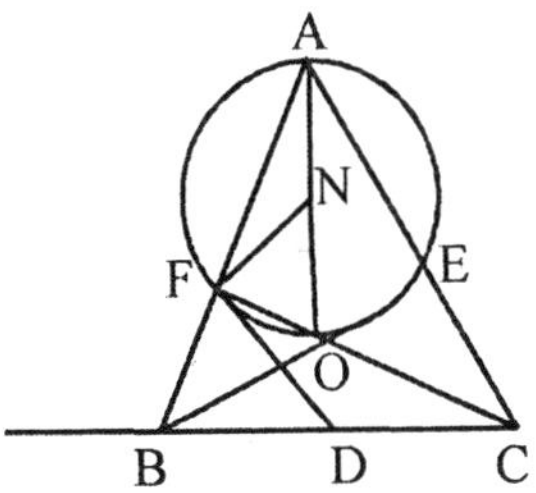

(2) Let the centre of AO be N.

(3) Centre of BFEC is the centre of BC. Let us call it D.

(4) DF is tangential at to AFOE

(5) The DF and NF are at rightangle.

Ex. 27 Similar to Ex 26.

Ex. 28 (A) $PA^2+PC^2=AC^2$

$$AC^2=MD^2$$

$$PB^2+PD^2=BD^2$$

$$\therefore\quad PA^2+PB^2+PC^2+PD^2=BD^2+MD^2$$

$$=BM^2\ (\text{diameter})^2$$

(B) $AB^2+CD^2=PA^2+PB^2+2PA.PB+PD^2+CP^2+2CP.PD$

$=(\text{diameter})^2+2EP.PF+2EP.PF$

$=(\text{diameter})^2+4EP.PF$

$=(\text{diameter})^2+4\,(EO-PO)\,(EO+PO)$

$=(\text{diameter})^2+4\,(r^2-PO^2)$

$=(\text{diameter})^2+(\text{diameter})^2-4(PO)^2$

$AB^2+CD^2+4(PO)^2=2\,(\text{diameter})^2$

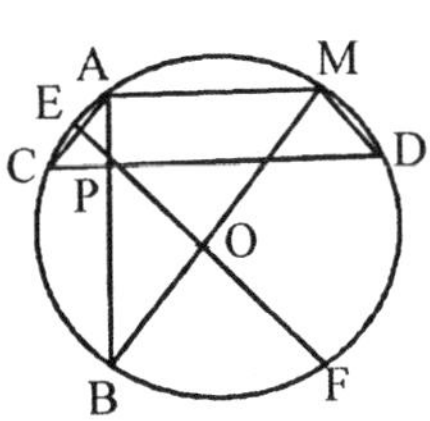

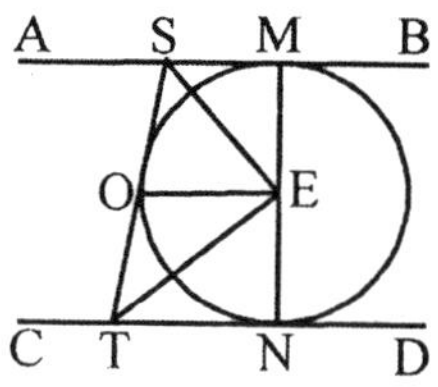

Ex. 29

(1) AB and CD are 2 parallel tangets to a circle, interceptped by a third tangent ST.

(2) To prove that SO. OT = (Radius)^2

(3) $\angle TES = \frac{1}{2}(\angle MEO + \angle NEO)$

$= \frac{1}{2} \times 180° = 90°$

(4) $\tan \angle TES = \tan 90° = \infty$

(5) $\therefore \tan \angle TEO = \cot \angle SEO$

(6) $\frac{OS}{OE} = \frac{OE}{OT}$

(7) $OS.OT = OE^2 = r^2$

Ex. 30 (1) PB and DF are 2 diameters of circles touching each other at A.

(2) P, A, F are in the same straight line as also B, A, D.

(3) $PF \perp BD$

(4) Line of centres of the 2 circles meet them in C and E.

(5) PBXF is a parallelogram.

(6) XD = CE = Sum of the 2 diameters.

(7) $XD^2 = BX^2 + BD^2$

$CE^2 = PF^2 + BD^2$

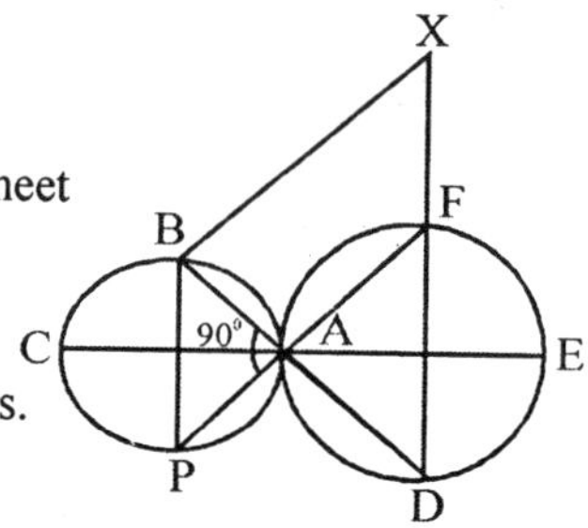

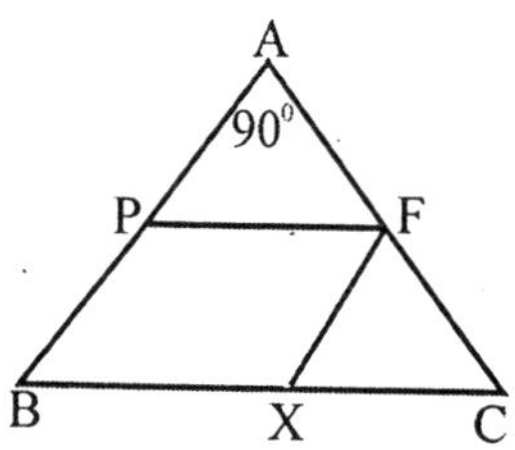

Ex. 31 $(BC - PF)^2 = (AB - AP)^2 + (AC - AF)^2$

$2BC.PF = 2AB.AP + 2AC.AF$

$BC.PF = AB.AP + AC.AF$

Ex. 32 h is the height of the flagstaff

$(h+30)30 = 54 \times 35$

$30h = 1890 - 900$

$30h = 900$

$h = 30$ m

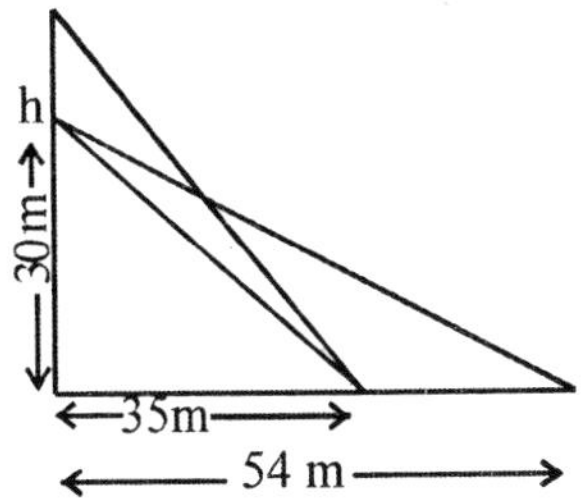

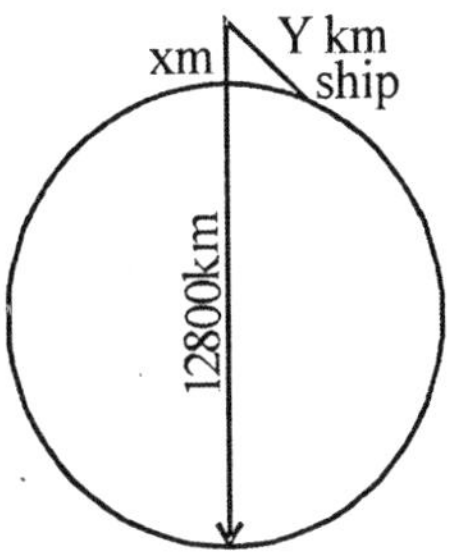

Ex. 33 $y = \sqrt{.03(.03+12800)}$

$= \sqrt{12800.03 \times .03}$

$= \sqrt{384.0009}$

$= 19.596$ km

Ex. 34 (1) ABCD is a cyclic quadrilateral

(2) BA and CD meet in F.

(3) DA and CB meet in P.

(4) Circle about ABP and circle about FDA meet in M.

(5) AMPB and ADFM are cyclic quadrilaterals.

(6) Square on tangent to AMPB= PF. FM.

(7) Square on tangent to ADFB = ADFM will be = PF.PM

(8) These 2 squares will be

PF (PM + MF) = PF.PF = PF^2

= Sum of squares to the circle for P and F.

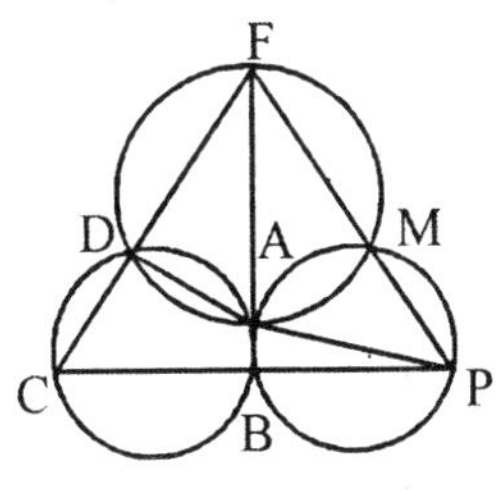

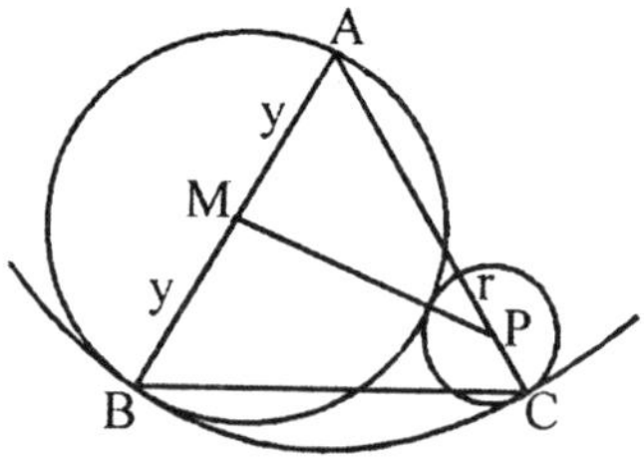

Ex. 35 (1) ABC is an equilateral Δ.

(2) With A as centre and AB as radius a circle is drawn.

(3) Another circle is drawn with AB as diameter. Let AB = 2y

(4) Another circle is drawn with P as centre (P on AC) such that it touches the circle with A as centre internally and the other with AB as diameter externally.

(5) Consider ΔMAP in which $\angle MAP = 60^{\circ}$

(6) $MP^2 = AM^2 + AP^2 - 2MA.AP \cos \angle MAP$

$$= AM^2 + AP^2 - 2MA.AP.\frac{1}{2}$$

$$= AM^2 + AP^2 - MA.AP$$

(7) Let the circle last drawn be = r

(8) $(y + r)^2 = y^2 + (2y-r)^2 - y(2y-r)$

$y^2 + 2yr + r^2 = y^2 + 4y^2 - 4yr + r^2 - 2y^2 + yr$

$2yr = 2y^2 - 3yr$

$5yr = 2y^2$

$5r = 2y$

$AP = 2y - r$

$$= 2y - \frac{2y}{5}$$

$$= \frac{8y}{5} = \frac{4}{5} AC$$

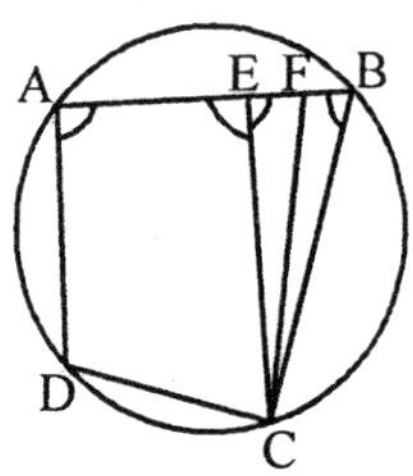

Ex. 36 (1) ABCD is a cyclic quadrilateral in which BC = AD.

(2) If $\angle D$ is obtuse, $\angle B$ is acute.

$\angle ADB > \angle ABC$

AB > AD.

(3) From AB is cut off AE = AD

(4) Let F be the midpoint of EB.

(5) Then $CF \perp AB$

(6) $\angle ADC = \angle AEC$

(7) $AC^2 = AB^2 + BC^2 - 2AB.FB$

$= BC^2 + AB^2 - AB.EB = BC^2 + AB.(AB - EB)$

$= BC^2 + AB.AE = BC^2 + AB.AD$

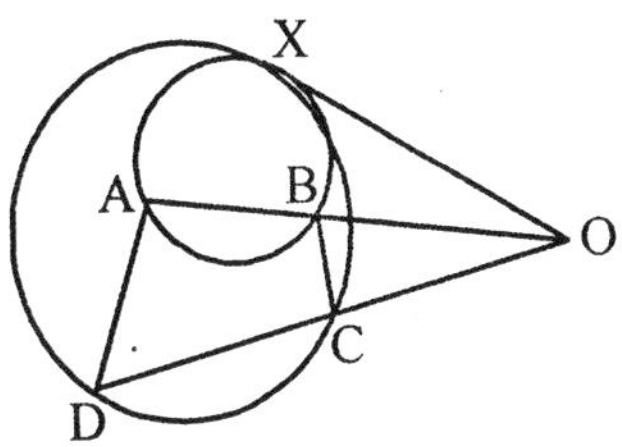

Ex. 37 (1) ABCD is a quadrilateral

(2) Circle AXB and CXD have a common tangent OX.

(3) Locus of X will be a circle with O as centre and OX as radius.

Ex. 38 $XA^2 + XB^2$ $\qquad$ $XC^2 + XD^2$

$= 2XE^2 + 2AE^2$ $\qquad$ $= (OC - OX)^2 + (OD + OX)^2$

$= 2XO^2 + 2OE^2 + 2AE^2$ $\qquad$ $= (OC - OX)^2 + (OC + OX)^2$

$= 2XO^2 + 2OA^2$ $\qquad$ $= 2OC^2 + 2OX^2$

$= 2XO^2 + 2OD^2$ $\qquad$ $= 2OM^2 + 2OX^2$

$= 2XO^2 + 2OM^2$ $\qquad$ $= 2MX^2$

$= 2MX^2$

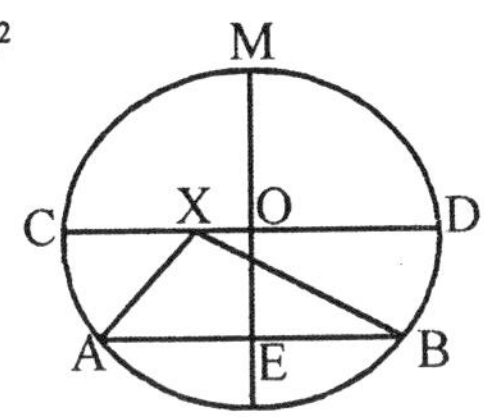

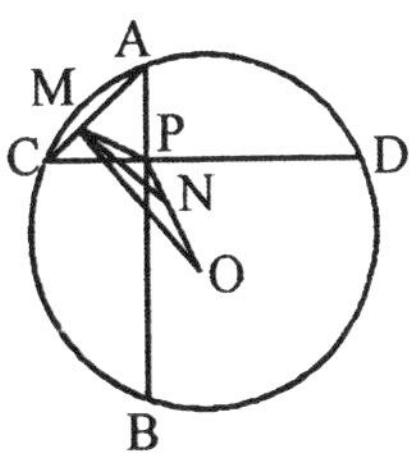

Ex. 39 (1) AB and CD meet at right angles at P.

(2) M is the centre of AC.

(3) N is the centre of OP.

(4) $2MN^2 + 2PN^2$

$= MP^2 + MO^2$ (Th. 63)

$= MC^2 + MO^2$

$= (\text{Radius of the circle})^2$

Ex. 40 (1) AB is the diameter of the circle. M is its centre MB = x, AB = 2x.

(2) N is the centre of the smaller circle of which NC = y the radius

(3) PTF is the chord of the bigger circle meeting it at rt. angles MT=r.

(4) It touches the smaller circle at the chord PTF at D.

(5) $\angle TPB = \angle PAT = 90^\circ - \angle TBP$.

(6) $APB = 90^\circ$

$\therefore PB^2 = BT.BA = (x + r).2x$

$= 2x^2 + 2xr$

(7) $NM = (x - y)$, $NC = y$ and $MC = CT + MT = (y + r)$

(8) $\therefore (x - y)^2 = (y + r)^2 + y^2$

$x^2 - 2xy = (y + r)^2$

(9) $BC^2 = (x + y + r)^2$

$= x^2 + 2x(y + r) + (y + r)^2$

$= x^2 + 2xy + 2xr + (x^2 - 2xy)$

$= 2x^2 + 2xr$

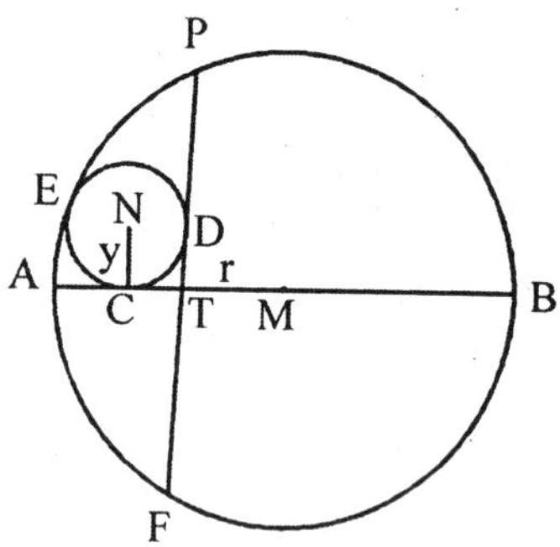

(10) From (9) and (6) PB = BC

$\therefore \angle PCB = \angle CPB$

(11) But $\angle PAT = \angle TPB$

(12) $\therefore \angle PCB - \angle PAT = \angle CPB - \angle TPB$

$\therefore \angle APC = \angle CPT$

$\therefore$ PC bisect $\angle APT$

(13) From (8) above

$x^2 - 2xy = (y + r)^2$

$x^2 - (y + r)^2 = 2xy$

$(x + y + r)(x - y - r) = 2xy$

$BC \times AC = 2xy$

$\frac{1}{2} BC \times AC = xy$

Ex. 41 $\frac{AB}{BC}=\frac{BD}{AB}$

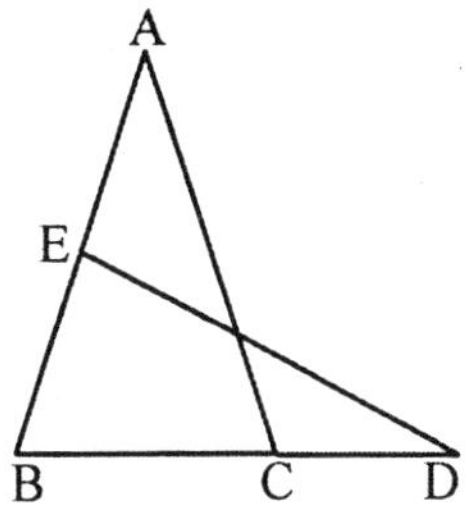

$AB^2 = BC.BD$

$BD^2 - AB^2 = BD^2 - BD.BC$

$= BD(BD - BC)$

$= BD.CD$

$\frac{AB}{BC}=\frac{BD}{AB}$

$AB^2 = BD.BC$

$AC^2 = BD.BC$

Ex. 42 ABCD is a square

$2PA.\ PC = 2$ Area of ΔAPC

= Area of quad ARCP

$= AC.RP$

$PB^2 - PD^2 = BE^2 - ED^2$

$= (BE + ED)(BE - ED)$

$= BD.BM$

$= AC.RP$

$\therefore \quad PB^2 - PD^2 = 2PA.\ PC$

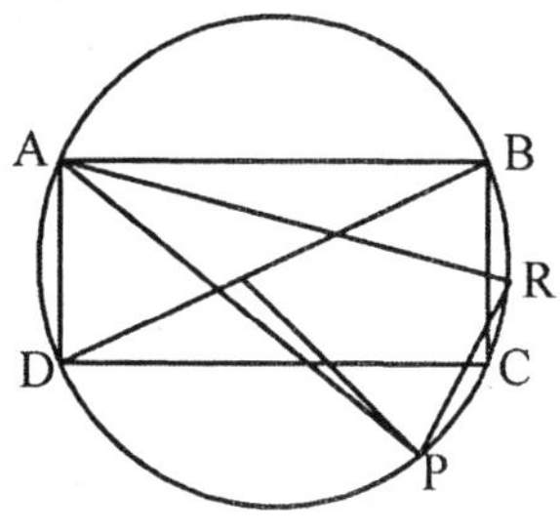

Ex. 43 $AB = x, \qquad AP = a$

$a^2 = x(x - a)$

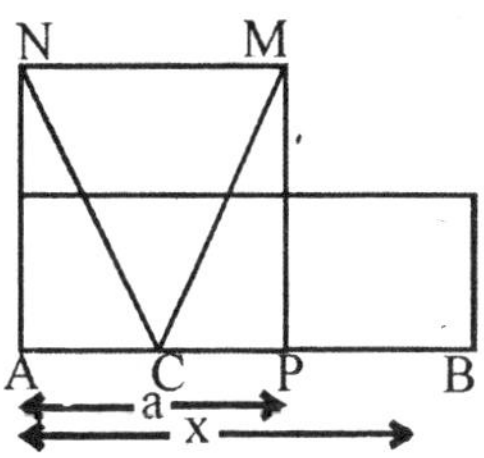

$CM = CN = \frac{\sqrt{5}}{2}a$

$x(x - a) = a^2$

$x^2 - ax - a^2 = 0$

$x = \frac{+a \pm \sqrt{a^2 + 4a^2}}{2} = \frac{a\left(1 \pm \sqrt{5}\right)}{2}$

$CB = x - \frac{a}{2} = \frac{\sqrt{5}}{2}a$

Ex. 44 Extend HB to K so that BK = BH.

Join IK.

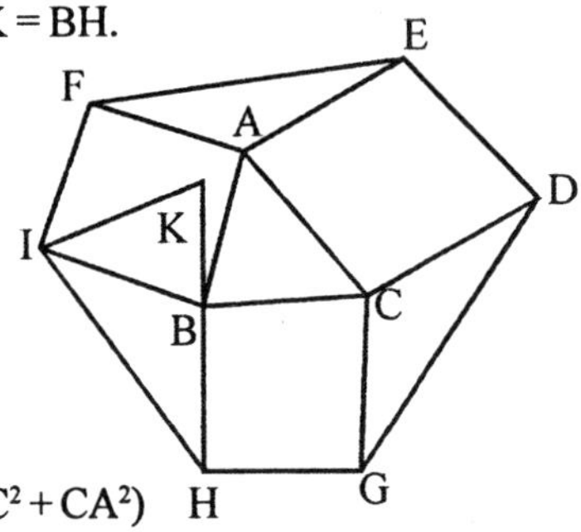

$$IH^2 + IK^2 = 2IB^2 + 2BH^2$$

$$IH^2 + AB^2 = 2AB^2 + 2BC^2$$

$$IH^2 = AB^2 + 2BC^2$$

$$GF^2 = BC^2 + 2CA^2$$

$$ED^2 = CA^2 + 2AB^2$$

$$\therefore \quad IH^2 + GF^2 + ED^2 = 3(AB^2 + BC^2 + CA^2)$$

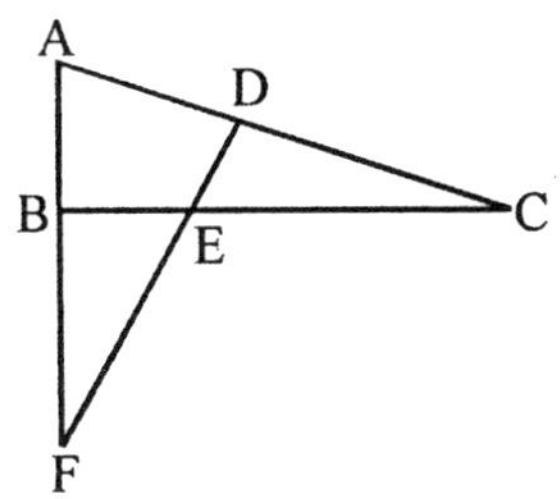

Ex. 45 (1) $\frac{AD}{AF} = \frac{AB}{AC}$

$$\frac{AC}{2AF} = \frac{AB}{AC}$$

$$AC^2 = 2AB.AF$$

Ex. 47 (1) In ΔMCP and MGC because

$$\angle MGC = 90^\circ = \angle MCP$$

$\angle M$ is common

$\therefore$ The Δs are equiangular

$\therefore \quad MC^2 = MP.MG$

(2) But $\quad MC^2 = MX.MY$

$\therefore \quad MX.MY = MP.MG$

$\therefore$ XYMG is cyclic.

(3) Now

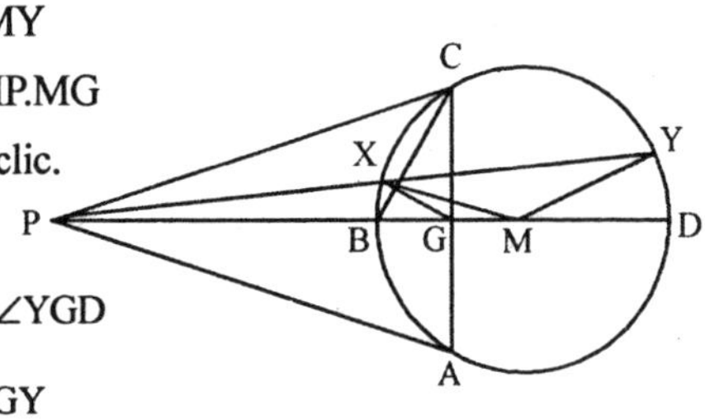

$$\angle XGP = \angle XYM = \angle MXY = \angle YGD$$

$\therefore \quad \angle CGX = \angle CGY$

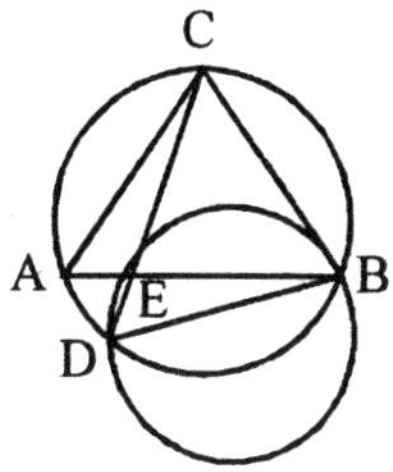

Ex. 48 $\angle CAB = \angle CBA = \angle CDB$

$\therefore$ CB is a tangent to the circle EBD

$CB^2 = CE.CD$

$a^2 = x(x+b)$

$a^2 = x^2 + xb$

$x^2 + xb - a^2 = 0$

$$x = \frac{-b \pm \sqrt{b^2 + 4a^2}}{2}$$

Thus x = CE can be found out.

Ex. 49 $AB^2 = AC^2 + CB^2 + 2AC.CB \cos(ACB)$

$AC^2 = AD^2 + CD^2 + 2AD.CD \cos(ADB)$

Since AC = CD, BC = AD,

$\angle C + \angle D = 180^\circ$

$AB^2 + AC^2 = AD^2 + CB^2 + AC^2 + CD^2$

$AB^2 = AD^2 + CB^2 + CD^2$

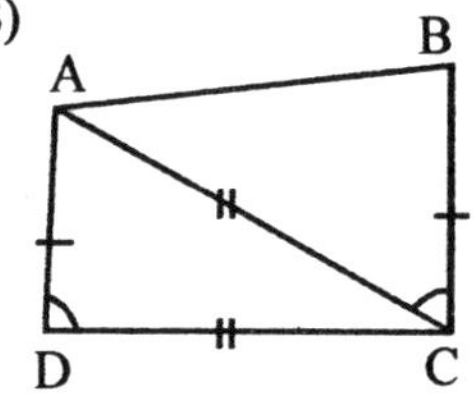

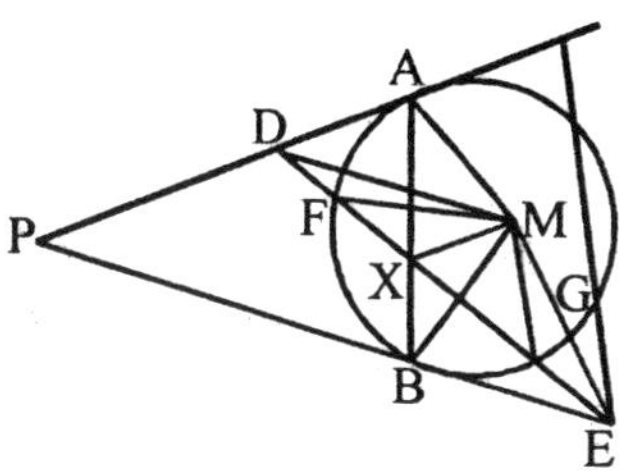

Ex. 50 (1) M is the centre of a circle. PA and PB are 2 tangents to the circle from P.

(2) On PA any point D is taken and on PB extended to a point E is taken so that BE = AD.

(3) DE is joined and it cuts the circle in point F and G.

(4) We have to show that AB cuts DE at point X so that point X is centre point of DE as well as FG.

(5) $\Delta MAD = \Delta MBE$ $\therefore$ MD = ME

(6) $\Delta MDF = \Delta MGE$ (Theorem 18 corollary)

(7) DF = GE are DE and FG have the same centre that is X.

(8) Let DE and AB meet in the point X.

(9) ΔAMB is equiangular ΔDME.

(10) $\angle XBM = \angle XGM$ $\therefore$ EMBX is cyclic quadrilateral.

(11) $\therefore$ $\angle MXE = \angle MBE = 90^\circ$

(12) $\therefore$ X is a common mid point.

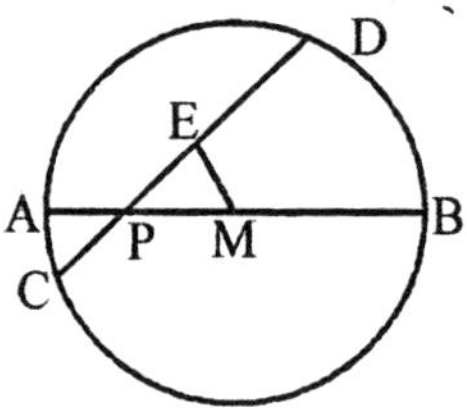

Ex. 51 $2CP^2 + 2DP^2$

$= 2(CE - EP)^2 + 2(DE + EP)^2$

$= 2(CE - EP)^2 + 2(CE + EP)^2$

$= 4CE^2 + 4EP^2 = 4(CE^2 + EM^2)$

$= 4CM^2 = AB^2$

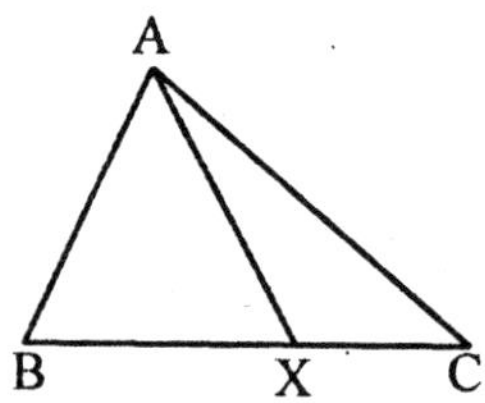

Ex. 52 $AX^2 + BX^2 = 2AF^2 + 2FX^2$ ...(1)

$BX^2 + CX^2 = 2BD^2 + 2DX^2$ ...(2)

$CX^2 + AX^2 = 2CE^2 + 2EX^2$ ...(3)

Adding (1), (2) and (3)

$2(AX^2 + BX^2 + CX^2) = 2(AF^2 + BD^2 + CE^2) + 2(DX^2 + 2FX^2 + EX^2)$ ...(4)

$(AX^2 + BX^2 + CX^2) = (AF^2 + BD^2 + CE^2) + (DX^2 + FX^2 + EX^2)$ ... (4)

In a ΔABC the base BC is divided at X so that m BX = n CX then

$m\,AB^2 + n\,AC^2 = m\,BX^2 + n\,CX^2 + (m+n)AX^2$

(Ex I ex. 13)

Applying this example to ΔXAD, ΔXBE and ΔXCF

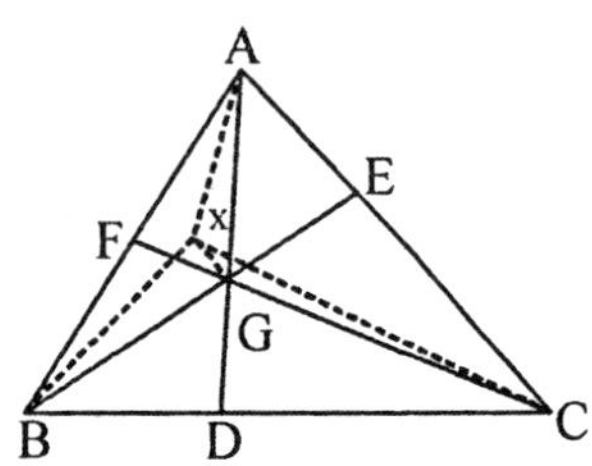

we get

$AX^2 + 2XD^2 = AG^2 + 2DG^2 + 3XG^2$ ...(5)

$BX^2 + 2XE^2 = BG^2 + 2EG^2 + 3XG^2$ ...(6)

$CX^2 + 2XF^2 = CG^2 + 2FG^2 + 3XG^2$ ...(7)

Adding (5), (6) and (7), we get

$(AX^2 + BX^2 + CX^2) + 2(XD^2 + XE^2 + XF^2) = (AG^2 + BG^2 + CG^2) + 2(DG^2 + EG^2 + FG^2) + 9XG^2$ ...(8)

$$\frac{(AX^2+BX^2+CX^2)}{2}+(XD^2+EX^2+FX^2)=\frac{AD^2+BE^2+CF^2}{3}+\frac{9}{2}XG^2 \quad ...(9)$$

Adding (4) + (9)

$$\frac{3}{2}(AX^2+BX^2+CX^2)+(XD^2+XE^2+XF^2)=\frac{AD^2+BE^2+CF^2}{3}+\frac{9}{2}XG^2$$

$$+\frac{AB^2+BC^2+CA^2}{4}+(XD^2+XE^2+XF^2)$$

$$\frac{3}{2}(AX^2+BX^2+CX^2)=\frac{3(AB^2+BC^2+CA^2)+4(AD^2+BE^2+CF^2)}{12}$$

$$+\frac{9}{2}\times G^2$$

$$(AX^2+BX^2+CX^2)=\frac{AB^2+BC^2+CA^2}{2}+3\times G^2$$

Ex. 53. Apply **Ex. 52.**

Ex. 54 Obviously

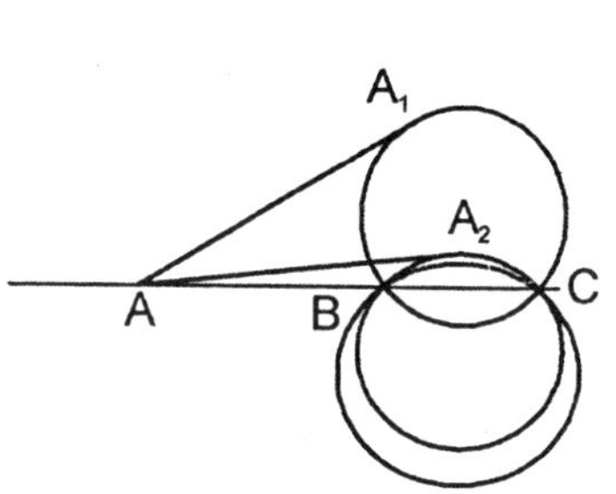

OA_1, OA_2, OA_3, OA_4................. will be equal.

Hence locus will be a circle with O as centre and OA_1 as radius.

Ex. 55 PF is always || to BC. That means.

AP : AB = AF : FC

That means M is on a st. line || to BC.

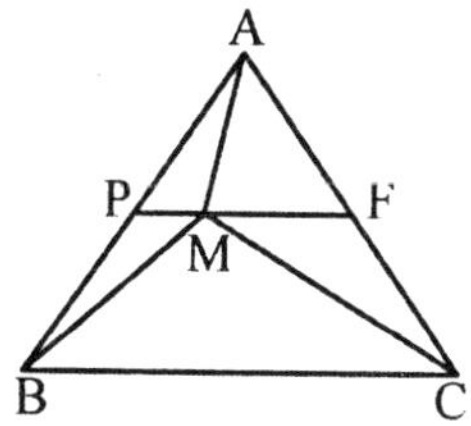

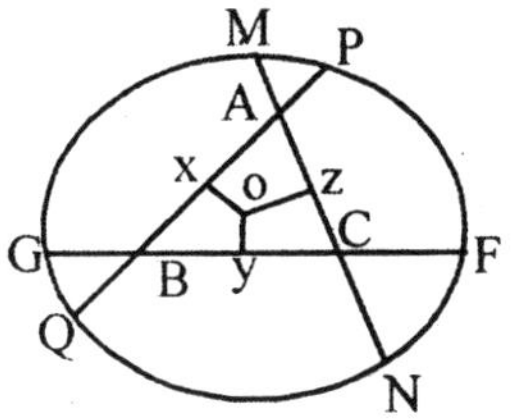

Ex. 56 If O is the centre of the circle and ABC is the equilateral Δ.

$$PX+GY+NZ=QX+FY+MZ$$

$$AX+BY+CZ=BX+CY+AZ$$

Subtracting

$$AP+GB+CN=BQ+CF+AM$$

Ex. 57 $\frac{AI}{IK}=\frac{\sqrt{3}+1}{\sqrt{3}-1}=\frac{2+\sqrt{3}}{1}=\frac{AB}{BK}$

Let $BK=1, KC=x$

$\therefore$ $AB=(2+\sqrt{3})\times BK$

$$\frac{AB}{BK}=\frac{AI}{IK}=\frac{AC}{KC}$$

$\therefore$ $AB=(2+\sqrt{3})\times BK$

$$\frac{AB}{BK}=\frac{AI}{IK}=\frac{AC}{KC}$$

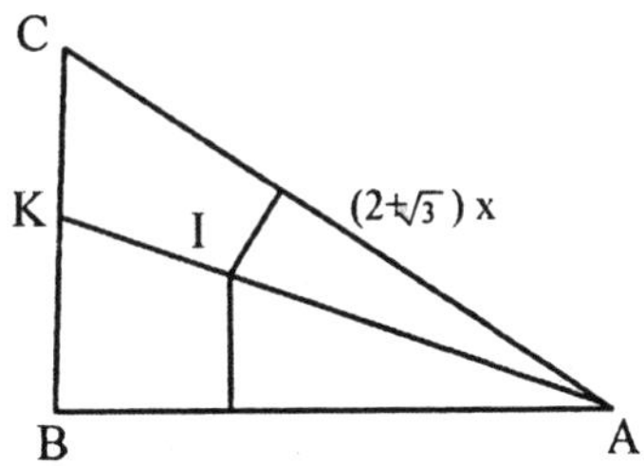

$$\frac{2+\sqrt{3}}{1}=\frac{AC}{X}$$

$$AC=\left(2+\sqrt{3}\right)x$$

By pythagoras theorem

$$AC^2=AB^2+BC^2$$

$$x^2\left(2+\sqrt{3}\right)^2=\left(2+\sqrt{3}\right)^2+(1+x)^2$$

$$(x^2-1)\left(2+\sqrt{3}\right)^2=(x+1)^2$$

$$\frac{x+1}{x-1}=\frac{7+4\sqrt{3}}{1}$$

$$\frac{2x}{2}=\frac{8+4\sqrt{3}}{6+2\sqrt{3}}$$

$$x=\frac{4+2\sqrt{3}}{3+2\sqrt{3}}$$

$$AC=\left(2+\sqrt{3}\right)x$$

$$=\left(2+\sqrt{3}\right)\frac{\left(4+2\sqrt{3}\right)}{3+2\sqrt{3}}=\frac{\left(2+\sqrt{3}\right)\left(4+2\sqrt{3}\right)}{3+2\sqrt{3}}$$

$$\frac{BC}{AC}=\frac{1}{2}\ \therefore \angle A=30°\text{and } \angle C=60°$$

Ex. 58 Given: (i) O is the circumcentre of ΔABC.

(ii) Lines AO, BO, CO intersect the sides BC, CA, AB in P, Q, R respectively.

(iii) $\angle BAC=\angle AQR=\angle RQB$

To Prove: $\angle BAQ=50$

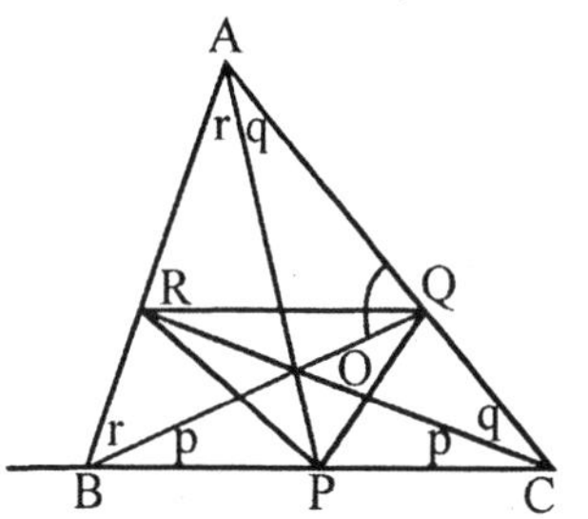

Proof: Let $\angle OBC=\angle OCB=p$

$\angle OCA=\angle OAC=q$

$\angle OAB=\angle OBA=r$

Segment QR bisects $\angle AQB$ ...(given)

$$\therefore \quad \frac{AQ}{QB}=\frac{AR}{RB} \quad ...(1)$$

$$\frac{BP}{PC}\cdot\frac{CQ}{QA}\cdot\frac{AR}{RB}=1$$

$$\frac{BP}{PC} \cdot \frac{CQ}{QA} \cdot \frac{AQ}{QB} = 1$$

or $$\frac{BP}{PC} = \frac{QB}{QC}$$

Segment QP bisects $\angle BQC$

$$\angle BQC = 180° - \angle AQB = 180° - 2\ \angle AQR$$

$$= 180°\ \angle BAC$$

$$(\because \angle BAC = \angle AQR = \angle R.QB)$$

$$= 180° - 2(\angle BAO + \angle OAC)$$

$$= 180° - 2(\angle r + \angle q) = 2q$$

$$\angle A + \angle B + \angle C = 180°$$

$$q + r + r + p + p + q = 180°$$

$$2p + 2q + 2r = 180° \quad ...(2)$$

From (2) and (3)

$$\angle BQC = 2p$$

$\angle BQP = \angle PQC = p$ ($\because$ segment QP bisects $\angle BQC$)

$\angle OBC + \angle OCB = 2p$

OPCQ is a cyclic quad.

$\angle POC = \angle PQC$ ...(5)

$2q = p$

$\therefore$ $\angle POQ + \angle PCQ = 180°$

$\therefore$ $\angle AQB + \angle PCQ = 180°$

$2(p + q) + (p + q) = 180°$ ...(6)

$\therefore$ $3q = 60 \quad q = 20$

$p = 40 \quad \angle BAQ = 50°$

$p + q = 60°$

From (5) and (6)

$3q = 60$

$q = 20$

$p = 40$ $r = 30°$

$\angle BAC = r + q = 50°$

Ex. 59 (1) Two circles intersect at C and A. Their centres are M and N.

(2) P moves on the first circle.

(3) PC and PA cut the second circle with centre at N in F and B.

(4) As P moves round the first Circle, A and B move round the second circle.

(5) $\angle P$ is fixed also the angle CFA.

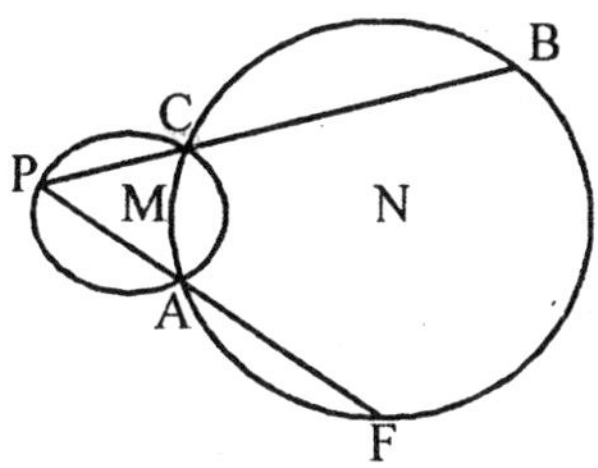

(6) $\therefore$ $\angle FCB$ is fixed.

(7) $\therefore$ length of FB is also fixed.

(8) $\therefore$ ΔFNB fixed. As far as size and shape.

(9) Let X be the circumcentre PBF

(10) $\therefore$ ΔXBF moves round with ΔNBF.

(11) But NX is fixed.

(12) $\therefore$ x moves round with N in a as centre circle.

Exercise Q

Ex. 1 (1) Two points P and Q are given inside a circle with centre O.

(2) On PQ draw a segment of a circle which will contain $\angle A$.

(3) Two points are possible at A and A'.

(4) Hence 2 Δs are available ΔACD and $\Delta A^1C^1D^1$.

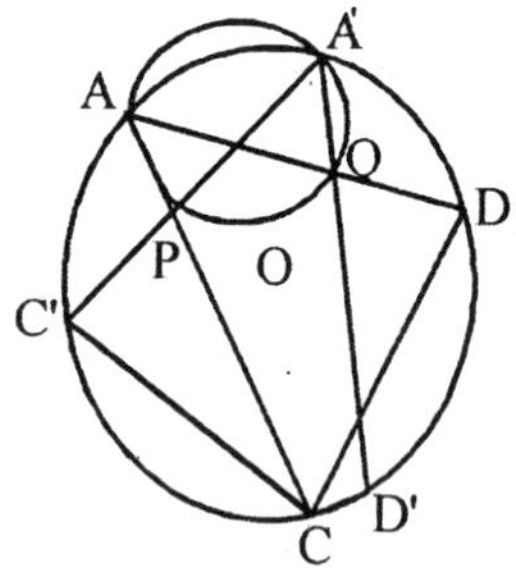

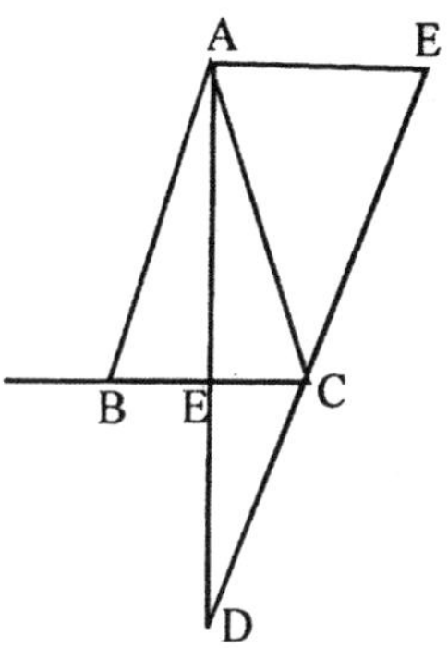

Ex. 2 (1) Draw AD = perp + base (given)

(2) At A draw AF $\perp$ AD= $\frac{\text{perp} + \text{base}}{2}$

(3) Join DF.

(4) From A cut off AC = equal distance (given)

(5) From C draw a perp. CE on AD and produce to B making EB = EC.

(6) ABC is the Δ.

Ex. 3 (1) Take a distance AD = perp. – base = (given)

(2) Produce AD to E.

(3) With DE make an angle = $\tan^{-1}\frac{1}{2}$

(4) From DC cut off AC= equal distance given.

(5) From C drop CE$\perp$AD.

(6) Produce CE = EB.

(7) Join ABC which is the Δ req.

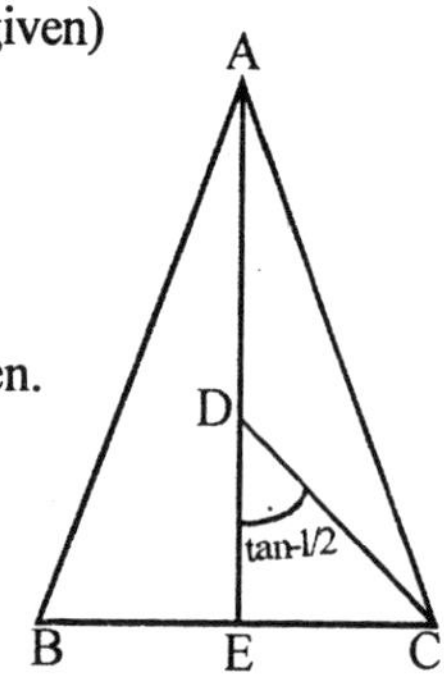

Ex. 4 First we draw the base. Then the base is divided internally and externally in the rafio of the other two sides which is given. We get the locus of the vertex by drawing a semicircle on these 2 points of division. This is common to all the five cases given below.

(a) *Difference of the base angles is given i.e.* $\angle B - \angle C$.

Let AD be the bisector

To calculate $\angle ADC$

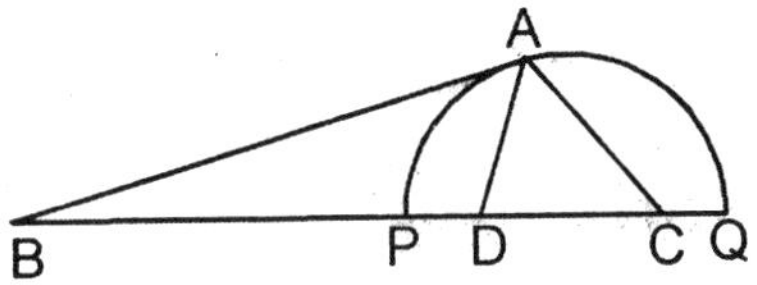

$$\angle ADC = \frac{A}{2} + B$$

$$= \frac{A}{2} + \frac{B}{2} + \frac{B}{2} = \frac{A}{2} + \frac{B}{2} + \frac{C}{C} + \frac{B}{2} - \frac{C}{2}$$

$$= 90 + \frac{B-C}{2}$$

It is easy to find A when (B-C) is known.

(b) *A Median bisecting one of the other sides than the base is given* First to draw the base and bisect it at D.

Then on DC draw the semi circle as above.

That is on half the base draw the circle. Then cut off the circle by median BE. Join CE and produce it by an equal length upto EA. ABC is the Δ.

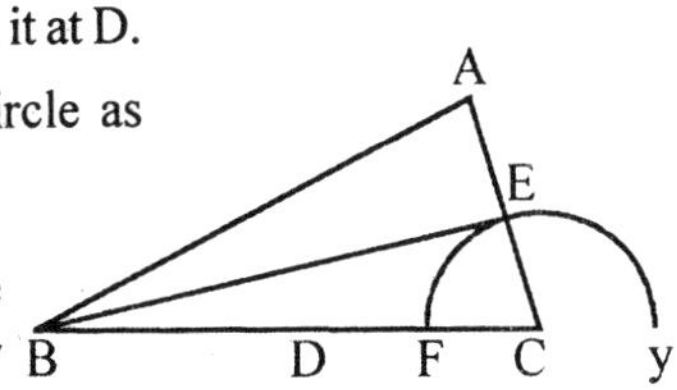

(c) *Difference of the squares of the other sides is given.*

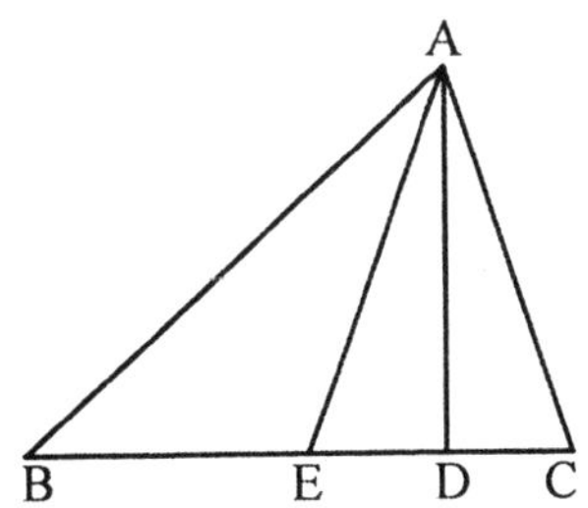

$$AB^2 - AC^2$$

$$= BD^2 - DC^2$$

$$= (BD + DC)(BD - DC)$$

$$= BC.BE$$

Thus E can be found out

$\therefore$ A can be found out.

(d) *Radius of the circum-circle in given*

Base is drawn.

Its centre is found out at D. Knowing BO we can find out point O at the ⊥ bisector of BC. Knowing O we can draw the circumcentre.

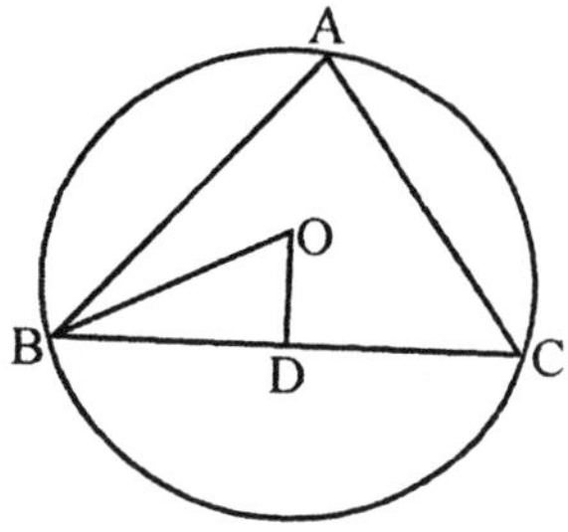

(e) Point where the In-circle touches the base is given.

There both (s–b), (s–c) are known. A = base is already known.

∴ (b–c) = (s–c) – (s–b).

All sides are known and base is know. Hence Δ can be drawn.

Ex. 5 When $(b^2 + c^2)$ and $(b^2 - c^2)$ is known both b and c can be found out. All sides are known. Hence Δ can be drawn.

Ex. 6 Diagonal of the square is more than the side by 2 cm.

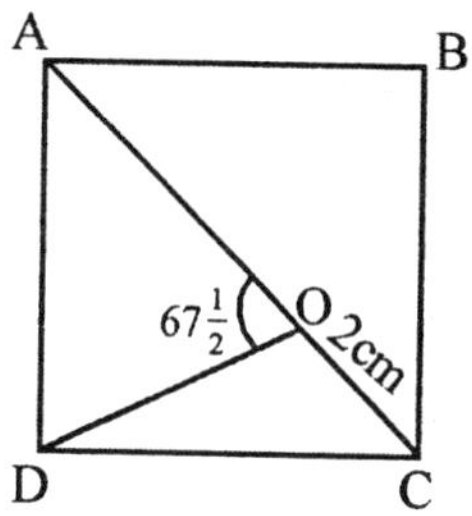

The problem reduces to Drawing the ΔDOC

Where OC = 2 cm. ∠C=45°

$$\angle DOC = 180° - 67\frac{1}{2}^{\circ} = 112\frac{1}{2}^{\circ}$$

This can be drawn.

Ex. 7 In a cyclic quadrilateral

∠B+∠D=180°

∴ cos B = – cos D

$AC^2 = a^2 + b^2 - 2ab \cos B$

also $AC^2 = d^2 + c^2 + 2d \cos D$

equating the 2 equations

$a^2 + b^2 - 2ab \cos B = d^2 + c^2 + 2dc \cos D$

$a^2 + b^2 - d^2 - c^2 = 2ab \cos B + 2dc \cos B$

$a^2 - b^2 - d^2 - c^2 = \cos B\,(2ab + 2dc)$

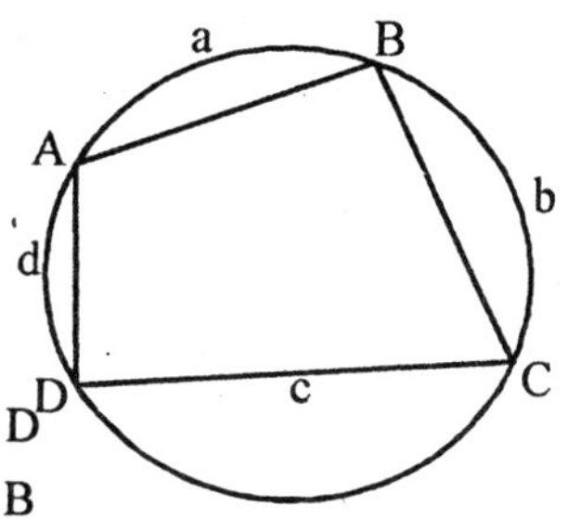

$$\cos B = \frac{a^2 + b^2 - c^2 - d^2}{(2ab + 2ac)}$$

$\therefore$ cos B can be calculated

Hence $\angle B$ can be drawn.

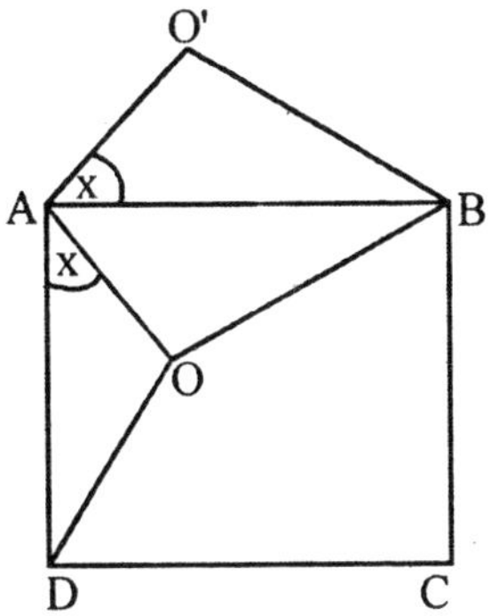

Ex. 8 OA, OB, OD are given.

To draw the square ABCD

Make $\angle BAO' = \angle DAO$

Make $O^1A = OA$ (known)

$\angle O'AO = 90°$

$\therefore$ quadrilateral O. AOB can be drawn

Knowing BA draw the square.

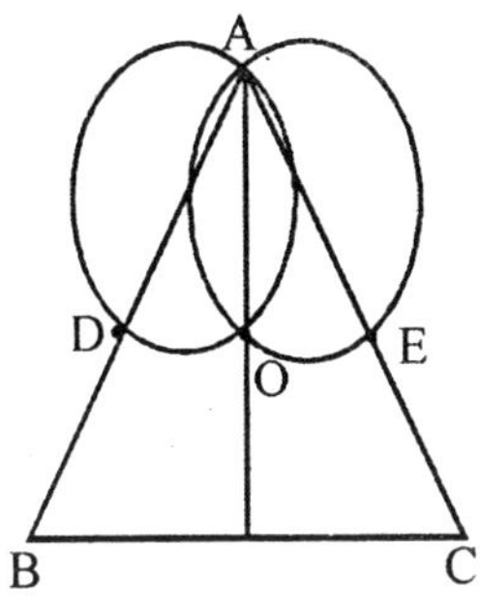

Ex. 9 Points D, E and O are given On DO draw a circle which can subtend 30°. Similarly draw a semi ⊙ on OE than can subtend $\angle$ of 30. We get point A. Then we can draw the Δ. A^1 and O are known.